NASA SP-4020

# ASTRONAUTICS AND AERONAUTICS, 1975

A Chronology

by
Nancy L. Brun
and
Eleanor H. Ritchie

The NASA History Series

*Scientific and Technical Information Branch*  1979
NATIONAL AERONAUTICS AND SPACE ADMINISTRATION
*Washington, D.C.*

# Preface

Users of this series will notice that the current volume, the 15th, is more condensed than earlier volumes in the series, a trend started with the 1974 volume. This is attributable more to staff limitations than to any reduction in the historical significance of the subject matter. Fewer non-NASA activities have been included this year. The annual summary has been eliminated, not without regret. More material has been presented in tabular form for economy in both preparation and use. The staff of the History Office expects to produce one volume per year of *Astronautics and Aeronautics* with approximately the coverage and detail of the present volume— and with greater regularity than in the past few years.

Note that the errata section, introduced in the 1974 volume, has been continued. We want this series to be a reliable reference work, and the reader can help us in this. Let the History Office staff know of any errors that you detect in this or previous volumes, so that corrections can be published.

Text for the first ten months of the present volume's coverage was written by Nancy L. Brun, who has since transferred to the National Institutes of Health. The volume was completed by her successor, Eleanor H. Ritchie, formerly of the Goddard Space Flight Center.

May 1979
Arnold W. Frutkin
*Associate Administrator for External Relations*

# Contents

| | PAGE |
|---|---|
| Preface | iii |
| NASA Associate Administrator Arnold W. Frutkin | |
| January | 1 |
| February | 17 |
| March | 41 |
| April | 57 |
| May | 73 |
| June | 99 |
| July | 125 |
| August | 157 |
| September | 181 |
| October | 201 |
| November | 217 |
| December | 235 |
| Appendix A: Satellites, Space Probes, and Manned Space Flights, 1975 | 251 |
| Appendix B: Major NASA Launches, 1975 | 277 |
| Appendix C: Manned Space Flights, 1975 | 281 |
| Appendix D: NASA Sounding Rocket Launches, 1975 | 283 |
| Appendix E: X-24B Lifting-Body Flights, 1975 | 293 |
| Appendix F: Abbreviations of References | 297 |
| Index and List of Abbreviations and Acronyms | 301 |
| Errata in Earlier Volumes | 327 |

## January 1975

*4 January*: Funding in the FY 1976 budget for a third Earth Resources Technology Satellite (ERTS) was in doubt, Walter Sullivan said in a *New York Times* article. Despite worldwide enthusiasm for continuing the program, which was developing remote-sensing techniques for monitoring and managing earth's resources, doubts about funding approval were appearing within the Office of Management and Budget. Factors that would influence OMB's decision included the reported existence of far more detailed information from secretly launched intelligence satellites. Although scientists would probably prefer a system free from intelligence links, some argued that the information needed by agencies such as the Bureau of Census and Dept. of Agriculture could be extracted from imagery from these satellites. Another reason to delay immediate funding was that a more effective system would eventually become feasible. Scientists argued against this reasoning, saying that the lack of fine detail in ERTS images was not a serious impediment but that lack of continuity would be. (Sullivan, *NYT*, 4 Jan 75, 17)

*8 January*: Twenty-five spacecraft on twenty-four vehicles were scheduled for launch by NASA during 1975, NASA announced. The 25 spacecraft included 14 for NASA programs, 10 for other organizations, and 1 as a cooperative program with a foreign government. Highlighting the year would be the 15 July launch of an Apollo spacecraft carrying Thomas P. Stafford, Vance D. Brand, and Donald K. Slayton as the U.S. contribution to the joint U.S.–U.S.S.R. Apollo-Soyuz Test Project. Apollo would rendezvous and dock with *Soyuz 19*—launched 7 hr earlier than Apollo and carrying Aleksey A. Leonov and Valery N. Kubasov—allowing both crews to exchange visits and perform joint scientific experiments.

The two sets of Viking spacecraft, scheduled for 11 and 21 August launches, would journey through space for a year before arriving in the vicinity of Mars. There each set of spacecraft would separate into an orbiter and a lander to make orbital and surface investigations of the Martian environment.

Other NASA spacecraft would include ERTS-B Earth Resources Technology Satellite and SMS-B Synchronous Meteorological Satellite, in January; GOES-C Geodynamic Experimental Ocean Satellite, in March; OSO-I Orbiting Solar Observatory, Nimbus-F experimental meteorological satellite, and Explorer 53 (SAS-C Small Astronomy Satellite), in May; GOES-A Geostationary Operational Environmental Satellite, in June; Explorers 56 and 57 Dual Air Density Explorers, on a single booster in July or August; Explorer 54 (AE-D Atmospheric Explorer), in September; and Explorer 55 (AE-E), in December.

The 10 spacecraft to be launched by NASA for other organizations included 5 comsats for Communications Satellite Corp.: Intelsat-

IV F−1 in February, Intelsat-IVA F−1 and F−2 in July and October, and Marisat A and B in April and July. Other reimbursable launches would include Telesat−C (Anik 3) for Canada in March, COS−B Celestial Observation Satellite in July for the European Space Agency, Symphonie−B experimental comsat in September for France and West Germany, RCA−A (RCA−*Satcom 1*, in orbit) comsat in December for RCA Corp., and ITOS E−2 Improved TIROS Operational Satellite in late 1975 for the National Oceanic and Atmospheric Administration.

Helios−B would be launched late in 1975 as a cooperative effort with West Germany. The schedule called for 18 of the 24 launches to be made from Kennedy Space Center, 5 from Western Test Range, and 1 from San Marco. For the launches NASA would use a Saturn IB, 3 Titan-Centaurs, 5 Atlas-Centaurs, and 15 Thor-Delta launch vehicles. (NASA Release 75−5; KSC Release 180−74)

- The Department of Defense announced the award of a $1 500 000 cost-plus-incentive-fee contract to General Dynamics Corp. for launch services for the Atlas E and F vehicles at Vandenberg Air Force Base. (DOD Release 8−75)

*9 January:* Sea Satellite (SEASAT)—a new program, approved in the FY 1975 NASA authorization, to monitor the oceans and provide continuous weather and sea condition reports—was announced by NASA. SEASAT−A, a proof-of-concept mission scheduled for 1978 launch, would carry sensors to measure wave heights, current directions, surface wind directions, and surface temperatures. Objectives of the initial mission were to disseminate these data to users concerned about weather predictions; to route shipping to avoid storms, adverse currents, and ice fields; and to provide coastal disaster warnings.

SEASAT−A would also accumulate scientific data on the curvature of the oceans; ocean circulation; transport of mass, heat, and nutrients by surface currents; and the interaction between air and sea.

SEASAT would be managed by Jet Propulsion Laboratory with Langley Research Center, Goddard Space Flight Center, Wallops Flight Center, and JPL each having responsibility for one of the four sensors-off-the-shelf NASA or Air Force spacecraft to which a sensor module would be attached. Existing tracking facilities and support hardware would also be used. Cost of the SEASAT−A mission, as currently planned, was $58.2 million. (NASA Release 75−1)

- McDonnell Douglas Corp., working under contract to NASA, began flight and static tests of the refanned engine to demonstrate NASA-developed noise-reduction techniques. More than 2 yr of work at Lewis Research Center; Pratt & Whitney Aircraft Div. of United Aircraft Corp.; Boeing Co.; United Air Lines, Inc.; and American Airlines, Inc., had been spent in modifying a JT8D engine to reduce irritating high-frequency noise. When installed on the McDonnell DC−9, the new engine was expected to reduce by 60% the ground area exposed to excessive noise levels. Similar results were expected with the engine installed on Boeing's 727 and 737 aircraft. During the initial test phase, McDonnell Douglas would fly for 90 hr a DC−9 outfitted with the modified engine, to evaluate noise characteristics and flight performance. The second phase of testing would begin in mid-January when Boeing would flight-test it in a 727. (NASA Release 75−4; LeRC Release 75−2)

*11 January:* The 38 000-kg second stage from the Saturn V booster that placed the *Skylab 1* orbital workshop in orbit 14 May 1973 reentered the atmosphere over the Atlantic Ocean just before 3 am EST. NASA reported that radar tracked one large chunk that survived the fiery plunge through the atmosphere and fell into the ocean at 34° north latitude, 19° west longitude, about 1600 km west of Gibraltar. Smaller pieces might have scattered over an area several kilometers to the northwest and southeast of that point and some charred debris might have fallen on the Sahara Desert, but NASA received no report that any fragments had caused damage or injury. (NASA PAO, interview, 30 June 1976; Reuter, *W Post*, 12 Jan 75, A14; AP, *W Star-News*, 12 Jan 75, A1; UPI, *NYT*, 12 Jan 75, 55)

*11 January–9 February:* The U.S.S.R. launched *Soyuz 17*, carrying cosmonauts Lt. Col. Aleksey A. Gubarev and Georgy M. Grechko, from Baykonur cosmodrome at 2:43 am local time (4:43 pm EST 10 Jan.) to rendezvous and dock with the *Salyut 4* space station launched 26 Dec. 1974. *Soyuz 17* entered orbit with a 249-km apogee, 186-km perigee, 88.9-min period, and 51.6° inclination. Tass announced that the mission would carry joint experiments with *Salyut 4*, including a comprehensive checkout of the spacecraft's onboard systems in various flight conditions.

After an orbital correction that raised the spacecraft's orbit to 354-km apogee, 90.7-min period, and 51.6° inclination, *Soyuz 17* docked with *Salyut 4* on 12 Jan. Gubarev and Grechko entered the station, switched on the power and radio transmitters, and inspected the scientific equipment.

During their nearly 30 days aboard *Salyut 4*, the cosmonauts studied solar phenomena, x-radiation from celestial bodies, and earth's radiation. They also studied the effects of weightlessness on the human body, made earth-resources observations, and studied the earth's upper atmosphere. They resprayed two telescope mirrors dulled by exposure to space; carried out biological "Oasis" experiments using insects, microorganisms, tissue cultures, and plants; and recycled water, condensed from the cabin's atmosphere, for drinking and food preparation.

The crew began preparations to return to earth on 9 Feb., reboarding *Soyuz 17* and undocking from the space station at 11:08 am Baykonur time (1:08 am EST). *Soyuz 17* softlanded in the U.S.S.R. 110 km northeast of Tselinograd, Kazakhstan, "in complex meteorological conditions" after 29 da 13 hr 20 min in space.

On-the-spot and subsequent medical checks showed the cosmonauts to be in good health. The *Soyuz 17* cosmonauts broke the previous 23-day 18-hr 22-min Soviet record for time in space, set by the *Soyuz 11* crew (6–30 June 1971) before they were killed during reentry. (GSFC SSR, 28 Feb 75; Tass, FBIS–Sov, 13 Jan–13 Feb 75; *SF*, April 75, 144–5, June 75, 235; UPI, *NYT*, 16 Jan 75, 14; *SBD*, 14 Jan 75, 62; *LC S&T News Alert* 2645)

*12 January:* The communications satellite business was becoming one of the fastest growing businesses on earth, Thomas O'Toole said in a *Washington Post* article. The huge dish-shaped antennas used as telephone and television links between earth and the dozen or so orbiting comsats had sprouted up in the suburbs of Moscow and Peking, in Brazilian jungles, in the mountains of Iran, and in the deserts of Algeria. In addition, Algeria was building 14 antennas;

Indonesia, 60; and Brazil, 2000. More than 90 countries were currently communicating by satellite, most using the seven Intelsat satellites orbited by the 89-country consortium, the International Telecommunications Satellite Organization. Worldwide Intelsat traffic was growing by 20% a year. In addition to the Intelsat comsats, the Soviet Union, Canada, France, and West Germany, and Western Union already owned their own comsats. Japan would have one by 1977, and Iran, Saudi Arabia, and Indonesia were each planning to orbit their own.

Besides making it easier to communicate around the world, satellites were responsible for cutting the cost of overseas conversations. In 1947, a 3-min phone call from New York to London cost $12. The same call in 1975 cost $5.40. (O'Toole, *W Post*, 12 Jan 75, 1)

*13 January:* Dr. John L. McLucas, Secretary of the Air Force, announced the selection of the General Dynamics Corp. YF−16 prototype for full-scale engineering development as the Air Force's F−16 air combat fighter. A $417 904 758 fixed-price-incentive contract to fabricate 15 engineering development F−16 aircraft was awarded to General Dynamics, which had been in competition with Northrop Corp. and its YF−17 lightweight fighter prototype during flight-test evaluations at the Air Force Flight Test Center. Dr. McLucas said the decision in favor of General Dynamics had been based on cost and technical engineering proposals submitted by the companies. The Air Force planned to introduce a minimum of 650 F−16 aircraft into the active inventory by the early 1980s.

The Air Force also awarded a $55 500 000 fixed-price-incentive contract to United Aircraft Corp's Pratt & Whitney Div. to produce the F100 engine used in the F−16. (DOD Release 16−75)

- Marshall Space Flight Center plans for the FY 1975 reduction-in-force had been revised, MSFC announced. The target date for issuance of RIF notices to MSFC employees had been delayed from mid-January to 29 Jan. with an effective date of 14 Mar. The 2-wk delay resulted from the need for an additional decrease in the end-of-FY 1975 personnel ceiling, from 4145 to 4113. This further reduction was MSFC's share of the government-wide reduction of 40 000 employees called for by President Ford in 1974. (MSFC Release 75−6)
- Marshall Space Flight Center announced the award of a $59 950 contract to Abbott Laboratories for living human kidney cells to be used by U.S. astronauts in the German electrophoresis experiment during the July U.S.−U.S.S.R. Apollo-Soyuz Test Project. Abbott Laboratories would provide the cells as part of an experiment to find a way to isolate effectively the one kidney cell in 20 that produced the enzyme urokinase, an enzyme capable of dissolving blood clots. Because gravity made such separations difficult on earth, scientists hoped to develop new technology in the zero-g environment of space. The experiment was being developed, built, and tested by West Germany's Messerschmitt-Boelkow-Blohm Gmbtt. (MSFC Releases 75−7, 75−13)

*13−21 January:* Preparations for U.S. participation in the joint U.S.−U.S.S.R. Apollo-Soyuz Test Project mission in July continued at Kennedy Space Center. After removal from environmentally protected shrouds, poststorage inspection, and installation of eight stabilizing fins, the Saturn IB booster, SA−210, was stacked on the mobile launcher inside the Vehicle Assembly Building on 13 Jan. Engineers began electrical and mechanical systems tests on the first stage and installed flame curtains and panels to shroud the outboard engines. The Saturn IVB second stage was mated to the booster 14 Jan. and the instrument unit added to the stack 16 Jan. A boilerplate unit simulating the Apollo command module was added 17 Jan. The boilerplate would be removed and replaced with the actual spacecraft, and the rollout of the entire vehicle to the launch pad was scheduled for March.

The docking system and docking module were mated 17 Jan. after combined systems tests. The complete unit would be placed into the spacecraft adapter in February. Swing-arm launch-control center integration was completed 21 Jan. (KSC Release 165–74; MSFC Release 75–19; *Spaceport News*, 23 Jan 75, 3; 6 Feb 75, 3; *Marshall Star*, 2 Jan 75, 2)

*14 January:* NASA's Earth Resources Technology Satellite, ERTS, had been renamed Landsat, NASA Associate Administrator for Applications Charles W. Mathews announced at a Landsat–B prelaunch press briefing in Washington, D.C. *Erts 1*, launched 23 July 1972, would be called *Landsat 1*; ERTS–B, scheduled for launch 19 Jan., would be designated *Landsat 2* once in orbit. Mathews said that, since NASA planned a SEASAT to study the oceans (see 9 Jan.), Landsat seemed a more appropriate name for a satellite that studied the land.

Landsat Program Manager Harry Mannheimer said that Landsat–B would be launched into the same orbital plane as *Landsat 1* but 180° out of phase. Between them the two satellites would retrace the identical ground track every 9 days rather than the 18-day interval of one satellite. The additional coverage would enable scientists to study dynamic phenomena more effectively. Mannheimer said that a significant improvement over *Landsat 1* was Landsat–B's increase in command storage, adding flexibility in commanding the various instruments while the spacecraft was out of view of a ground station.

James R. Morrison, Landsat Resources Survey Program Manager, said that, whereas the emphasis of the *Landsat 1* mission was on experimentation, the emphasis for Landsat–B would be on fewer but larger scale applications of remote-sensing information to real resource problems. Of the 120 investigators—compared with 320 investigators during the *Landsat 1* mission—57 were U.S.-sponsored and 52 were from foreign countries. During the rigorous selection, preference had been given to investigations that included direct participation and cost-sharing by bona fide users in both the public and private sectors.

New with Landsat–B were the Applications Systems Verification Tests (ASVT), based on results of *Landsat 1*, to further develop, demonstrate, and document Landsat results. The ASVTs were larger in design and scope than earlier investigations, essentially beyond the capability of any one investigator to manage. ASVTs for Landsat–B included a large-area crop inventory experiment (LACIE), ice-warn program, snow-cover and snow-melt prediction program, natural resources information system, and environmental information system. (Text)

- Apollo-Soyuz Test Project crew members Thomas P. Stafford, Donald K. Slayton, and Vance D. Brand participated in a 9-hr exercise at

Kennedy Space Center to test the Apollo spacecraft's internal systems, earth-landing systems, and instrument packages. The astronauts climbed into the Apollo, which was inside a huge altitude chamber, to test the systems at simulated altitudes of up to 61 000 m while cabin pressure was maintained at 34 474 newtons per sq m. Pressure inside the cabin was also lowered to zero to test the integrity of the astronauts' space suits. On 16 Jan. the U.S. backup crew for the mission, Alan L. Bean, Ronald E. Evans, and Jack R. Lousma, participated in a similar exercise. (*NASA Activities*, Feb 75, 11; KSC Release 5–75)

- Flight Research Center was investigating the use of TV to provide visual information to pilots of future aircraft and spacecraft, NASA announced. TV might enhance target visibility for pilots of high-performance aircraft that might have to approach and land at high angles of attack with reduced window area, and for unmanned remotely piloted research vehicles that would augment flight instrumentation with visual. For testing, FRC had installed a TV camera on top of a Piper PA–30 aircraft with a 12.5- by 17.5-m TV screen installed on the instrument panel. A curtain would restrict the test pilot's view through cockpit windows with a safety pilot riding in the right seat of the aircraft. Besides demonstrating use of TV for approach and landings, the first phase of the program would evaluate the pilot's ability to judge altitude using TV. The second phase would aim at optimizing the TV system for landing maneuvers. (NASA Release 75–11)

*14–17 January:* NASA and Air Force Cambridge Research Laboratory launched 10 rocket-borne experiments to study the upper atmosphere as a follow-on to the June 1974 Atmospheric Layering and Density Distribution of Ions and Neutrals program (Project ALADDIN '74). Two experiments on the Ute-Tomahawk rockets released chemical clouds into the atmosphere to measure winds and temperature and diffusion coefficients. The first, launched 15 Jan., released aluminum vapor at 160-km altitude, creating a blue-green glowing cloud visible along much of the eastern seaboard. A second, launched 17 Jan. after a 1-day postponement, released trimethylaluminum in puffs 10 sec apart from 180 km down to 90 km. Five single-stage Super-Lokis carried instruments to measure density, temperature, and wind, and instrumentation aboard two Paiute-Tomahawks measured the atomic oxygen density profile. A prelaunch performance test of a Ute-Tomahawk was made 14 Jan. (WFC Releases 75–1, 75–2; *NYT*, 16 June 75, 29)

*15 January:* President Ford signed Executive Order 11834 establishing the Energy Research and Development Administration—under Administrator Dr. Robert C. Seamans, Jr.—and the Nuclear Regulatory Commission—under Chairman William A. Anders—effective 19 Jan. Establishment of the two new agencies was provided for in the Energy Reorganization Act of 1974, signed into law on 11 Oct. 1974 by President Ford, which abolished the Atomic Energy Commission and reassigned its responsibilities. ERDA would take over AEC's research and development programs including the development of fossil, nuclear, solar, and geothermal energy to meet present and future needs. ERDA would also take the lead in energy R&D programs transferred from the Dept. of the Interior, National Science

Foundation, and Environmental Protection Agency. NRC was assigned AEC's licensing and regulatory responsibilities. (*PD,* 20 Jan 75, 54; ERDA Release 75−1; ERDA *Fact Sheet*)

*16 January−8 February:* U.S. and U.S.S.R. Apollo-Soyuz Test Project working groups held their last major meeting in the U.S. before the scheduled July launches. The five working groups included 80 Soviet aerospace specialists led by Academician Boris N. Petrov and ASTP Technical Director for the U.S.S.R. Konstantin D. Bushuyev. Working Group 3, Soviet specialists in docking systems, arrived 16 Jan. and went directly to Rockwell International Corp. in California for joint tests of the docking system alignment pin and socket. Following the tests, the group returned to JSC to review the results of those tests and others made earlier in Moscow.

NASTRAN to analyze Space Shuttle structures. Recipients of the awards included 12 NASA and 12 contractor employees. (NASA Release 75−13)

*16 January−8 February:* U.S. and U.S.S.R. Apollo-Soyuz Test Project working groups held their last major meeting in the U.S. before the scheduled July launches. The five working groups included 80 Soviet aerospace specialists led by Academician Boris N. Petrov and ASTP Technical Director for the U.S.S.R. Konstantin D. Bushuyev. Working Group 3, Soviet specialists in docking systems, arrived 16 Jan. and went directly to Rockwell International Corp. in California for joint tests of the docking system alignment pin and socket. Following the tests, the group returned to JSC to review the results of those tests and others made earlier in Moscow.

On 20 Jan. a portion of Working Group 1—for experiments, trajectories, and contingencies plans—arrived at JSC to prepare for joint simulations scheduled for March. Group 4, for communications systems, also arrived to review the results of independent U.S. and Soviet tests of the flight communications system. The latter group—accompanied by Dr. Glynn S. Lunney, U.S. ASTP Technical Director, and Prof. Bushuyev—then traveled to Kennedy Space Center 3 Feb. to make electromagnetic radio and TV compatibility tests.

The remaining Soviet specialists arrived in the U.S. on 27 Jan. Working Group 2, for control systems and docking targets, prepared contingency-control modes for the mission, and Working Group 5 prepared the final report on the compatibility of ASTP life-support systems. Group 5 also certified the equipment to be transferred between the two spacecraft.

Although a portion of Group 1 would remain in the U.S. through crew training scheduled for 11 Feb., most of the Soviet contingent returned to the Soviet Union 8 Feb. The next major meeting for the working groups was scheduled for mid-May in Moscow. (JSC Release 75−02; *Spaceport News,* 6 Feb 75, 3)

*17 January:* A request for proposals for integrated electronic assemblies (IEA) for the Space Shuttle's solid-rocket boosters had been issued by Marshall Space Flight Center, MSFC announced. The proposals would be for design, development, test and evaluation, fabrication, and assembly of 33 IEAs. The Space Shuttle would carry two refurbishable IEAs, one forward and one aft, to route commands from the Orbiter to the thrust-vector control system to release the nosecap and frustum, jettison the solid-rocket motor nozzle, detach parachutes, and turn on the recovery aids. (MSFC Release 75–11)

*18 January:* New members of the House of Representatives Committee on Science and Technology, led by Committee Chairman Olin E. Teague (D–Tex.), made an orientation tour of Kennedy Space Center. In addition to visits to Launch Complexes 36 and 39, the KSC industrial area, and the Visitors Information Center, the committee members viewed the Apollo spacecraft being readied for the July U.S.–U.S.S.R. Apollo-Soyuz Test Project and the Viking lander being prepared for the August twin mission to Mars. The members also received special briefings on Center activities. (KSC Release 11–75)

*20 January:* The appointments of four senior European Space Research Organization directors became effective. The appointments were Professor of Engineering Massimo Trella (Italy) as Technical Operator, Bernard Deloffre (France) as Director of the Spacelab program, Walter Luksch (West Germany) as Director of the Communications Satellite Programme, and John Hawkes (United Kingdom) as Head of Development and Technology at ESRO's Space Research and Technology Center (ESTEC) in the Netherlands. (ESRO Release, 20 Jan 75)

• An interagency agreement between NASA and Department of Interior to use NASA technology for mineral extraction was announced by Secretary of the Interior Rogers C. B. Morton. Funded and directed by DOI and combining efforts of NASA and contractor scientists, the project would apply NASA experience in developing of systems for manned and automated operations in hostile environments to demonstrate new coal-mining technology. NASA has designated Marshall Space Flight Center as the lead NASA Center. (NASA Release 75–17)

*21 January:* The current fleet of supersonic transports—including the 16 Anglo-French Concordes and 14 Soviet Tu–144s flying or scheduled for service—would cause minimal damage to the ozone layer, Dr. Alan J. Grobecker, Director of the Dept. of Transportation's Climatic Impact Assessment Program, announced at a press briefing. Citing a 3-yr DOT study, "The Effects of Stratospheric Pollution by Aircraft," Dr. Grobecker said it would require 125 Concordes flying 4.5 hr daily to cause a minimally detectable change in the ozone layer. However, the study concluded that future expansion of stratospheric jet fleets should be carefully monitored.

Recommendations made by the study included development of engines to meet specific nitrogen oxide-emission standards, development of low-sulfur aviation fuels, and establishment of stratospheric air-quality standards and engine-emission regulations. (AP, *NYT*, 22 Jan 75, 33; Robinson, *Av Wk*, 27 Jan 75, 16–17)

*22–25 January:* NASA launched *Landsat 2* (Landsat–B), formerly called Earth Resources Technology Satellite (ERTS–B), from Western Test Range at 9:56 am PST on a two-stage Thor-Delta booster supplemented by nine strap-on rockets. The satellite entered a near-polar orbit with a 918.23-km apogee, 912.89-km perigee, 103.32-min period, and 99.09° inclination. Primary objective of the mission was to acquire multispectral imagery over the U.S. and foreign countries in quantity sufficient to improve remote-sensing interpretative techniques and to further demonstrate the practical application of Landsat data. Secondary objectives were to acquire sufficient multispectral coverage over the U.S. to supply data requirements for the Applications System Verification Tests for at least 1 yr, to acquire multispectral coverage for at least 2 yr over the major agricultural areas of the world to illustrate further applications of Landsat imagery for crop inventory, and to demonstrate successful operation of the Data Collection System.

A 3-day delay in the launch changed the initial orbital phasing of *Landsat 2* relative to *Landsat 1* (launched as *Erts 1* on 23 July 1972) into a 12- to 6-day repeat cycle instead of the planned 9-day cycle. However, an orbital correction maneuver beginning 27 Jan. and ending 6 Feb. altered phasing to a 9-day repeat cycle. The payload separated from the launch vehicle 50 min 3 sec after launch, followed by solar-panel deployment and earth acquisition using the attitude-control system. The command system was turned on automatically, permitting activation of the payload by ground stations. During the night of 22–23 Jan. the mechanical integrity of the wideband video tape recorder was verified, Data-Collection System (DCS) activated, and data-collection platform experiment data transmitted. By 12:00 pm EST 23 Jan. all spacecraft systems were operating and by 25 Jan. the wideband transmitter, wideband video tape recorder, multispectral scanner (MSS), and return-beam vidicon (RBV) were turned on, and operating normally.

Teaming up with *Landsat 1*, the 953-kg *Landsat 2* would provide repetitive coverage of almost the entire earth to demonstrate the practical benefits of resources management from space for NASA in cooperation with the Dept. of Agriculture, National Oceanic and Atmospheric Administration, Dept. of the Interior, Environmental Protection Agency, U.S. Army Corps of Engineers, and various state, local, and foreign organizations. Experiments included the cooperative NASA–DOA–NOAA Large Area Crop Inventory (LACIE), which combined crop-acreage measurements derived from Landsat data with meteorological information from ground stations and NOAA satellites to assess crop yields and make production forecasts. During the first yr of operation, LACIE would concentrate on U.S.-grown wheat, but the experiment would later be expanded to other crops and other regions.

Other experiments included land-use survey and mapping; mineral resources, geological, structural, and land-form surveys; water resources studies; marine and ocean surveys; meteorological and environmental studies; and interpretative techniques development.

Data were being transmitted to three NASA tracking and data-acquisition facilities at Fairbanks, Al.; Goldstone, Calif.; and Greenbelt, Md. In addition, Canada and Brazil operated Landsat ground data-acquisition stations and Italy and Iran were constructing similar facilities. Data received from the satellite were sent to the NASA Data Processing Facility at Goddard Space Flight Center where 1300 images, covering 45 million sq km, could be processed each week. The data would then be forwarded to the Federal Data Center and made available to the public.

The Landsat mission was part of a U.S. program to develop remote-sensing methods for improved management of earth's resources. In addition to *Landsat 1* and *2*, the program included remote-sensor instrument development; data-analysis research using data from spacecraft, aircraft, and ground-truth sites; low-, intermediate-, and high-altitude aircraft flights; and Earth Resources Experiment Package (EREP) experiments completed during the 1973 manned Skylab missions. (NASA MORs, 9, 23 Jan 75; NASA Releases 74−329, 75−31)

*22 January:* NASA's *Oso 5* Orbiting Solar Observatory began its 7th yr in earth orbit. Launched 22 Jan. 1969 to study the sun and its influence on earth's atmosphere, *Oso 5* had completed more than 34 000 orbits. It was the fifth of eight spacecraft launched in the OSO program to observe the sun during most of its 11-yr solar cycle. *Oso 5* had been shut down 3 yr after launch but was reactivated in July 1974 after decay of the orbit of *Oso 7* (launched 29 Sept. 1971). *Oso 5*, which had a planned lifetime of 6 mo, was obtaining data on the frequency and extent of eruptions on the sun's surface. (NASA Release 75−46)

*23 January*: The Air Force's prototype B−1 strategic bomber successfully completed its second flight test at Air Force Flight Test Center. Primary objectives of the 3-hr 21-min flight included evaluation of the aircraft's wing sweep control system, flying qualities in the initial climb, and power approach configuration. The auxiliary power unit was also evaluated and air-start tests were made.

Following takeoff, the B−1 climbed to 3050 m, where its landing gear, flaps, and slats were retracted. After flying at low speeds with the wings in the full forward 15° position, the wings were swept 25°, and the aircraft climbed to 4900 m and conducted air-start tests at mach 0.7. The prototype then slowed to 463 km per hr and its speed brake was checked in a descent to traffic-pattern altitude. The B−1 then landed safely at AFFTC. (AFSC *Newsreview*, March 75)

- The Federal Communications Commission approved a plan for International Business Machines Corp. and Communications Satellite Corp.'s Comsat General subsidiary to form a domestic satellite company with a third company as long as no partner owned less than 10% or more than 49% of the stock. However, a plan in which IBM would replace Lockheed Aircraft Corp. and MCI Communications Corp. as Comsat's partner and own 55% was rejected. The FCC stated that this plan would give IBM too much power over the specialized communications business.

  Other alternatives suggested by the FCC included a plan whereby IBM and Comsat could independently enter the communications satellite business or Comsat could join another consortium without IBM. Another alternative was that Comsat, which was not permitted to deal directly with the public, could lease space communications circuits to IBM under an arrangement similar to the one Comsat had with American Telephone & Telegraph Co.

  Whatever the arrangement, FCC ruled, IBM must establish a separate corporation for its satellite operations. (FCC Memorandum Opinion and Order 75−156; FCC Release 46281; DJ, *W Post*, 24 Jan 75, D7; *W Star-News*, 24 Jan 75, F21)

- Ground was broken for Marshall Space Flight Center's x-ray telescope test facility for the High Energy Astronomy Observatory (HEAO) program. The facility, which would cost $4 million including equipment, would be used to test instruments for the HEAO program as well as to calibrate rocket payloads for x-ray stellar studies and make advance telescope calibrations. The HEAO program included three unmanned scientific satellites to be launched into low earth orbit between 1977 and 1979 to study the invisible light of x-rays, gamma rays, and cosmic rays. (*Marshall Star*, 29 Jan 75, 4)

*24 January*: Studies of the heavy particle cosmic-ray exposure received by the astronauts during the nine Apollo lunar orbital missions revealed a significant variation in exposure as a function of the amount of shielding and the phase of the solar cycle, *Science* magazine reported. The command module pilot, who remained in the spacecraft while the two other crew members landed on the moon, had received the least exposure. The ankle, the least shielded part of the body, received a higher exposure than the chest, which had greater shielding. A sharp rise in exposure had been evident beginning with *Apollo 14* when mission times increased. Observed variations in flux from mission to mission—the flux during *Apollo 16* and *17* was 2.4 times higher than *Apollo 8* through *12*—was attributed to solar modulation of the primary cosmic ray beam.

Data obtained during the Apollo missions would be used for planning long-range missions and for estimating expected biological damage. (Benton *et al.*, *Science*, Vol. 187, 263–5)

- The U.S.S.R. commanded the retrorockets aboard *Salyut 3* (launched 25 June 1974) to fire, putting the orbital workshop on a descent trajectory. After 7 mo in space, it reentered and burned up over the western part of the Pacific Ocean. After a 16-day visit to *Salyut 3* in July 1974 by a two-man crew aboard *Soyuz 14*, a second crew launched 26 Aug. 1974 aboard *Soyuz 15* had failed, in several attempts, to dock with the station and returned to earth after 48 hr in space. (Tass, FBIS–Sov, 24 Jan 75, U1)

*27 January*: NASA Black Brant VC sounding rockets would be used to carry materials-processing experiments as a low-cost way of expanding observations made during the 1973–1974 Skylab missions, NASA announced. Three flights a year aboard the sounding rockets were planned from 1975 through 1980 to provide a better understanding of material behavior during melting, solidification, and heat treatment without the effects of gravity. Although the sounding rockets would provide only 6 min of low gravity—one ten-thousandth of the gravity on earth—rocket missions were the only way to get low-gravity materials-processing data between the July 1975 manned Apollo-Soyuz Test Project and the Space Shuttle flights in the 1980s. (NASA Release 75–12)

- Dr. James C. Fletcher, NASA Administrator, and New York City mayor Abraham D. Beame met in New York to review NASA-developed safety standards for handling liquefied natural gas. Following a 1973 gas storage tank fire that killed 40 persons, New York officials had asked NASA to use its experience in handling highly volatile rocket fuels to help the city establish a comprehensive risk-management plan to design, construct, and operate natural gas facilities. The plan, developed at Kennedy Space Center and incorporated into New York City Fire Department operating procedures, identified liquid gas risks, controlled the risks through redundant fail-safe techniques, and set up criteria for discussions to eliminate or accept certain risks.

Also discussed was the new fiberglass and aluminum breathing system developed at Johnson Space Center and being tested by firefighters in New York, Houston, and Los Angeles. Results to date indicated that the new unit—which was 30% lighter than conventional systems and featured a high-pressure, longer duration air tank and a face mask allowing better vision—was superior to old-style units. The new system was expected to be introduced commercially later in the year. (NASA Release 75−25)

- Sen. Edward M. Kennedy (D−Mass.) introduced S. 397 to authorize the Secretary of the Interior to acquire and maintain for future generations the site in Auburn, Mass., on which Dr. Robert H. Goddard launched the first liquid-propelled rocket on 16 March 1926. (*CR*, 27 Jan 75, S884)

*27 January−13 February*: A joint conference of technical directors for the U.S.−U.S.S.R. Apollo-Soyuz Test Project met at Johnson Space Center. Participants in the conference, held while U.S. and Soviet working groups also met at JSC (see 16 Jan.−8 Feb.), reviewed the progress of ASTP preparations and published the necessary documentation.

A joint communique issued 13 Feb. stated that discussions and agreements were completed on technical questions relating to the development of new spacecraft equipment and design improvement, mission ground-support equipment, and publication of technical documentation.

The communique also stated that communication lines between U.S. and U.S.S.R. centers fully guaranteed the flow of information necessary to conduct the joint mission. Both docking systems were carefully checked and were ready for flight, with final checkouts scheduled at the respective launch sites. Crew training was proceeding satisfactorily (see 7−26 Feb.) with the final joint training session scheduled for April in the U.S.S.R.

Soviet ASTP Technical Director, Professor Konstantin D. Bushuyev, had informed U.S. officials that *Soyuz 16* (launched 2 Dec. 1974) had successfully completed a full test of basic flight phases planned for ASTP as well as performance of spacecraft systems and interaction with U.S.S.R. ground control systems. Joint U.S.−U.S.S.R. tracking activity during the flight had shown that the precision of tracking information was satisfactory.

During a 13 Feb. press conference at JSC, Professor Bushuyev said that, in addition to changes in the Soyuz life-support systems and the addition of the docking apparatus, other design changes on the Soyuz spacecraft included the installation of a transporter for the Apollo and related antennas, a new interspacecraft communications system, an optical target to permit Apollo to target the approach and rendezvous, and some flashing beacons and orientation lights. (Joint communique text; JSC press briefing transcript, 13 Feb 75)

*28 January*: Lee R. Scherer assumed the duties of Director of Kennedy Space Center, succeeding Dr. Kurt H. Debus, who retired in October 1974. Scherer had been Flight Research Center Director since 1971 and, before that, Assistant Director for Lunar Programs and manager of the Lunar Orbiter program at NASA Hq. (KSC Release 17−75)

- Marshall Space Flight Center announced the award of three contracts, part of a program to design, fabricate, test, and demonstrate a low-cost aesthetically appealing solar collector for residential heating and cooling. Chamberlain Manufacturing Corp. was awarded $72 621 and

Honeywell, Inc., was awarded $104 255 to deliver a low-cost collector based on existing technology and the requirements of the solar heating and cooling demonstration. A third contract was awarded to PPG Industries which would furnish collector panels from off-the-shelf hardware at essentially no cost. (MSFC Release 75-22)

*29 January:* NASA's basic goal in the U.S. stratospheric research program was to determine the normal makeup of the upper atmosphere with emphasis on understanding the dynamic processes occurring and the perturbations caused by natural and man-made events, Dr. James C. Fletcher, NASA Administrator, said at a hearing before the Senate Committee on Aeronautical and Space Sciences. NASA's current high-altitude program was being carried out by conventional and special purpose aircraft such as the U-2, instrumented packages placed on commercial aircraft, instrumented balloons, sounding rockets, and computer modeling techniques. Satellites still provided the best potential for continuing repetitive global observations of the upper atmosphere. The Nimbus satellites had measured properties affecting the ozone and had made vertical profiles of the upper atmosphere. Atmospheric Explorer (AE) satellites had made coordinated investigations of photochemical processes in the upper atmosphere caused by solar ultraviolet radiation.

Dr. Fletcher said that NASA would continue these efforts and had plans for new atmospheric studies. In addition to payloads planned for launch on the Space Shuttle, NASA would use instrumented aircraft, balloons, and sounding rockets to study the effects of chlorofluoromethanes, or Freon, on atmospheric ozone. (Transcript)

- Reduction-in-force and reduction-in-grade notices were sent to 213 Civil Service employees at Marshall Space Flight Center as a result of the reduction in the end-of-FY 1975 personnel ceiling. In addition to the 93 persons separated and the 120 persons reduced in grade, 191 MSFC employees received reassignments as part of a reclassification survey conducted during the summer of 1974. (MSFC Release 75-24)

*30 January:* NASA's Office of Space Science announced assignment to Marshall Space Flight Center of overall responsibility for a definition study of the Atmospheric, Magnetospheric and Plasmas-in-Space (AMPS) Spacelab payload, to fly on the Space Shuttle. AMPS would be a manned laboratory equipped to study the dynamic process of the atmosphere and magnetosphere, using active and passive probing techniques. (NASA Release 75-28)

*31 January:* An unexpected and sudden drop in the total number of ionospheric electrons within 1000 km of the burning Saturn V second-stage engines had been noted during the 15 May 1973 launch of *Skylab 1, Science* magazine reported. The probable cause of the electron loss was the large number of hydrogen ($H_2$) and water vapor molecules added by the engines to the ionospheric F-region of the atmosphere. This initiated a recombination process between ionospheric positive ions and ambient electrons, causing the removal of ion-electron pairs. The Saturn's first-stage engines, which burned kerosene in an oxygen environment, cut off at 88 km and, therefore, had little effect on the ionosphere F-region.

Similar changes in ionospheric chemistry had not been noted in association with Saturn V vehicles launched before *Skylab 1* because all their final parking orbits (and therefore their second-stage burns)

had been below 190 km, where the ionospheric chemistry is different. (Mendillo et al., *Science*, 31 Jan 75, 343–5)
- Marshall Space Flight Center announced the award of two contracts to Martin Marietta Corp. for further studies on the Earth Orbiting Teleoperator System (EOTS). Under a $246 570 contract Martin would design EOTS, an unmanned remotely controlled spacecraft, to operate in earth orbit near the Space Shuttle Orbiter. The spacecraft would carry a TV camera for close-up inspection, have docking capability, and be propelled by cold-gas expulsion. Under a $64 000 contract Martin would develop a complete set of requirements for the control and display station, a unit in the Orbiter that would contain control equipment common to all experiments requiring a teleoperator system. (MSFC Release 75–29)
- NASA had awarded a $152 565 000 cost-plus-award-fee contract to Martin Marietta Corp. for the design, development, and test of the Space Shuttle external tank, Marshall Space Flight Center announced. The contract, which covered the first increment of the external tank project for 1 Sept. 1973 to 30 June 1980, called for a maximum production rate of 24 tanks per year and delivery of major ground-test articles and 6 flight-model tanks. The work would be done at Michoud Assembly Facility under the direction of MSFC. (MSFC Release 75–28)

*During January: Opportunities and Choices in Space Science, 1974*, a report published by the Space Science Board of the National Research Council, recommended that NASA undertake the Large Space Telescope as the only new space research start in FY 1976. The telescope should be followed by a series of permanent national and international observatories in orbit. The report also recommended that, budget permitting, NASA undertake in 1977 a lunar–polar mission, a Pioneer–Jupiter orbiter, a Mariner–Jupiter–Uranus mission, and a solar maximum mission. However, NASA should compare the importance of the Pioneer–Jupiter orbiter with the Mariner–Jupiter–Uranus mission if current FY 1977 budget estimates would not support both missions. Other recommendations included an immediate reevaluation of strategy for missions to explore the outer solar system during the next decade and for returning a Martian surface sample to earth instead of landing an unmanned laboratory to perform analyses as current plans propose.

The report strongly endorsed the High Energy Astronomical Observatories (HEAO), Pioneer–Venus mission, and Mariner–Jupiter–Saturn mission as vital to the nation's space science efforts. (Text)
- Contracts involving $25 000 or more awarded by Marshall Space Flight Center during the month totaled nearly $200 million. Among them were awards to Lockheed Missiles and Space Co., Inc., for $67 500 to develop the capability of predicting radiant heating at the base region of the Space Shuttle, and for $49 971 to continue a two-phase study of flow effects on Space Shuttle plume simulation. Lockheed also had received $59 740 to continue analyses of Apollo Telescope Mount data obtained during the Skylab mission. A $29 945 contract had gone to Northrop Services, Inc., to continue studies on Space Tug recovery of a spinning satellite.

The U.S. Army Engineer Div., Ala., had received two contracts. A $2-million contract had been awarded for construction at MSFC of the structural test facility for the Solid Rocket Booster and a $2.93-million contract has been awarded for construction and modifications to the Dynamic Test Facility for vibration testing of the Space Shuttle in the vertical positions. The contracts also included a $152.6-million award to Martin Marietta Corp. for the design, development, test, and evaluation of the Space Shuttle external tank, including six flight units and test hardware.

MSFC had awarded the Dept. of Commerce $60 000 for an atmospheric measuring program. Bendix Corp. had received $99 736 for a Skylab control-moment gyro anomaly investigation, and the University of New York had received $36 253 to continue data analysis of a Skylab zodiacal light experiment. (MSFC Release 75−39)

# February 1975

*2 February*: High-speed computer techniques were producing profiles of giant red stars, in a NASA-sponsored research project conducted by California Institute of Technology scientist Dr. Juliana Christy-Sackmann at the Jet Propulsion Laboratory.

Analyzing high-resolution infrared data on successive star flashes representing from a fraction of a year to a million years in the evolution of a red giant, Dr. Christy-Sackmann found that the violent combustion of helium in the heart of the stars, releasing heat up to 260 million°C, produced a predominance of carbon. She stressed the convective role of helium and hydrogen in fueling the stellar fires: A convective tongue of helium was driven through layers of burning shells toward the outer envelope of hydrogen. Hydrogen might also drop down to overlap, giving extra energy to an old star. The release of energy from the interior of the stars varied, apparently because of convective zones that seemed to come and go. Although each flash produced extreme interior disruption, surface changes in most cases were minor.

The length of time required for computation had prevented previous investigators from following and analyzing more than a few of the hundreds of successive flashes occurring in a star; using the new computer technique, Dr. Christy-Sackmann planned to analyze hundreds. (JPL Release, 2 Feb 75)

*3 February:* President Ford sent a $349.4 billion FY 1976 budget request to Congress, an increase of $45.0 billion over the FY 1975 request. The recommended deficit of $51.9 billion was a peacetime record high designed to help revive the Nation's sagging economy. In his budget message to Congress, the President said that his recommendations provided for fiscal policy actions to increase purchasing power and stimulate economic revival; a major new energy program to hold down energy use, accelerate development of domestic energy resources, and promote energy research and development; an increase in outlays for defense to maintain preparedness and preserve force levels in the face of rising costs; a 1-yr moratorium on new Federal spending programs other than energy; and a temporary 5% ceiling on increases in pay for Federal employees and on individual benefits tied to changes in consumer prices. Proposals included a one-time $16-billion tax cut— $12 billion for individual taxpayers and $4 billion for business—to stimulate economic recovery.

*Energy:* The President specifically requested that no new Federal spending programs be initiated in FY 1976 except in the field of energy, which had a total recommended budget authority set at $2.491 billion. Federal energy functions were divided into three broad categories: general operating programs (analysis and development of energy policy), regulatory programs, and research and development programs. General operating programs, with a recommended budget authority of $548 million, would be administered by the Federal

Energy Administration (FEA), Dept. of the Interior, and the new Energy Research and Development Administration (ERDA). Federal energy regulation, with a recommended budget authority of $178 million, was managed by FEA, Nuclear Regulatory Commission (NRC), and Federal Power Commission. Although the new budget called for support from NASA and the National Science Foundation (NSF) for energy R&D, ERDA would have the major responsibility.

The proposed energy program called for an increased fee on imported oil and an excise tax on domestically produced petroleum and natural gas. The proposal also called for decontrol of oil prices—coupled with a windfall profits tax—and deregulation of prices on new natural gas.

**Energy**

(in millions of dollars)

| Program or agency | Outlays | | | Recommended budget authority for 1976 |
|---|---|---|---|---|
| | 1974 actual | 1975 estimate | 1976 estimate | |
| Energy: | | | | |
| General operating programs | 223. | 152. | 498. | 548. |
| Regulation | 90. | 171. | 164. | 178. |
| Research and development | 739. | 1131. | 1577. | 1764. |
| Total energy | 606. | 1454. | 2240. | 2491. |

The Federal government had further expanded its research and development program to provide new and improved technologies necessary to increase domestic energy resources. Outlays for energy R&D would be $1.7 billion in FY 1976, an increase of 36% over FY 1975 and 102% over 1974. Budget recommendations continued a vigorous nuclear R&D program and accelerated non-nuclear energy R&D, particularly in coal and solar energy.

**General Science, Space, and Technology**

(in millions of dollars)

| Program or agency | Outlays | | | Recommended budget authority for 1976* |
|---|---|---|---|---|
| | 1974 actual | 1975 estimate | 1976 estimate | |
| Space research and technology: | | | | |
| Manned space flight | 1473 | 1538 | 1705 | 1782 |
| Space science, applications, and technology | 1168 | 1040 | 1127 | 1119 |
| Supporting space activities | 322 | 327 | 351 | 324 |
| Subtotal | 2963 | 2905 | 3183 | 3225 |
| General science and basic research: | | | | |
| National Science Foundation | 647 | 649 | 720 | 757 |
| Energy Research and Development Administration | 369 | 393 | 414 | 438 |
| Earth Sciences: | | | | |
| Geological Survey | 178 | 238 | 266 | 268 |
| Deductions for offsetting receipts | −3 | −3 | −3 | −3 |
| Total | 4154 | 4183 | 4581 | 4686 |

*Compares with budget authority of $3874 million in 1974 and $4299 million in 1975.

*General Science and Technology:* Requested funding totaling $4.686 billion for general science, space, and technology—not including energy or Dept. of Defense programs—was highlighted by the buildup in the development and production of the Space Shuttle; continued development of spacecraft to explore the sun, planets, and universe; continued research, development, and experimentation in the application of space technology for surveying natural resources and improving weather forecasting; and increased support for basic science.

Agencies receiving the major portions of the general science, space, and technology funds were NASA, NSF, ERDA, National Oceanic and Atmospheric Administration, and Geological Survey.

*Air Transportation and Nonmilitary Aeronautical Research:* In the field of transportation, for which the recommended FY 1976 budget authority was $5.568 billion, air transportation accounted for $2.660 billion. The Administration would propose major legislation in aviation development, and revenues providing for continued long-term Federal development of the airway system, substantial restructuring of the airport grant program, and more equitable structuring of air-transportation user fees.

NASA would spend $316 million in FY 1976 on its broad program of research and technology to support civilian and military aeronautical objectives. Major aims of the program: to reduce aircraft noise and exhaust pollution, reduce fuel consumption, and improve aircraft performance, reliability, and safety.

**Air Transportation**

(in millions of dollars)

| Program or agency | Outlays | | | Recommended budget authority for 1976 |
| --- | --- | --- | --- | --- |
| | 1974 actual | 1975 estimate | 1976 estimate | |
| Air transportation: | | | | |
| Airways and airports | 1870 | 2092 | 2288 | 2285 |
| Air carrier subsidies | 73 | 67 | 66 | 61 |
| Aeronautical research and technology | 292 | 304 | 316 | 314 |
| Total, air transportation | 2236 | 2464 | 2670 | 2660 |

*Defense Research, Development, Test, and Evaluation:* Total obligational authority of $103 billion for military defense included a $10.236-billion estimate for RDT&E to permit continued engineering development of the B−1 strategic bomber, of the Trident submarine missile system, of long-range and intercontinental ballistic missiles, and of the command, control, and communications of strategic forces. Also funded under the RDT&E function were military astronautics programs to improve space technology for military applications and to develop space vehicles for specific military missions. Efforts would be increased on development of a missile early warning system, communications satellite system, NAVSTAR global positioning system, and uses of NASA's Space Shuttle for launching military space payloads.

## Military Research, Development, Test and Evaluation Programs

(in millions of dollars)

|  | Budget plan | | | Obligations | | |
| --- | --- | --- | --- | --- | --- | --- |
|  | 1974 actual | 1975 estimate | 1976 estimate | 1974 actual | 1975 estimate | 1976 estimate |
| Summary of programs by activities: | | | | | | |
| 1. Military sciences | 427 346 | 432 586 | 485 899 | 428 092 | 432 700 | 484 393 |
| 2. Aircraft and related equipment | 1 667 463 | 1 615 499 | 2 123 267 | 1 660 347 | 1 620 520 | 2 113 520 |
| 3. Missiles and related equipment | 2 123 007 | 2 107 534 | 2 488 503 | 2 071 245 | 2 126 370 | 2 477 253 |
| 4. Military astronautics and related equipment | 590 028 | 523 757 | 625 451 | 585 187 | 524 130 | 622 214 |
| 5. Ships, small craft, and related equipment | 633 990 | 665 301 | 664 346 | 647 832 | 668 026 | 664 146 |
| 6. Ordnance, combat vehicles, and related equipment | 426 630 | 464 607 | 569 941 | 418 264 | 469 429 | 564 245 |
| 7. Other equipment | 1 594 571 | 1 844 160 | 2 195 680 | 1 611 604 | 1 850 172 | 2 186 443 |
| 8. Programwide management and support | 731 938 | 962 068 | 1 083 513 | 727 287 | 960 709 | 1 081 407 |
| Total direct | 8 194 973 | 8 615 513 | 10 236 600 | 8 149 858 | 8 652 056 | 10 193 621 |

*Defense Procurement:* Out of a total FY 1976 recommended defense procurement budget authority of $24.420 billion, $8.015 billion was designated for aircraft and $3.306 billion was designated for missiles.

## Defense Procurement

(in millions of dollars)

|  | 1974 actual | 1975 estimate | 1976 estimate |
| --- | --- | --- | --- |
| Aircraft | 5 938.2 | 6 081.9 | 8 014.8 |
| Missiles | 2 681.8 | 2 544.8 | 3 305.6 |
| Ships | 3 508.4 | 3 174.6 | 5 446.0 |
| Combat vehicles, weapons, and torpedoes | 575.5 | 708.4 | 1 314.6 |
| Other | 4 763.5 | 4 846.0 | 6 338.5 |
| Total | 17 467.4 | 17 355.7 | 24 419.5 |

(*PD*, 10 Feb. 75, 128–136; OMB, *Budget of the U.S. Govt.*, FY 1976 and appendixes)

- Dr. James C. Fletcher, NASA Administrator, released the budget statement he had given at a 1 Feb. press briefing. The FY 1976 NASA budget called for the authorization of $3.539 billion and net outlays of $3.498 billion for the 12-mo period beginning 1 July, and additional

## NASA Budget Plan, FY 76

(in millions of dollars)

|  | FY 1974 | FY 1975 | FY 1976 |
| --- | --- | --- | --- |
| Research and Development | 2310.9 | 2323.3 | 2678.4 |
| Construction of Facilities | 101.1 | 142.7 | 84.6 |
| Research and Program Management | 744.0 | 765.2 | 776.0 |
| Total | 3156.0 | 3231.2 | 3539.0 |

amounts of $958.9 million in authorizations and $905.6 in net outlays for the 3-mo transition period beginning 1 July 1976.

Dr. Fletcher said the new budget was "lean but manageable." For the first time since 1961 the budget provided no new program starts. In accordance with President Ford's request to hold FY 1975 and FY 1976 expenditures to a minimum, NASA, in FY 1975, had deferred obligations of $72 million and outlays of $70 million, reduced civil service employment by 300, and made program adjustments to stay within the total FY 1975 and 1976 budgets. Further adjustments might be necessary if the inflation rate continued to grow.

Although NASA's FY 1976 budget was up $300 million from the previous year, $200 million of this increase was specifically anticipated in last year's budget to carry forward commitments built into programs approved in FY 1975. Therefore, NASA had received an effective increase of only $100 million, or about 3%. This, along with the nation's current 9% inflation rate, was a "good indicator of the leanness of NASA's FY 1976 budget."

The budget would permit NASA to proceed with current major programs as planned: The Space Shuttle, scheduled for a first manned orbital flight in mid-1979, and Landsat−C, scheduled for a fall 1977 launch, would continue on schedule. Reduction of fuel use in aircraft remained a principal focus in NASA's aeronautics research and technology. NASA was also continuing stratospheric studies, including the effects of pollutants on the environment, with significant contributions coming from NASA meteorological and atmospheric satellites, sounding rockets, and balloons and high-flying aircraft.

Although there would be no new starts, NASA would continue its study and advanced technical development for science, applications, and aeronautics projects to be started in future years, including preliminary work for payloads to fly on the Space Shuttle and Spacelab.

The 1975 calendar year launch schedule was one of the most ambitious in U.S. history with 28 launches scheduled, including the joint U.S.−U.S.S.R. Apollo-Soyuz Test Project in July and the two Viking missions to Mars scheduled for August.

Although NASA's legal responsibilities for energy research and development—designated under the Solar Heating and Cooling Act of 1974—had been transferred to the Energy Research and Development Administration, NASA would work closely with ERDA in various aspects of solar research and development.

Dr. George M. Low, NASA Deputy Administrator, said at the press briefing that the largest budget increase was for manned spaceflight, where the increase for the Shuttle was partially offset by reduced spending for the rest of manned spaceflight. The small increase for space science did not tell the whole story: If funds for the Viking program were subtracted, the budget for the rest of science had increased $170 million from FY 1974 to FY 1976. Although applications, aeronautics, and space technology programs each showed small increases, funds for tracking and data acquisition decreased slightly despite increased maintenance costs for tracking stations, because NASA was beginning to close down tracking stations, preparing for the time when tracking and data relay satellites would be available.

## NASA Budget Plan

(in millions of dollars)

|  | FY 1974 | FY 1975 | FY 1976 |
|---|---|---|---|
| Manned space flight | 999.9 | 1110.3 | 1414.6 |
| Space science | 664.5 | 542.8 | 582.6 |
| Applications | 159.0 | 174.5 | 175.0 |
| Aeronautics and space technology | 234.3 | 237.8 | 250.3 |
| Tracking and data acquisition | 244.0 | 248.0 | 243.0 |
| Energy programs | 4.7 | 4.4 | 5.9 |
| Technology utilization | 4.5 | 5.5 | 7.0 |
| Total Research and Development | 2310.9 | 2323.3 | 2678.4 |
| Construction of Facilities | 101.1 | 142.7 | 84.6 |
| Research and Program Management | 744.0 | 765.2 | 776.0 |
| Total budget plan | 3156.0 | 3231.2 | 3539.0 |

William E. Lilly, NASA Comptroller, said that, of the $84.6 million budgeted for construction of facilities, $47.2 million was to continue the buildup of Space Shuttle facilities, mostly for launch and landing facilities at Kennedy Space Center. The remaining funds were for normal rehabilitation and modification of existing facilities, with smaller amounts for minor construction and for facilities planning and design.

Civil service manpower was expected to continue at the same level for FY 1976 with 24 316 employees. However, the number of support service contractors—excluding mission or stage contractors—would decrease from 18 637 in FY 1975 to 17 811 in FY 1976. The estimated 125 000 persons working on NASA programs during FY 1975 was expected to increase by 6000 to 7000 by the end of FY 1976.

Distribution of funds to the Centers would change: An increase of $224.3 million would go to Johnson Space Center for the Shuttle program. A decrease of $37.3 million for KSC reflected a decrease in funding for the construction program. Goddard Space Flight Center

## Estimated Distribution of Total NASA Budget

(in millions of dollars)

|  | FY 1975 | FY 1976 | Increase or Decrease |
|---|---|---|---|
| Johnson Space Center | 922.6 | 1146.9 | 224.3 |
| John F. Kennedy Space Center, NASA | 275.9 | 238.6 | -37.3 |
| Marshall Space Flight Center | 428.7 | 541.2 | 112.5 |
| National Space Technology Laboratories | 10.0 | 10.9 | 0.9 |
| Goddard Space Flight Center | 493.7 | 461.1 | -32.6 |
| Wallops Flight Center | 31.9 | 31.3 | -0.6 |
| Jet Propulsion Laboratory | 180.9 | 188.1 | 7.2 |
| Ames Research Center | 168.2 | 193.0 | 32.0 |
| Flight Research Center | 34.6 | 40.3 | 5.7 |
| Langley Research Center | 285.5 | 249.4 | -36.1 |
| Lewis Research Center | 218.1 | 253.3 | 35.2 |
| NASA Headquarters | 159.1 | 166.4 | 7.3 |
| Undetermined | 22.0 | 18.5 | -3.5 |
| Total | 3231.2 | 3539.0 | 307.8 |

was down $32.6 million, largely because of a decrease in funding for Delta vehicles. (OMB, *Budget of U.S. Govt., FY 76* and Appendix; NASA FY 76 budget briefing, 1 Feb 76, transcript; NASA Release, FY 76 budget briefing background material; Fletcher budget statement, 3 Feb 76, text)

- The Federal Council for Science and Technology, a committee of officials representing major research and development agencies, released "Report on the Federal R&D Program—FY 1976," a summary of 17 R&D agencies' efforts in applying science and technology to national issues. The report stated that the FY 1976 budget included $21 billion for R&D programs, a 15% increase over the FY 1975 R&D budget of $18.8 billion. Of this, approximately $7.4 billion was planned for civilian programs including energy, health, education, agriculture, environment, urban problems, and transportation. Approximately one-fourth the civilian R&D budget was planned for energy, with the largest share of that—$1.55 million—going to the Energy Research and Development Administration. Nine other government agencies would receive increases in R&D budgets, with NASA's FY 1976 budget estimate increasing to $3.539 billion, up from $3.231 billion in FY 1975, an increase of more than $300 million. Space R&D would continue with projects to explore Mars, Venus, and outer planets, as well as the sun. Satellites would continue to be used for earth resources assessment, weather prediction, and surveys of potential geothermal energy sources.

    The Federal Aviation Administration would receive increases for air traffic safety, and Dept. of Defense increases would support development of the B−1 manned strategic aircraft, Trident missile systems, and Navy and Air Force air combat fighters. (Text; OMB, *Budget of the U.S. Govt., FY 1976)*

- NASA management had permitted inflight use of "undesirable components" in the electrical system of the Thor-Delta launch vehicle to preserve the program's schedule, *Aviation Week & Space Technology* reported. At least two Thor-Deltas had been launched with the same electrical-system contamination failure potential that "probably caused the loss" of the U.K.'s *Skynet II A* military communications satellite 18 Jan. 1974. *Av Wk* reported that, although NASA officials considered the policy undesirable, they had used the components until they could be recycled or replaced, in the belief that the risk had been acceptable. On 18 Dec. 1974 the French−West German *Symphonie 1* was almost lost when an apparent electrical contamination problem occurred in the Thor-Delta's second stage. *Av Wk* said that Symphonie program management had not been aware of the vehicle's electrical-system failure potential. (*Av Wk*, 3 Feb 75, 44)

- A 9.6-gram finely powdered lunar sample arrived at Marshall Space Flight Center where scientists would measure in a vacuum the sample's thermal conductivity, a fundamental property governing heat flow through the surface of the moon. The data, correlated with data obtained from previous studies on simulated lunar samples, would be used to develop a thermophysical model of the moon, helping scientists better to understand the moon's origin, history, and relationship to the earth. (MSFC Release 75−41)

- Kennedy Space Center awarded a $79 991 contract to the Univ. of Florida's Institute of Food and Agriculture Sciences for research on freeze-temperature prediction. Under the terms of the contract, a follow-on of a 1973 contract to study the feasibility of using remote-sensing devices for temperature measurement and evaluation, the Univ. of Florida would receive satellite and aircraft data from KSC and the National Oceanic and Atmospheric Administration that would provide actual measurements of leaf and ground temperatures at specific times and places. Following calibration, the data would be combined into a temperature-prediction model. Then, using known variables such as current temperature, humidity, time, cloudcover, and wind velocity, the user could accurately forecast where freezing temperatures would occur. (KSC Release 21–75)

*3–11 February:* A delegation representing the U.S.S.R. Communications Administration met in Washington with representatives of the International Telecommunication Satellite Organization (INTELSAT) to discuss preliminary arrangements for the U.S.S.R. to use INTELSAT satellites for communications between the Soviet Union and the U.S., Canada, and Mexico. According to preliminary agreements, INTELSAT would provide the U.S.S.R. with "occasional use" of space segment capacity in the INTELSAT system for preassigned voice-grade services, television relays, and digital transmission. The U.S.S.R. Communications Administration was constructing an earth station to work with the INTELSAT system beginning in mid-1975. (INTELSAT release, 28 Feb 75)

*4 February:* Air Force Systems Command's Aeronautical Systems Div. announced plans to begin flight evaluation of fan blades constructed of advanced composite materials. The new blades, made of an aluminum alloy matrix enforced by silicon carbide-coated boron filaments, would be flown on an F–111 aircraft over a 2- to 3-yr period at the Air Force Flight Test Center in California. The composite blades, 40% lighter than conventional titanium blades, were expected to increase operating efficiency by reducing rotating mass, increasing tip speeds, and eliminating part-span shrouds. (AFSC Release OIP 301.74)

- Edward N. Cole, retired president of General Motors Corp., was designing a freight carrier, called the Huskie, twice the size of the Boeing 747, the *Christian Science Monitor* reported. Cole, whose plan was still on the drawing board, envisioned a fleet of 300 Huskies carrying freight around the world, each aircraft flying 15 hr a day. Because fast loading and unloading was critical to the plane's economic success, an automated modular loading and unloading system was also being designed to permit the craft to take off again within 30 min of landing. With the price tag for the fleet estimated at $13 billion, Cole was counting on government interest to bolster his plans. *CSM* reported that Dept. of Defense officials were interested in a fleet of privately built cargo planes—with government-insured mortgages—that could be commandeered by the military in an emergency. (AP, *CSM*, 4 Feb 75, 5B)

- NASA announced that General Counsel R. Tenney Johnson would leave NASA to become general counsel of the Energy Research and Development Administration, effective 16 Feb. Johnson had been with NASA since May 1973. NASA Deputy General Counsel S. Neil Hosenball had been named acting General Counsel. (NASA Release 75–32)

- Academician Anatoly A. Blagonravov, one of the Soviet Union's leading space scientists, died in Moscow at the age of 80. Blagonravov, chairman of the Soviet Academy of Sciences' Commission for Space Research and head of the State Research Institute of Engineering Studies, was one of the key scientists responsible for launching the world's first satellite, *Sputnik 1*, 4 Oct. 1957. Blagonravov had begun his career in the military and much of his early work was on the development of automatic infantry weapons. He later turned to the development of spacecraft, making major design contributions to the Soviet Union's *Lunik I*, launched 2 June 1959, as the first man-made object to fly by the moon. Blagonravov also represented the U.S.S.R. in many international organizations: He had been vice-president of the International Space Research Committee, permanent U.S.S.R. representative on the United Nations Scientific and Technical Subcommittee for the Peaceful Use of Space, and permanent deputy Soviet representative on the U.N. Space Committee.

  In 1962 Blagonravov held discussions with Dr. Hugh L. Dryden, Deputy Administrator of NASA, during which details for exchange of satellite data were worked out. (Moscow Domestic News Service, FBIS-Sov, 7 Feb 75, U1; *W Post,* 7 Feb 75, C8; McElheny, *NYT*, 6 Feb 75, 34)

*5 February:* Bills were introduced in the Senate (S. 573) by Sen. Frank E. Moss (D-Utah), and in the House of Representatives (H.R. 2931) by Rep. Olin E. Teague (D-Tex.), "to authorize appropriations to the National Aeronautics and Space Administration for research and development, construction of facilities, and research and program management, and for other purposes." Bill S. 573 was referred to the Committee on Aeronautical and Space Sciences. Bill H.R. 2931 was referred to the Committee on Science and Technology. (*CR*, 5 Feb 75, H659, S1430)

- NASA's new Low Cost Modular Spacecraft (LCMS), a recoverable compartmented spacecraft being considered for use aboard the Space Shuttle, was demonstrated at a news conference at Rockwell International Corp.'s Space Division. The LCMS consisted of an instrument and spacecraft assembly joined by a transition ring. Scientific instruments or earth-resources sensors could be attached to the LCMS for specific missions. Designed by Goddard Space Flight Center engineers, the LCMS would fit into the 5- by 18-m payload bay of the Orbiter and also could serve as a platform for as many as 14 satellites. The system could be placed in orbit by the Shuttle and later serviced or recovered and returned to earth for needed repairs. Dr. Robert G. Wilson, Director of Advanced Payload Analysis at NASA Headquarters, said that a recoverable modular satellite system such as LCMS could save as much as $100 million annually over the cost of nonrecoverable satellites. (JSC *Roundup*, 14 Feb 75, 3; *Goddard News*, Feb 75, 8)

*6 February:* NASA launched *Sms 2* (SMS-B Synchronous Meteorological Satellite) into transfer orbit from the Eastern Test Range at 5:04 pm EST. Launched for the National Oceanic and Atmospheric Administration, the three-stage inertially guided Thor-Delta booster placed the satellite in an orbit with a 36 810-km apogee, 197-km perigee, 651.0-min period, and 23.9° inclination. Originally scheduled for 30

Jan., the launch had been delayed to replace the Inertial Measurement Unit of the launch vehicle's flight-control system with a newer more reliable unit. The spacecraft performed normally during the transfer orbit, and attitude and reorientation maneuvers were performed as planned. The apogee boost motor was fired on second apogee at 9:43 am EST 7 Feb. placing *Sms 2* in a synchronous orbit; orbital parameters were 36 685-km apogee, 35 680-km perigee, 1456.4-min period, and 1.1° inclination. By 10 Feb. the spacecraft was drifting westward at 6° per day toward the final operational station of 115° west longitude. Over several days beginning 10 Feb., the onboard auxiliary propulsion system was fired to place the spacecraft into proper attitude for operational use. First photos were transmitted 11 Feb.

NASA objectives—to launch the spacecraft into a synchronous orbit of sufficient accuracy to enable the spacecraft to accomplish its operational mission requirements, conduct an in-orbit evaluation and checkout of the spacecraft, and turn operational control over to NOAA—were met, and the mission was adjudged successful 22 April. NASA turned over control of *Sms 2* to NOAA 10 March.

The 628-kg cylindrical spacecraft carried five instruments to provide high-quality day and night cloudcover data; take radiance temperatures of the earth's atmosphere; measure proton, electron, and solar x-ray fluxes, and magnetic fields; transmit processed data from central facilities to regional stations; and transmit environmental information to NOAA from the thousands of manned and unmanned data-collection platforms on land, rivers, lakes, and the sea.

*Sms 2*, together with its sister craft *Sms 1* (launched 17 May 1974), provided continuous coverage of the Western Hemisphere, transmitting weather photos every 30 min. SMS photos would be made into film loops of clouds moving over oceans and land masses to help meteorologists determine what types of cloud formations and weather conditions might cause the destructive tornadoes and hurricanes that frequently occur over the eastern U.S.

*Sms 2* was the second in a series of two operational prototypes and one operational spacecraft (SMS-C, to be called *Goes 1* in orbit) developed and funded by NASA to meet the requirements of the Dept. of Commerce's Geostationary Operational Environmental Satellite (GOES) system. Under a NASA-DOC agreement, NASA was conducting a program to develop improved sensors and techniques—based on technology developed in its Tiros, Nimbus, and Applications Technology Satellite programs—for DOC's operational program. Follow-on GOES spacecraft were being planned. Goddard Space Flight Center, which managed the SMS program under the direction of NASA's Office of Applications, also had responsibility for the Thor-Delta launch vehicle. (NASA MORs, 9 Jan, 10 Feb, 23 April 75; NASA Releases 75-6, 75-27; GSFC *Wkly SSR*, 6-12 Feb 75)

- DOD announced the award of a $3.6-million contract to Aerojet Electrosystems Co. for design, development, and delivery of one prototype and one flight microwave temperature sounder to be used on military meteorological satellites between 1977-1980. Under the contract, Aerojet would build and test one mechanical-thermal simulator and associated ground equipment for calibration and

- testing, and would develop a software data package to convert satellite data to temperature profiles.

  The microwave sounder would be designed to measure atmospheric temperature from the surface of the earth to altitudes above 30 km without regard to cloudcover. (AFSC Release OIP 028.75)

- France launched *Starlette*, a passive geodetic satellite, from Kourou, French Guiana, on a Diamant B/P.4 launcher. The satellite entered orbit with a 1137-km apogee, 804-km perigee, 104.5-min period, and 49.8° inclination. *Starlette*, with an inner core of uranium 238 to give it a high mass and reduce gravitational effects on the spacecraft, would reflect laser emissions directed at it from ground stations to study the earth's gravitational field, elasticity of the earth, and moments of the earth's poles.

  The launch was the first for the Diamant booster developed by the Centre National de Recherches Scientifiques. (GSFC *Wkly SSR*, 6–12 Feb 75; *Av Wk*, 9 Dec 75, 55; 17 Feb 75, 19)

*7 February:* NASA had issued a request for proposals for the procurement of telecommunications services provided by a Tracking and Data Relay Satellite System (TDRSS) developed and operated by industry to meet NASA requirements, NASA announced. TDRSS, consisting of two specialized relay satellites in synchronous earth orbit and a U.S.-based ground terminal, would relay nearly continuous data and voice commands to and from mission spacecraft and the ground control center. The network could support all earth-orbiting spacecraft below 5000 km, including the Space Shuttle and Spacelab, and provide 85% coverage, compared with the current 15% coverage by the conventional ground tracking system. (NASA Release 75–37)

- An A–7D aircraft with a DIGITAC digital flight-control system—first of its kind to be flight-tested by the Air Force—made its first flight at Air Force Flight Test Center. DIGITAC was part of a program to investigate the feasibility of installing digital flight-control systems in current and future tactical fighter aircraft.

  The system, featuring a pilot-selectable control mode, would increase flight-control reliability with its component self-test capability, improve combat effectiveness by permitting multimode aircraft-handling qualities, and cost less because of the system's inherent flexibility and the general availability of its digital components. (AFSC Release OIP 057.75)

- Dr. Hertha Firnberg, Austrian Minister of Science and Research, and Roy Gibson, Director General of the European Space Research Organization (ESRO), signed an agreement in Vienna permitting Austria to participate in individual European space programs—including Spacelab—without obligation of membership. (*Spacelab Newsletter* 75–1)

- A decrease in the amount of ozone in the stratosphere would cause an increase in skin cancer in white (the most susceptible) Americans, Walter Sullivan wrote in a *New York Times* article. The findings were based on the combined results of a 1974 NASA study that used instrumented aircraft to profile ozone and nitrogen oxide in the stratosphere, the National Cancer Institute's 1974 Third National Cancer Survey, and the Dept. of Transportation's 3-yr study "The Effects of Stratospheric Pollution by Aircraft" [see 21 Jan.].

The NASA study, which used high-altitude U-2 aircraft, confirmed that atmospheric ozone increased from the equator to the poles and was also influenced by regional and seasonal variations. Ozone molecules above the latitude of Houston were only three fifths as plentiful as above the latitude of Minneapolis. Correspondingly, the Cancer Institute survey showed an annual incidence of nearly 400 skin-cancer cases per 100 000 whites in Dallas, Tex., and less than half that figure in the Minneapolis area.

These data, when combined with the DOT study, indicated that the number of skin-cancer cases in the U.S. would increase by 6000, or 1%, a year if 120 supersonic transports operated in the stratosphere for 4 hr daily, thus depleting the zone by 0.5%. (Lowenstein et al., *Journal of Atmospheric Sciences*, Nov 75, 2185– 90; Sullivan, *NYT*, 7 Feb 75, 11)

*7 February–1 March:* The eight prime and backup Soviet cosmonauts training for the July Apollo-Soyuz Test Project joined with U.S. prime and backup astronauts in the U.S. to continue training for the joint mission. The cosmonauts—Aleksey A. Leonov, Valery N. Kubasov, Anatoly V. Filipchenko, Nikolay N. Rukavishnikov, Vladimir A. Dzhanebekov, Boris D. Andreyev, Yury V. Romanenko, and Aleksandr S. Ivanchenkov—arrived in Washington, D.C., 7 Feb. and traveled to Kennedy Space Center the following morning.

The group toured KSC 8 Feb. giving the cosmonauts their first glimpse of the U.S. launch complex. Col. Leonov commented that the pad from which Alan B. Shepard was launched in 1961 "is a very simple launch stand compared to the ones we use now."

Following a tour of Disney World in Florida 9 Feb., the cosmonauts spent 10 Feb. inspecting the Vehicle Assembly Building, where the Apollo would be stacked on the Saturn IB launch vehicle; the Launch Control Center firing room, from which the rocket would be controlled; and Launch Complex 39 Pad B, from which the Apollo spacecraft would be launched. Press coverage of the first Soviet visit to KSC noted that previous invitations for Soviet officials to visit KSC had been turned down, apparently because of Soviet unwillingness to reciprocate with invitations to Baykonur Cosmodrome. But in May U.S. astronauts would return the visit by going to the Soviet launch site.

The astronauts and cosmonauts flew to Johnson Space Center to begin a 2-wk training period that included exercises in command and docking module simulators and mockups; joint language-training sessions; briefings on experiments, contingencies, and mission rules; and practice run-throughs of the planned joint activity days. Following a press conference at which U.S. and Soviet ASTP crew members were introduced to JSC employees, the Soviet cosmonauts returned to the U.S.S.R. in two groups, on 28 Feb. and 1 March. (NASA Release 75– 34; JSC Release 75– 07; NASA ASTP Status Report No. 7; *Spaceport News*, 21 Feb 75; JSC *Roundup*, 28 Feb 75, 1; McElheny, *NYT*, 10– 17 Feb 75)

*10 February:* More than 19 000 aerospace workers struck McDonnell Douglas Corp. facilities in Missouri, California, and Florida, saying that their demands for increased wages and benefits had not been met. In McDonnell's St. Louis plant, where the Defense Dept.'s F– 4 Phantom jets and F– 15 Eagle fighters were built, 12 000 employees walked off the job. The press quoted company officials as saying that the strike would slow, and perhaps stop, production. Only the few aircraft ready for final assembly and flight testing would be completed. About 7000 workers walked out of California facilities, including Vandenberg Air Force Base. At Kennedy Space Center, 200 McDonnell Douglas employees struck, causing postponement of the Telesat−C launch scheduled 6 March on a McDonnell-built Delta launch vehicle. (*C Trib*, 11 Feb 75; NASA PIO, interview, 14 Feb 75)

- Formation of a Federal interagency group—the Federal Interagency Task Force on Inadvertent Modification of the Stratosphere (IMOS)—to investigate the relationship of freons (fluorochlorohydrocarbons) to ozone reduction in the stratosphere was announced by the Chairman of the Council on Environmental Quality, Dr. Russell W. Peterson, and the Chairman of the Federal Council for Science and Technology, Dr. H. Guyford Stever. The group included representatives from NASA; the Departments of Agriculture, Commerce, Defense, Justice, and Transportation; Dept. of Health, Education, and Welfare; Energy Research and Development Administration; Environmental Protection Agency; National Science Foundation; and Interdepartmental Committee for Atmospheric Sciences.

  IMOS would prepare a report summarizing atmospheric, medical, and ecological information on the freon/ozone relationship. The report would also evaluate possible economic impacts and alternatives to industry, consider what Federal action could be taken, and propose a Federal program to resolve the issues. (Fed Council for Sci and Tech Release 1)

*11 February:* Dr. John F. Clark, Goddard Space Flight Center Director, and Dr. James W. Smith, president of Greenbelt Homes, Inc., signed a cooperative agreement to install experimental solar heating units in four four-family housing units in Greenbelt, Md. During the first phase of the project, all four of the units would be instrumented to measure heat loss and fuel-consumption characteristics, and two of the four would be fully insulated. Based on the cost-effectiveness of the first phase, the two insulated buildings would be equipped with solar heating units designed to augment existing hot-water radiator systems. The remaining two units would serve as control for comparison. The solar units, which would be designed at GSFC, would be installed in time for the 1975−1976 heating season. (*Goddard News*, March 75, 1−2; GHI manager's ofc, interview, 5 Oct 1976)

- The first airborne terminal for an Air Force communications satellite system was being installed in a C−141 aircraft as part of a flight-test program to develop a system for transmitting high-priority messages throughout the world via Air Force repeater satellites. In addition to the C−141 terminal, the system included six additional airborne and four ground terminals. Results of the 6-mo test would help DOD decide whether the terminals should be produced in quantity. (AFSC Release OIP 303.74)

*12 February:* Water vapor was detected for the first time deep in the atmosphere of Jupiter in October 1974 by a group of Univ. of Arizona scientists aboard NASA's C−141 Airborne Infrared Observatory, Ames Research Center announced. The data, combined with data gathered by the *Pioneer 10* and *11* flybys of the planet in 1973 and 1974, added to the speculation that organic compounds being formed in Jupiter's atmosphere might account for the red, brown, and orange coloring of the planet's clouds.

The discovery was made with a 91.5-cm infrared telescope, newly installed aboard the C−141, which was flying above virtually all the water vapor in the earth's atmosphere. Scientists would use the new data to predict Jupiter's temperature and pressure, develop a structural model of the planet's atmosphere, and determine Jupiter's character at the time of its formation. (ARC Release 75−3; ARC Airborne Astronomy Prog Mgr, interview, 26 July 76; Larson et al., *Astrophysical Journal*, 1 May 75, L137−40)

- Communications Satellite Corp. released its 1974 annual report: earnings during 1974 had increased to a record $44 918 000, or $4.49 per share, up from $36 299 000 or $3.63 per share in 1973.

    In his "Message for the Shareholders," ComSatCorp President Joseph V. Charyk attributed the growth to the continued increase in communications traffic over ComSatCorp's global network. Excluding the U.S. mainland-Hawaii service, the number of half-circuits leased by ComSat to U.S. international carriers increased during the year by about 20%, from 2933 at the end of 1973 to 3510 at the end of 1974.

    Responding to the increased demand for service, ComSatCorp and its partners in the International Telecommunications Satellite Organization had continued to improve and expand the system during 1974. A sixth Intelsat IV satellite, *Intelsat IV F8*, had been launched 21 Nov. Three higher capacity Intelsat IV−As were scheduled for launch during 1975, and three additional Intelsat IV−As were scheduled for launches beginning in 1977. Even higher capacity Intelsat V satellites were being planned for use in the 1980s.

    The ground network had been expanded to a year-end total of 104 antennas at 84 earth station sites in 60 countries. Construction of additional antennas was begun at ComSatCorp-operated stations in Maine and West Virginia. (ComSat Release 75−7; ComSatCorp 1974 Annual Rpt, 3 March 75)

*14 February:* High-temperature core reactors, fueled by fissioning uranium plasmas, could be used to "burn up" or eliminate radioactive wastes produced in the fission process, NASA announced. In a NASA-sponsored program to develop advanced nuclear-powered rocket propulsion, scientists had found that long-lived radioactive wastes could be transformed into harmless materials by bombarding them with neutrons. The uranium-compound fuel could be circulated and returned to the reactor to burn up the materials produced in the fuel. Computer analyses showed that a gaseous fuel reactor could establish a balance in rates of production and elimination of the waste produced after 3 yr of operation; once this equilibrium was established, no additional long-lived radioactive wastes would be produced. The gaseous fuel reactor also could be designed to burn wastes from conventional nuclear-fission reactors. (NASA Release 75−44; Schneider et al., *Nuclear Technology*, Sept 75, 34−50)

- Marshall Space Flight Center announced the award of a 10-mo $149 325 contract to Econ, Inc., to study the technical and economic feasibility of two different solar power plants, both in space and on the ground, and compare them with conventional terrestrial plants expected to exist in the future. The first system was an orbiting system which would generate electrical power from the sun and transmit it to industries and cities on earth. A second kind of solar-power generating system, located on a remote area of the earth, could transfer power via an earth-orbiting satellite to users thousands of kilometers away. Econ, Inc., would use results from earlier concept studies to define requirements for payload packaging, development and checkout, and resupply for launch on the Space Shuttle. (MSFC Release 75−37)

- The government of Indonesia selected Hughes Aircraft Corp. to build a satellite communications system linking 120 million residents of the Indonesian islands. Under the $23.6-million contract signed in Jakarta, Hughes would build two satellites, a master control system, and nine earth stations. Thirty additional stations would be built to Hughes specifications by two other U.S. firms. The satellites were tentatively scheduled for launch in fall 1976. (*LA Times*, 16 Feb 75)

*14 February−17 June:* Flight Research Center flew 16 flights using an instrumented B−57 to gather detailed information on jet streams, thermal turbulence, and mountain waves over the western U.S. as part of the Measurement for Atmospheric Turbulence program. Onboard instruments measured velocity and acceleration in the various samples as the B−57 flew through the turbulence for 10 min on a straight-line course at altitudes up to 15 000 m. Langley Research Center, which had been making similar flights over the eastern U.S. since March 1974, would analyze the data for use in making future aircraft safer in turbulent air. (FRC Release 4−75)

*15 February:* The Indonesian government concluded a $120.6-million (4.6 billion Belgian francs) agreement with the General Bank of Belgium to finance construction of two communications satellites, a satellite control station in Jakarta, and 39 comsat network stations throughout the Indonesian islands. (Jakarta Domestic Service, FBIS−Indonesia, 19 Feb 75, N1)

- The U.S. could land men on Mars in 10 yr if a commitment similar to President Kennedy's pledge in 1961 to make a moon landing were made, Dr. Wernher von Braun, former NASA Deputy Associate Administrator, said in a speech at the Johns Hopkins Univ. Applied Physics Laboratory. Dr. von Braun, vice president for engineering and development at Fairchild Industries, Inc., said that nuclear-powered command ships could be sent to orbit the planet while a lander traveled to the Martian surface. The cost of the program would approximate the near-$25 billion cost of the Apollo program. (B *Sun*, 16 Feb 75, B16)

- Soviet and American fraternizing at Cape Canaveral and Florida's Disney World in recent weeks were evidence of how much the world had changed since the 1950s, a *New York Times* editorial said. "What was once a ferociously pursued race in space has given way to détente in space," with the preparations for the joint Apollo-Soyuz Test Project in July. Skeptics, complaining that Moscow would be the "big net gainer" in the project, were correct only from a narrow intelligence point of view. They missed the main point for détente in space. Neither the U.S. nor the Soviet Union had resources sufficient for fruitful human activity in space for decades ahead. "It is simply enlightened selfishness for Washington and Moscow to work together in space." And this was only a start. What was needed even more was "broadly based cooperation in space efforts embracing the personnel and talents of all nations for the benefit of all humanity." *(NYT,* 15 Feb 75, 28)

*18 February:* U.S.S.R. and India signed a 5-yr agreement on scientific exchange and cooperation between the Soviet Academy of Sciences and India's Dept. of Science and Technology. The agreement provided for wider cooperation between the two countries in various fields of science and technology, and for the exchange of delegations and information. (Tass, FBIS—Sov, 6 March 75, J1)

*19 February–19 March:* Hairline cracks were discovered in the holddown fittings on the fins of the Saturn IB launch vehicle scheduled to launch the Apollo spacecraft in the Apollo-Soyuz Test Project in July. After an earlier routine inspection at Michoud Assembly Facility had turned up one crack, a more detailed inspection at Kennedy Space Center 19 Feb. revealed 25-cm cracks on the fittings of two of the eight fins. A decision was made to replace the two fins. The cracks would not affect flight, but were in an area carrying the weight of the vehicle as it sat atop the mobile launcher. During a later inspection, cracks also were found in the remaining six fins, making it necessary to replace all eight.

Replacement work began 11 March. Engineers strengthened the cracked areas by placing compressive stresses in the surface of the material, by pneumatically pounding it with a small bundle of rods. Six of the fins were replaced by 15 March.

The spacecraft was placed atop the Saturn IB on 17 March and replacement of the remaining two fins was completed by 19 March. (KSC Release 27–75; NASA Release 75–50, 75–57; *Marshall Star,* 19 March 75)

- The Air Force System Command's Materials Laboratory had successfully static-tested a full-scale F–15 aircraft wing constructed of advanced composite materials, AFSC announced. The wing had withstood a maximum load 45% greater than expected during actual flight, before giving way at the point where the titanium rear spur flange and upper skin were bolted together. The composite wing, 18% lighter than an all-metal wing, was made of a brown epoxy thin honeycomb sandwich material reinforced with graphite-epoxy stiffeners. The ribs were graphite-epoxy and the spurs were metal. The wing, the largest aircraft wing ever built of advanced composite materials, was 4.9 m wide at the root chord and 1.5 m wide at the tip chord, and had a 4.3-m span.

  The program had begun in May 1971 with an $8.2-million contract award to McDonnell Douglas Corp. for development of the wing, which was expected to reduce fuel requirements by increasing range, speed, and maneuverability. (AFSC Release OIP 021.75)

- Despite the immediate need for aircraft noise-abatement regulations, little or nothing had been done, Sen. John V. Tunney (D–Calif.) said in a speech on the floor of the Senate. The Federal Aviation Administration was considering proposals to curb jet-fleet noise by retrofitting engine nacelles of JT–3D and JT–8D aircraft with sound-absorbing material. Delay had come in part from industry, which favored an alternate NASA refan program costing five times that of retrofit and not technologically feasible for several years. In addition, the NASA program could not quiet the JT–3D aircraft—the Boeing 707s and McDonnell Douglas DC–8s—which made up the noisiest fourth of the nation's jet fleet. Sen. Tunney said that the potential benefits of the NASA program did not justify waiting until the technology was ready at a cost five times greater than retrofitting. If FAA was to prove responsible in the area of aircraft noise abatement, it could not "succumb to industry pressure against retrofit." A retrofit regulation should be put into final form at the first available opportunity. (*CR*, 19 Feb 75, S2160)

*20 February:* Some "very heavy" influential politicians had been rallying around NASA's "little known and unglamorous" Earth Resources Technology Satellite [renamed Landsat 14 Jan.] program, inducing President Ford to overrule his budget advisers and provide money for a third spacecraft in 1977, Arlen J. Large said in the *Wall Street Journal*. It wasn't only NASA but "the ERTS community—geographers, foresters, pollution fighters and land-use planners" that lobbied for a third satellite to continue the work of *Landsat 1* and *2* (launched 23 July 1972 and 22 Jan. 1975). The Office of Management and Budget had wanted to delay funding until NASA developed cameras with better resolution, but the "lobbying blitz" convinced President Ford to overrule OMB; money for the satellite was included in the FY 1976 budget. (Large, *WSJ*, 20 Feb 75, 44)

• *Intelsat IV F-6* communications satellite was destroyed by a NASA range safety officer 415 sec after launch from Eastern Test Range when the Atlas-Centaur test vehicle tumbled out of control. Lifting off at 6:35 pm EST, the mission had been normal to Atlas booster-engine cutoff at 138 sec. Then telemetry data indicated several malfunctions, including electrical problems and loss of altitude. After destruction, the vehicle impacted in the Atlantic Ocean 800 km downrange from the launch site. An investigation of the failure was begun under the direction of Lewis Research Center and the mission was officially judged unsuccessful 6 March. The failure was the first for an Atlas-Centaur since May 1971, when a circuit problem in the Centaur ended the launch of the Mariner 8 Mars probe.

The 1400-kg *Intelsat IV F-6*, launched by NASA on a cost-reimbursable basis for the Communications Satellite Corp. on behalf of the 89-member-nation International Telecommunications Satellite Consortium, was intended to be the primary spacecraft over the Indian Ocean region, with *Intelsat IV F-5*, launched 13 June 1972, becoming the backup. The failure of Intelsat IV F-6 would have no immediate effect on Intelsat global communications; *Intelsat IV F-5* could provide the primary function and *Intelsat III F-3*, launched 6 Feb. 1969, was still capable of providing limited backup.

The *Intelsat IV F-6* mission had been scheduled originally for launch in February 1973 but experienced a series of delays from problems in both the satellite and the launch vehicle. (NASA MORs, 26 Feb 75, 13 March 74; PAO, interview, 21 Feb 75; NASA Releases 73-83, 75-7, 75-47; NASA "Note to Editors," 3 May 73, 27 June 75; *SBD*, 24 Jan 75, 127)

*21 February:* The unexplained loss of calcium by astronauts during prolonged manned missions "seems to be a slow persistent loss," Dr. David L. Winter, NASA Director for Life Sciences, said during a press briefing. Data indicated calcium had been lost in space at the rate of 0.5% a month, a rate that showed "no evidence of leveling off." Bones tended to lose calcium at varying rates, weakening selective bones.

Scientists were studying various ways, including diet management, to compensate for the calcium loss.

Dr. Winter said that the observed decrease in red blood cells during the Skylab missions was not because of red-cell destruction but rather a slowdown of the red-cell production. "We seem to have stumbled on a way in which red blood cell production is suppressed." This mechanism, unknown before, would be a very interesting and important process to study.

Physical criteria for the selection of astronauts for the Space Shuttle would vary according to function, Dr. Winter said. Pilots would be selected using very rigid criteria; these would not be imposed upon astronauts, i.e. mission specialists, whose function would be primarily scientific. Because the objective for Shuttle missions would be to get the best qualified scientists, it would be impossible to predict what size, shape, age, or sex these scientists would be. The purpose of the Shuttle was to permit anyone with a legitimate need to go into space; therefore, NASA's approach would be to broaden the selection criteria as much as possible. (Transcript)

- Dr. Philip Handler, 18th president of the National Academy of Sciences, had been elected for a second 6-yr term, NAS announced. Dr. Handler had served as NAS president since 1 July 1969. Before that he had been chairman of the Department of Biochemistry at Duke Univ. Medical Center.

  NAS, established in 1863 by an act of Congress, was a private organization of over 1000 scientists and engineers frequently called upon to advise the Federal government on policy questions related to science and technology. (NAS Release, 21 Feb 75)

- DOD announced the award of a $3 500 000 firm-fixed-price contract to TRW, Inc., for software support of the Space Shuttle. (DOD Release 83-75)

*24 February*: Design and fabrication of the first Space Shuttle Orbiter was progressing well, with all major structural elements—forward fuselage, tail, wings, mid fuselage—scheduled for final assembly in mid-1975, NASA Deputy Director of the Space Shuttle LeRoy E. Day told the Canadian Science Writers Association in Toronto, Canada. The main rocket engine, the most difficult development item, was being built on a 10-yr technology base in which government and industry had invested nearly $100 million. During 1974, initial component engine firings had been made, and firings of the first integrated subsystems test bed engine were scheduled for July.

The emphasis for the expendable external tank had been on simplified low-cost production. An inexpensive spray-on insulation had been developed to protect the aluminum skin from excessive temperatures during ascent.

In summary, NASA and its contractors were making steady progress, holding within cost and schedule commitments. The first Orbiter would be rolled out in 1976, with its first captive flight on the back of the Boeing 747 aircraft scheduled for the second quarter of 1977. The first approach and landing tests, in 1977, would be made by separating the Orbiter from the 747 at about 8500 m, permitting the Orbiter to glide down and land on the runway at Edwards Air Force Base. The second Orbiter would be used for the first orbital flight in 1979, with six developmental flights scheduled before the Space Shuttle became operational in 1980. (Transcript)

- The recent detection of water vapor, and therefore oxygen, on Jupiter by Univ. of Arizona scientists aboard NASA's C−141 Airborne Infrared Observatory [see 12 Feb.] could stimulate additional support for a probe mission to obtain more detailed information, *Aviation Week and Space Technology* reported. Scientists had already known that the proportions of helium and hydrogen on Jupiter were approximately the same as on the sun. The additional information that oxygen existed on Jupiter would strengthen the belief that the planet's composition was similar to that of the sun, and that the sun and the planet had been formed from the same raw material. The conclusion was important to many scientists who believed that earth's atmospheric composition had evolved from one similar to Jupiter's. If this were true and if Jupiter's atmosphere had changed little from its primitive state, then the Jovian atmosphere could serve as a laboratory for studying the evolution of life on earth. This new information could influence NASA scientists charged with making recommendations for new starts. (*Av Wk*, 24 Feb 75, 58)
- Tokyo Univ.'s Institute of Space and Aeronautics launched *Srats* (Taiyo)—Solar Radiation and Thermospheric Structure Satellite—from Uchinoura Space Center at 2:25 pm local time after a 10-day delay because of faulty instrumentation. The three-stage MU3C liquid-fueled launch vehicle placed the 86-kg spacecraft into an orbit with a 3130-km apogee, 247-km perigee, 120.2-min period, and 31.5° inclination. The purpose of the satellite was to observe solar and cosmic phenomena. By 25 Feb. the satellite was reported to be functioning normally, with two NASA tracking stations—one in Ecuador and one in Chile—reporting signals from *Srats*. (Kyodo, FBIS−Japan, 18−26 Feb 75)

*24–27 February*: The American Institute of Aeronautics and Astronautics held its 11th annual meeting and technical display in Washington, D.C. "Frontier Technology and Shuttle Country" was the theme of the session chaired by Grant L. Hansen of General Dynamics Corp. An Aerospace Day was observed 27 Feb.

Sessions were held in four areas: transportation efficiency, national defense, space applications, and frontier technologies. The more than 500 attendees of the technical sessions listened to papers on computers in aerospace, short-haul aircraft systems, advanced communications, and satellite technology.

Harris M. Schurmeir, Jet Propulsion Laboratory's manager of the Mariner Jupiter−Saturn 1977 project, delivered the von Kármán Lecture 24 Feb., "Planetary Exploration: Earth's New Horizon." Schurmeir said that, despite American virtuosity in the unfolding drama of solar-system exploration, the never-abating high launch rate of the Soviet Union would make them competitors if the U.S. failed to exercise its mission options in the late 1970s and 1980s.

On 25 Feb. Hans E. W. Hoffman, managing director of West Germany's ERNO Raumfahrttechnik GmbH, said that the development of Spacelab was exactly on schedule, with a complete engineering model to be delivered to NASA in April 1978. The first flight unit would be delivered 1 yr later.

The first mission of the Spacelab would be a joint U.S.−European venture. A call for ideas for experiments had netted 243 proposals, from which a selection would be made; the selection process for European payload specialists would begin in 1975.

Dr. James C. Fletcher, NASA Administrator, addressed the gathering 26 Feb. He noted that, although aerospace technology had been one of the "key drivers" in retaining the U.S. lead in world technology, a feeling of the importance of technology was "not universally shared." A decline in the national emphasis on technology "has much to do with the decline in productivity in this country." The increasing cost and complexity of applying new technology had become a vexing national problem. Finding new ways to reduce costs was a key management task that the aerospace business must face.

Caspar W. Weinberger, Secretary of Health, Education, and Welfare, said in an address 26 Feb. that satellite-delivered services in Alaska had been so successful that the Alaskan Federation of Natives —a highly representative broad-based grassroots organization—had made them the organization's top priority. The goal was to achieve an operational satellite service for health, education, and other communications needs.

The formation of the Public Service Satellite Users Consortium, consisting of health and educational organizations, represented the first effort to organize a market of small independent users sufficient to attract private investment capital for large-scale technology to meet their collective needs. HEW was investigating the possibility of a "social service satellite" to meet the needs of various public and social service communities by means of an inexpensive ground receiver. The satellite could dramatically reduce the cost of such services while greatly increasing the quality.

At the Honors Night Banquet 26 Feb., Chairman Hansen was installed as the 13th president of the AIAA. The 19 newly elected fellows included Hans M. Mark, Director of Ames Research Center, and William H. Phillips of Langley Research Center. Other awards presented at the banquet included:

| Award (purpose) | Recipient (Affiliation) | Citation |
| --- | --- | --- |
| Goddard Award (for contributions in propulsion energy conservation) | Gordon F. Holbrook (General Motors Corp.) George Rosen (United Aircraft Corp.) | For leadership in development of turbo-propeller propulsion systems |
| Robert M. Losey Award (for contributions to meteorology as applied to aeronautics) | Paul W. Kadlec (Continental Air Lines, Inc.) | For identifying factors indicating clear air turbulence and publications on meteorological features of high atmosphere |
| G. Edward Pendray Award (for contributions to aeronautical and astronautical literature) | Prof. William R. Sears (U. of Arizona) | For contributions to aeronautical and astronautical literature since 1938 |
| Sylvanus Albert Reed Award (for contributions to aeronautical engineering design or aeronautical sciences) | Prof. Antonio Ferri (New York U.) | For advances in air-breathing propulsion and airplane design |
| deFlorez Award (for improvement in aerospace, aviation, or astronaut training) | Prof. Hugh H. Hurt, Jr. (U. of Southern Calif.) | For contributions to flight safety by development of military and civilian texts and training programs |
| Louis W. Hill Award (for enterprise and ingenuity in art and science of space flight) | Dr. Rocco A. Petrone (NASA) | For contributions to space flight technology while Director of KSC during last 6 Apollo missions |

| Award (purpose) | Recipient (Affiliation) | Citation |
|---|---|---|
| John Jeffries Award (for contributions to aerospace medical research) | Col. Malcolm C. Lancaster (USAF) | For clinical research in aerospace medicine |
| Spacecraft Design Award (for advancement in spacecraft, launch vehicle, reentry vehicle, or missile design) | Dr. Herbert A. Lassen (TRW, Inc.) | For conceptual design of spin-stabilized *Pioneers 6–11* using high-gain antennas and communications link for attitude reference |
| Space Science Award (for achievements in studies of physics of atmospheres of celestial bodies and dynamics and energy) | Dr. Murray Dryer (NOAA) | For contributions to understanding of interplanetary shock phenomena generated by solar disturbances |
| Lawrence Sperry Award (for contributions to advancement of aeronautics) | Jan R. Tulinius (Rockwell Int'l Corp.) | For development of advanced theoretical aerodynamic design methods |

(*A&A,* Feb 75, April 75; AIAA Releases 10–20 Feb 75; AIAA Bulletin, Aug 74; Fletcher speech, text)

*26 February:* Press briefings on the Apollo-Soyuz Test Project were held at Johnson Space Center. ASTP participants would be conducting experiments in space sciences, life sciences, and applications, ASTP Project Scientist Dr. Thomas Giuli said at an experiments briefing. The five space-science experiments included a soft x-ray experiment to study radiation by scanning across the celestial mode, pointing at known discrete sources in the soft part of the spectrum from 2 kev to 0.1 kev, and investigating 10 specific x-ray sources known to emit only in regions below 2 kev. A helium-glow experiment would investigate temperature and density of the interstellar medium in the vicinity of our solar system by observing two special frequencies of extreme ultraviolet light emitted from the sun after they traveled into interstellar space, hit the interstellar medium, and were reflected back again. Other space-science experiments included a study of the sun's outer atmosphere during an artificial eclipse created by maneuvering the Apollo spacecraft. During the "eclipse," the Soyuz crew would photograph the sun's corona.

Among the earth environmental and applications studies was an experiment to measure neutral atomic oxygen and neutral atomic nitrogen of the earth's upper atmosphere. Another experiment would try to devise a simple means of monitoring the aerosol content of the earth's atmosphere. A water-resources management experiment, planned and coordinated with the government of India, would photograph the Himalayan area to map drainage patterns and water-reservoir locations.

Life-science experiments included a study to observe optical sensations from cosmic rays experienced by the crew, an investigation of mutation in cells grown in space, and studies of the effect of space flight on the ability of microbes to infect humans and of changes in the human immunity system to resist infections under space flight conditions.

At a mission profile briefing, Kenneth A. Young of JSC's Mission Planning and Analysis Div. said that plans called for a 9-day Apollo mission beginning 7.5 hrs after the launch of Soyuz by the U.S.S.R. The planned rendezvous period would be 2 days, with docking scheduled for the 30th inertial orbit, or third day in flight, of the Apollo spacecraft. Following much negotiation, U.S. and Soviet mission planners determined that the two spacecraft would dock over Soviet territory, providing real-time telemetry coverage for the Soviet Union, but at a location where NASA's *Ats 6* communications satellite could provide communications coverage to the U.S. The first handshake would take place over Spain.

The spacecraft would undock on the 5th day and perform several joint experiments while separated. Soyuz was scheduled to land the following day. Apollo would continue unilateral experiments in orbit before landing at the end of the 9th day in space.

Astronaut Richard H. Truly said, during a joint crew activities briefing, that while the Apollo and Soyuz were docked, all communications would be conducted in the language of the listener regardless of the nationality of the speaker. (Transcript)

- Both the Soviet and U.S. Apollo-Soyuz Test Project crews had reached a level of language proficiency where "we can speak to each other," commander of the Soviet crew Aleksey A. Leonov said in an *Izvestiya* interview. Leonov said that the crews' first meeting had been "merely a glance" and "we wondered whether we would come to like each other." In preparing for the mission, the Soviets and the Americans had not only come to treat each other "as crew member to crew member" but had also developed a degree of comradeship. It would be impossible to work without this. In the event of some emergency, if "I have to carry him on my shoulders or he has to carry me, how could I do this unless I respect him and value him as a comrade? For we are virtually candidates of the whole world. Despite the difference of our social formations, despite the contradictions which have existed and continue to exist to this day, we have found points of contact." (Kondrashov, *Izvestiya*, FBIS—Sov, 28 Feb 75, U1)

- NASA announced the signing of a memorandum of understanding with Zaire to build the first African ground station designed to receive earth-resources data from *Landsat 1* (launched as *Erts 1* 23 July 1972) and *Landsat 2* (launched 22 Jan.). Zaire's new station would produce both computer tapes and photographic imagery that would include data on the African continent from Chad to South Africa and from Kenya to the Ivory Coast. Currently, data from that area had been stored by onboard tape recorders for transmission to U.S. ground stations. (NASA Release 75—53)

*28 February:* The preliminary design review of Space Shuttle Orbiter 102 for the first manned orbital flight was completed on schedule. (NASA Gen Mgmt Rev Rpt, 17 March 75, 39)

- Energy Research and Development Administration announced the award of a $2 612 000 contract to Hughes Aircraft Co.'s Hughes Research Laboratories to develop a new mercury valve for use in high-voltage direct-current transmission. The liquid-metal plasma valve was a spinoff from a spacecraft ion engine developed by Hughes for NASA. The valve, a single-anode device that fired a vacuum arc on signal, would be used to convert between alternating current and direct current at the terminals of high-voltage direct-current transmission lines. (ERDA Release 75—25)

- Soviet and French scientists had completed the "Araks" study of the earth's magnetically interlinked points, *Izvestiya* reported, by observing what happened when a bundle of electrons was artificially injected into the magnetosphere. French Eridan sounding rockets—equipped with electron accelerators, plasma generators, and instruments for measuring and recording wave radiation and particle streams—were launched on 26 Jan. and 15 Feb. 1975 from Kerguelen Island. During the ascent the accelerator emitted electron impluses at various angles to the line of magnetic force while the plasma generator emitted a stream of cesium plasma to compensate for the positive charge of the rocket.

  At the same time Soviet and French specialists in Arkhangelsk Oblast made optical observations using supersensitive TV installations, photometers, and electron optical-image intensifiers. Radar and radio spectrographs in Kostroma and Vologda Oblasts observed the dispersal of radio waves and radio emission in the area where the electrons penetrated the dense strata of the atmosphere.

  From the Araks data, scientists hoped to determine the true position of the magnetically interlinked spot and to learn more of the interaction between the electron bundle and its medium. (*Izvestiya*, FBIS−Sov., 13 March 75, E2)

- Aerodynamically designed wind-flow aids could reduce wind resistance to tractor-trailer trucks by up to 24%, NASA announced. In a series of tests conducted in 1974 at Flight Research Center, NASA and the Department of Transportation evaluated performance gains of a 15-m trailer outfitted with five different commercially available drag-reduction devices and two different trailer-cab spacings. The tests had been conducted without and then with the devices, at speeds from 45 to 30 km per hour, to evaluate changes in fuel economy resulting from reductions in air drag. (NASA Release 75−39)

*During February:* NASA biochemist B. C. Walverton began planting large quantities of water hyacinths as part of a study to determine the plant's ability to absorb and concentrate toxic metals and to metabolize various other chemical pollutants. Scientists at NASA's National Space Technology Laboratories had been experimenting since 1971 with vascular aquatic plants, which are equipped with a system of vessels to carry nourishment from the roots to the leaves. Results of the experiments were so promising that NASA, in cooperation with Mississippi state officials and the city of Bay St. Louis, had installed a special system using water hyacinths as a final filtration to remove nitrates, phosphates, and other chemical pollutants from the 245-sq-km Bay St. Louis, Miss. sewage lagoon. The water hyacinths would be harvested and analyzed to determine the amounts and kinds of impurities assimilated; saturated plants would then be recycled, using one of two methods being developed by NASA, into a mixed hydrocarbon fuel similar to natural gas. (NASA Release 74−332)

# March 1975

*1 March:* A *Baltimore Sun* editorial discussed "Distorted Priorities in the R&D Budget." The President's R&D budget for FY 1976 had little chance of passage in its existing form because of the hostility of much of the 98th Congress to some of its proposals. Defense-related R&D would be reduced, the energy proposals would be juggled to give more to solar and geothermal energy, and the proposed large cuts in biomedical research would be restored.

Congress might be more sympathetic to NASA. Once seen by congressional critics as a quasi-military organization or at least "a space circus that siphoned dollars away from more pressing needs," NASA was "beginning to prove itself a valuable scientific tool" in such valuable areas as environmental monitoring. That the Office of Management and Budget had not, as once threatened, scrapped funds for the Landsat–C satellite "may reflect an effort to make NASA more attractive by further pushing the agency into the areas that have gained humanist and environmentalist approval." (B*Sun*, 1 March 75, A18)

*2–7 March:* A meeting at Goddard Space Flight Center of U.S. and U.S.S.R. officials to discuss a cooperative meteorology program recommended that in May–July 1976 each country, within a single 27-day solar rotation, launch two series of sounding rockets—one under quiet magnetic conditions, the other under disturbed magnetic conditions. In 1977 the two countries should hold a rocket-data comparison test in the vicinity of Wallops Island, Va. Both countries should launch more meteorological rockets during atmospheric warming periods. In 1976–1978 both countries should research radiation transfer as a factor in temperature sensing from satellites. From 1975–1978 joint remote-sensing experiments using microwave were proposed, along with exchange of information on frequencies and data format to be used with direct-broadcast equipment. (NASA Gen Mgmt Rev Rpt, 17 March 75)

*3 March:* NASA's Remotely Piloted Research Vehicle (RPRV) was being equipped with landing gear to enable it to land under control of a pilot on the ground, Flight Research Center announced. Previously helicopters and parachutes had been used in midair recovery of the vehicles.

Developed by FRC, the RPRV—a scale-model airplane up to 9 m long—was an economical and safe way to test advanced aircraft. After air launch of the RPRV, a test pilot in a ground-based cockpit used flight controls and instruments to "fly" the RPRV through maneuvers, while a camera provided visual information. First application of the landing gear would be for a spin test of a 3/8-scale model of the Air Force F–15 fighter. (FRC Release 3–75)

- In NASA's 5–10 yr Global Air Sampling Program (GASP), a second instrumented Boeing 747 jumbo aircraft began sampling flights monitoring pollution in the earth's atmosphere. GASP data on carbon dioxide, ozone, water vapor, and particles larger than 0.3 microns would help determine whether the amount of ozone was being altered, whether vapor trails from jets contributed to cloud cover, and the amount of pollution caused by aircraft.

The first GASP-instrumented 747 began flights in December 1974. Four aircraft—one each from United Airlines, Pan American, Qantas Airways, and Scandinavian Airlines—would be instrumented for global sampling. (NASA Release 75-60)

- For coating solar collectors that heat and cool homes and buildings, black chrome had been demonstrated by a Lewis Research Center scientist to be some 20% more efficient than current coatings. Being "solar selective," black chrome allowed absorption of sunlight but prevented the ensuing heat from radiating out of the collector. Because of this, black chrome would raise the temperature of water in the system as much as 5°C (40°F) over the 95°C (200°F) provided by current coatings. A system with hotter water would require fewer collectors to heat or cool an area. (NASA Release 75-55)
- Sen. William Proxmire (D-Wis.) had said that NASA's Space Shuttle program "might be on the verge of a financial breakdown," the *Washington Post* reported. Citing a General Accounting Office report, Sen. Proxmire warned that the risk of cost overruns had increased and that Congress should look for "cheaper alternatives." (AP, *W Post*, 6 March 75)

*6 March:* NASA announced early findings of the 3 Dec. 1974 *Pioneer II* flyby of Jupiter. Data returned by *Pioneer II* (launched 5 April 1973, and renamed *Pioneer-Saturn* following the flyby) suggested that Jupiter's magnetic field might be created by a large ring current and many eddies deep within the planet. Earlier measurements from *Pioneer 10*, launched toward Jupiter on 2 March 1972, had indicated only a simple magnetic envelope with a single "ring current" much like earth's. The more complex magnetic field could explain the Jovian field's high-energy particle pattern as well as the bursts of intense radio energy observed emanating from the planet. Jupiter's magnetic field stretched across 14 million km in some places and shrank in volume by three-fourths or more in others; inside this pulsating field were belts of intense radiation with trapped electrons 10 000 times more intense than those in earth's Van Allen belts.

Pioneer had experienced peak intensities of high-energy electrons that were moderately higher than those found by *Pioneer 10*, encountering 10 times as many electrons as predicted by earth-based studies. Peak intensities of very high-energy protons were 100 times stronger than predicted but concentrated in small shallow regions.

Other early findings of *Pioneer 11* included the observation that Jupiter's cloud tops were substantially lower at the poles than at the equator and were covered by a thicker transparent atmosphere. Also, the clouds at the planet's south pole were lower than those at the north pole. Although *Pioneer 11* found much less evidence of rapid circulation at the poles than at the equator, the polar areas showed many small convective cells dwarfing similar earth disturbances such as thunderstorms. "Blue sky"—attributed to multiple molecule-scattering of light by gases of the transparent atmosphere—was visible at the poles.

*Pioneer 11* accurately determined, for the first time, that the mass of Jupiter's moon Callisto was 15 times that of earth's moon. Seen for the first time was an extensive white south polar cap on Callisto. (NASA Release 75−49)

- A fleet satellite communications-system simulator developed for the Air Force by TRW Systems Group had been delivered for testing, Air Force Systems Command announced. The simulator would be a low-cost means of testing compatibility between FLTSATCOM—an operational near-global satellite communications system for Air Force and Navy Communications—and preproduction terminals for the Air Force Satellite Communications (AFSATCOM) system, which provided reliable worldwide satellite communications for command and control of all DOD forces.

  The simulator contained all the communications channels that the actual satellites would have, but used cheaper off-the-shelf components instead of high-reliability electronic parts. (AFSC Release OIP 022.75)

- Rep. Olin E. Teague (D−Tex.) and Rep. Charles A. Mosher (R−Ohio) introduced H.R. 4461 "to establish a science and technology policy for the U.S. to provide for scientific and technological advice and assistance to the President, to provide adequate administrative organization to assure effective Federal support and utilization of research and development, to amend the National Aeronautics and Space Act of 1958, to amend the National Science Foundation Act of 1950, and for other purposes." The bill was sent jointly to the Committees of Science and Technology and on Government Operations.

  The legislation proposed the establishment of a new cabinet-level Dept. of Research and Technology Operations to bring together Federal research-related activities through a unified and efficient governmental structure; a five-member Council of Advisors on Science and Technology to help form and implement policy at the highest levels of government; and a Science and Technology Information and Utilization Corp. to assure the widest possible dissemination of scientific and technological information.

  The proposed legislation would administratively relocate NASA, Energy Research and Development Administration, National Bureau of Standards, National Science Foundation, and National Oceanic and Atmospheric Administration within the new department, but would not alter the basic mission and purpose of each agency. (*CR*, 6 March 75, H1464; House Com of Sci & Tech Print 48−694)

*7 March:* Approximately 31 000 contractor personnel in 47 states were current at work on the development of the Space Shuttle, NASA announced. That number was expected to increase to 34 000 by June and to reach a maximum of 50 000 persons within 2 yr. (NASA Release 75−63)

*9 March:* West German Federal Research Minister Hans Matthoefer said, in an interview with the West German news agency (Deutsche Presse Agentur), that West German and French officials had withdrawn their own candidates and agreed that United Kingdom's Roy Gibson would become secretary of the European Space Agency (ESA). This agreement removed substantial obstacles to the establishment of ESA. Gibson, who had been administrative director of the European Space

Research Organization (ESRO), was currently acting Secretary General until the organization merges into ESA, possibly in April.

Another important agreement specified that two of the five directorships would go to West Germans, two to French, and one to a Belgian. (DPA, FBIS—FRG, 11 March 75, J6)

*10 March:* NASA signed a contract with Italy's Consiglio Nazionale delle Ricerche (National Research Council) to launch the SIRIO experimental communications satellite. The contract—signed by Dr. James C. Fletcher, NASA Administrator, and Prof. Alessandro Faedo, CNR President—provided for the launch of the microwave propagation satellite aboard a Thor-Delta launch vehicle in the fall of 1976. (NASA Release 75—67)

- NASA announced the selection of Goodyear Aerospace Corp. for negotiations leading to a $1 000 000 contract to provide a special-purpose processor to support the Large Area Crop Inventory Experiment (LACIE) at Johnson Space Center. (JSC Release 75—17)
- Orders for U.S.-manufactured commercial jet transport aircraft at the end of 1974 totaled 564, valued at $7 587 000 000, the Aerospace Industries Association reported. The aircraft total was down from 573 aircraft, valued at $7 252 000 000, on order at the end of 1973. The orders included 210 aircraft from Boeing Co., 178 from Lockheed Aircraft Corp., and 176 from McDonnell Douglas Corp. (AIA Release, 10 March 75)

*11 March:* Kennedy Space Center announced the award of a 10-mo $30 000 contract to the Univ. of Miami for preliminary design of a laser system capable of penetrating water to a depth of 100 m to measure factors affecting water quality. The use of lasers to provide remote sensing of the ocean depths had been proved feasible during a NASA contract awarded to the school in 1974. The system, which would provide continuous information on temperature, salinity, and backscatter turbidity, would be tested first in a laboratory, then from a ship, and finally from an aircraft. (KSC Release 35—75)

*13 March:* The Soviet Union's Tu—144 supersonic transport made its inaugural scheduled flight from Moscow to Alma-Ata in a little more than 1.5 hr. Alma-Ata is capital of the Kazakh Republic, in South Central Asia bordering the Caspian Sea. The aircraft, designed by Aleksey A. Tupolev, flew at an altitude of 17 400 m and a speed of 2300 km per hr. (Alma-Ata Domestic Service, FBIS-Sov, 14 March 75, R1)

- A scientific conference of Soviet scientists and officials was being held in Zvenigorod, near Moscow, to discuss the study of the earth from outer space, Tass reported. Tass quoted Academician Roald Sagdeyev, Director of the Space Research Institute, as saying that the main task of the conference was to work out unified scientific principles and methods for exploring earth resources from outer space and for organizing a systematic control over the environment with the help of artificial earth satellites. To date, scientists had made progress in certain lines of research such as meteorology, but in other areas had received only bits of information. The task was to move from individual discoveries to a qualitative geological study of the planet as a whole. The task called for combined effort by geologists, soil scientists, foresters, and hydrologists.

Sagdeyev said that the study of earth from space was becoming an important aspect of international relations because man's active influences on the environment went beyond national boundaries. (Tass FBIS—Sov, 18 March 75, U1)

*14 March:* The House Committee on Science and Astronautics favorably reported out H.R. 4700 authorizing FY 1976 appropriations for NASA. Changes recommended by the Committee would increase the total NASA authorization by $46.9 million, to $3.586 billion. [See also 3 Feb.]

## Recommended Changes in the FY 1976 NASA Authorization by the House Committee

| Program | Difference from Submission (millions) | Recommended Authorization (millions) | Purpose for Recommended Change |
|---|---|---|---|
| *Research and Development* | | | |
| Space Shuttle | 0 | 1 206.0 | Accepted as presented. |
| Space Flight Operations | −4.0 | 203.1 | To decrease development, test, and mission-operation activities. |
| Advanced Missions | +1.5 | 3.0 | To increase studies of large space structures, manned orbital systems concepts, and in-orbit maintenance. |
| Physics and Astronomy | +1.0 | 156.8 | To increase stratospheric research. |
| Lunar and Planetary Exploration | −1.0 | 258.9 | To decrease funding for the Lunar Research Program. |
| Launch Vehicle Procurement | 0 | 166.9 | Accepted as presented. |
| Space Applications | +6.5 | 181.5 | To increase emphasis on weather and climate research, earth resources surveys, advanced communications research, data-management techniques, and Space Shuttle payload definition and early flight tests. |
| Aeronautical Research and Technology | 0 | 175.4 | Accepted as presented. |
| Space and Nuclear Research and Technology | +2.0 | 76.9 | To accelerate high pressure hydrogen propulsion systems tests and high temperature composites research. |
| Energy Technology Applications | 0 | 5.9 | Accepted as presented. |
| Tracking and Data Acquisition | −2.2 | 240.8 | To decrease operations and advanced missions activity. |
| Technology Utilization | +2.0 | 9.0 | To accelerate effort of NASA TU teams to work with the public and increase space benefits analysis documentation. |
| R&D Totals | +5.8 | 2 684.2 | |
| *Research and Program Management* | | | |
| Basic submission* | 0 | 776.0 | Accepted as presented. |
| R&PM Totals* | 0 | 776.0 | |
| *Construction of Facilities* | | | |
| Modification of 11- by 11-ft transonic wind tunnel at ARC | 0 | 2.7 | Accepted as presented. |
| Modification of 40- by 80-ft subsonic wind tunnel at ARC | +12.5 | 12.5 | To repower tunnel. Not in original request. |
| Addition to lunar sample curatorial facility at JSC | −2.5 | 0 | Deleted from funds. |
| Addition for composite model and metal finishing shops at LaRC | 0 | 1.9 | Accepted as presented. |
| Construction of transonic research tunnel at LaRC | +27.5 | 27.5 | Construction of high Reynolds number transonic wind tunnel. Not in original request. |

## Recommended Changes in the FY 1976 NASA Authorization by the House Committee (continued)

| Program | Difference from Submission (millions) | Recommended Authorization (millions) | Purpose for Recommended Change |
|---|---|---|---|
| *Construction of Facilities: (cont.)* | | | |
| Space Shuttle facilities | −0.9 | 46.3 | Deferment of modifications of two hypergolic checkout and refurbishment facilities and construction of facilities to house microwave scanning beam landing system. |
| Rehabilitation of facilities | 0 | 16.0 | Accepted as presented. |
| Minor construction of new facilities or additions to old | 0 | 5.0 | Accepted as presented. |
| Facility planning and design | +4.5 | 13.8 | For planning and design work associated with construction and modification of two wind tunnels mentioned above. |
| C of F Totals | +41.1 | 125.7 | |
| Grand Totals* | +43.7 | 3 585.9 | |

*Does not include a supplemental request for pay increases, considered separately as H.R. 13172 and signed into P.L. 94–303 1 June 1976.

In addition, for the transition period 1 July to 30 Sept. 1976, the committee recommended decreasing NASA's total budget request of $958 900 000 to $922 450 000. One specific recommended reduction was in the C of F request, from $14 500 000 to $8 050 000, which would allow construction to continue at the same rate as during FY 1976; the committee recommended reduction of the $730 600 000 request for R&D to $700 600 000. The committee allowed NASA management flexibility in apportioning the reductions. (H.R. Com of Sci & Tech Rpt 94−63)

*15 March: Helios 1*, the U.S.–West German solar probe launched 10 Dec. 1974, passed within 0.309 AU (46 million km) of the sun at 5:13 am EDT, closer than any previous spacecraft had come to the sun. During the perihelion, while traveling at 238 000 km per hr, *Helios 1* measured the solar wind, magnetic fields, solar and galactic cosmic rays, electromagnetic waves, micrometeoroids, and zodiacal light. Although the spacecraft encountered a heat load from the sun 10.5 times that encountered on earth, good data were obtained and the spacecraft continued to operate well. *Helios 1* was in a solar elliptical orbit with a 0.985–AU aphelion, 0.3095–AU perihelion, 190.15-day period, and 0.02° inclination. It would reach a second perihelion 21 Sept.

*Helios 1*, built by West Germany in cooperation with the U.S., carried 10 experiments, 7 of them West German. Spacecraft operations were controlled from the space operations center near Munich. Goddard Space Flight Center was responsible for U.S. participation. A second spacecraft, Helios–B, was scheduled for launch in early 1976. (NASA MORs 16 Jan 76, 11 April 76; Powers, GSFC Helios thermal control mgr, interview, 21 Sept 76)

- NASA announced establishment of the Office of Program Assurance under the Office of the Associate Administrator. Responsibilities and personnel of the former Reliability and Quality Assurance Div. and the Systems Safety Branch of the Office of Safety and Reliability and Quality Assurance were being transferred to the new office. George

C. White, who had been director of the Office of Safety and Reliability and Quality Assurance, had been appointed director of the new office. (NASA anno)

*16 March*: *Mariner 10* made its last and closest flyby of Mercury at a distance of 327 km at 6:39 pm EDT. To prepare for the flyby, the trajectory had been corrected on 30 Oct. 1974 and 13 Feb. 1975. An additional correction 7 March was to reduce the probability of *Mariner 10*'s impacting the surface of Mercury from 30% to 1%, targeting *Mariner 10* to approach the planet from the sunlit side, then loop around behind it.

One last-minute problem was navigational difficulty in locking onto the star Canopus so that the spacecraft would point in the proper direction for data collection. Because a commanded roll search for Canopus would expend all remaining attitude-control gas, *Mariner 10* had to be maneuvered using solar-radiation pressure on the solar panels and high-gain antenna. The problem was intensified because of the tandem demands of *Helios 1* and *Mariner 10* on the antennas of NASA's deep-space tracking network. However, the primary users—West German scientists tracking *Helios 1*—gave receiving time to the *Mariner 10* flight controllers, who obtained Canopus lock in time for the encounter.

Major objective of the third encounter was to determine whether Mercury had an intrinsic or induced magnetic field. In a press briefing 17 March, Dr. Norman F. Ness, principal investigator of the magnetic fields experiment, said that Mercury did have an intrinsic magnetic field, encountered on the incoming trajectory at precisely the time predicted from the first Mercury pass. *Mariner 10* had measured the field to be a maximum of 400 gamma. Plasma science experiment data supported the presence of the field; Dr. Ness observed that the interface between the solar wind and Mercury's magnetic field appeared to be a scaled-down version of the interaction between the solar wind and the magnetic field of earth.

The spacecraft returned about 300 TV pictures of the planet's surface between 12 and 17 March, with best resolution at 100 m obtained from quarter-frame near-encounter pictures. All photos were transmitted in real time and at 22 kilobits per sec, instead of the planned 117, because of overheating of the spacecraft. (NASA MORs, 10 March 75, 7 April 75; NASA Releases 75–58, 75–59; *NASA Activities*, April 75, 12–13; *Av Wk*, 24 March 75, 24–25)

*17 March*: *Vanguard 1*, oldest satellite still in orbit, completed its 17th year in space. The 1.5-kg 16.5-cm solar-powered spacecraft, which had revealed that the earth was slightly pear-shaped, had circled the earth more than 67 000 times since its launch in 1958. The satellite, fourth to orbit earth following two Soviet sputniks and *Explorer 1*, was one of 3304 man-made objects tracked by the North American Air Defense Command. (*A&A 1915–1960*; *A&A 1968*; Miles, *LA Times*, 18 March 75)

- NASA announced the award of a $334 100 contract for the design, fabrication, delivery, installation, and checkout of an optical bench and alignment system for an x-ray telescope test facility at Marshall Space Flight Center. The facility would be used to test x-ray verification and to calibrate x-ray mirrors, telescope systems, and instruments for NASA's High Energy Astronomy Observatory (HEAO) program. The HEAO program was to launch between 1977 and 1979 three unmanned scientific satellites into low circular earth orbit to study black holes, neutron stars, quasars, and supernovae. (MSFC Release 75−50)

- European spending for space programs would rise significantly during 1975 as major projects begun in the past reached a maximum pitch, *Aviation Week and Space Technology* reported. Most space spending was reflected in the European Space Research Organization (ESRO) budget which had jumped 50%, from $220 million in 1974 to $350 million in 1975. Two programs—the Ariane launch vehicle and the Spacelab for use on the Space Shuttle—accounted for half the ESRO budget for 1975.

    *Av Wk* reported that officials of France's Centre National d'Etudes Spatiales (CNES), prime contractor for the Ariane, had estimated that the booster would be used to orbit between 20 and 39 geostationary satellites during the 10-yr period from 1980 to 1990.

    Other European space programs under way included the Orbital Test Satellite (OTS), a test bed to evaluate equipment and concepts planned for a European comsat to be launched around 1980; Aerosat, a Canadian−ESRO−U.S. comsat program, with the first of two satellites scheduled for launch in 1978; Marots, a maritime comsat planned for launch in 1977; and Exosat, an x-ray astronomy satellite to be launched in 1979.

    All ESRO programs would be reviewed in the spring to resolve any pending disagreements, and to obtain approval of a 3-yr financial plan that called for a budget of $410 million in 1976 and $415 million in 1977. (*Av Wk*, 17 March 75, 65−68)

- "A significant influence on NASA's ability to maintain viable programs in the face of possibly unstable funding will be the extent to which commercial and foreign government users participate in the development and financial support of future payloads," *Aviation Week and Space Technology* reported. *Av Wk* interviewed the Director of Mission and Payload Integration in NASA's Office of Manned Space Flight, Philip E. Culbertson, who said that it would be better, in the long term, to move into a position where more organizations outside NASA were sponsoring and paying for missions and payloads. The agency was encouraged by the interest of industries in flying and funding Space Shuttle missions involving a wide range of basic scientific and technical operations.

    Foreign reimbursable payloads, already an important part of NASA launch operations, were seen as an even larger factor in Shuttle payloads development. Japan, already holding discussions with NASA to lay the groundwork for a viable national space program, was considered a serious Shuttle customer; Japan would like to develop both scientific and earth applications programs in the communications, navigation, and weather satellite fields. Canada's interest in positioning domestic communications satellites over its territory would continue into the Shuttle era. Even individual Soviet scientists had made inquiries into the possibilities of flying payloads on the U.S. Shuttle.

    Culbertson pointed out that, unlike earlier NASA manned space programs that included scientific funding in overall program budgets, NASA in-house Shuttle payload development was the sole responsibility of the specific NASA disciplines themselves. NASA's FY 1976 budget request contained from $50 to $100 million for various efforts that could evolve into Shuttle payloads, and the budget for FY 1977 would markedly increase payload funding. However, with the budget crunch, funding for Shuttle payloads would compete directly with other programs and the payload proposals "will have to stand on their own merit." Culbertson also said that major decisions would have to be made on how long to keep which expendable launch vehicles in the program.

In another interview, NASA Spacelab Program Director Douglas R. Lord said that upcoming major decisions concerning Spacelab included a follow-on procurement plan for Spacelab hardware and priorities on how often ESRO (European Space Research Organization) and other crews and experimental hardware would fly on Shuttle–Spacelab missions.

Lord said that operations of overall Spacelab functions would probably be maintained separately from payload operations: Johnson Space Center was the logical place to handle overall operations, and Marshall Space Flight Center was the leading candidate for the Spacelab payload operations center. Goddard Space Flight Center had expressed interest in handling operational needs of the high-energy experiments on Spacelab pallets. (Covault, *Av Wk*, 17 March 75, 59–60)

*18 March*: "The expertise and facilities of LeRC [Lewis Research Center], as well as other NASA centers, are being considered by ERDA [Energy Research and Development Administration] as part of its overall definition of a total institutional structure required to carry out the nation's energy R&D program," Dr. James C. Fletcher, NASA Administrator, and Dr. Robert C. Seamans, Jr., ERDA Administrator said in a letter to Representative Charles A. Mosher (R–Ohio). The letter was a reply to correspondence from members of the Ohio congressional delegation and Cleveland businessmen and industrialists to President Ford calling for increased use of LeRC talent to meet the nation's energy crisis.

Dr. Fletcher and Dr. Seamans noted that the newly established ERDA was the lead Federal agency for managing national energy R&D. ERDA intended to use the expertise and facilities of other Federal agencies where it was feasible and in the national interest with respect to cost, timeliness, and program management.

LeRC, NASA's lead center for aerospace propulsion and power, was already involved in ERDA-sponsored projects in solar heating and cooling, wind energy, and topping cycles to increase the efficiency of coal-fueled steam plants. But LeRC would continue its main task to work on propulsion technology, conducting energy research in support of ERDA "when it was consistent with the requirements of its principal mission within NASA." (Text, letter to Mosher from Fletcher and Seamans, 18 March 75)

- A *New York Times* editorial commented on *Mariner 10*. Men had become so blasé that even near-miracles were taken for granted. Consider *Mariner 10*, which had traveled nearly a billion miles since its launch in 1973: The spacecraft had taken the first close-up photos of Venus and provided us with sharp pictures of a large part of Mercury's surface during its three flybys of that planet. "In a world where so much that is manufactured is shoddy, faulty in conception, or the end result of workers who could not care less, it is worth remembering that men can also build durable, reliable and trustworthy mechanisms like *Mariner 10* whose historic contribution to knowledge has substantially enhanced man's understanding of the solar system." The Jet Propulsion Laboratory of the California Institute of Technology and its contractors, responsible for *Mariner 10* and its superlative performance, have every reason for pride in their creation, the *Times* said. (*NYT*, 18 March 75, 36)

- NASA announced establishment within the Office of Center Operations of the Office of Safety and Environmental Health under the acting directorship of Reuben P. Prichard. The new office, which would have agency-wide functional responsibility for all safety matters, except systems safety, and for environmental health, would also be the focal point for agency coordination on policy and program matters pertaining to the Occupational Safety and Health Act of 1970. Transferred to the new office were environmental health functions formerly assigned to the Office of Occupational Medicine and Environmental Health. (NASA anno, 18 March 75)

*19 March:* NASA announced that lunar samples would be available to colleges and universities for teaching purposes. Under a new NASA program, a "thin-section educational package," consisting of several 0.03-mm slices of representative lunar material plus brief descriptions, could be borrowed for several months by institutions offering undergraduate or graduate work in the geosciences. The purpose of the program was to broaden the use of NASA's lunar sample collection. (NASA Release 75—76)

*19—20 March:* A special conference at Lewis Research Center, "Selected Technology for the Gas Industry," discussed how NASA-developed technology could aid the U.S. gas industry in ways such as satellite search for fossil fuels, turbine machinery and compression technology, heat transfer, fluid flow and fluid properties, safety, technology for liquefied natural gas, instrumentation, materials and life prediction, and reliability and quality assurance.

Addressing a conference dinner 19 March, Dr. James C. Fletcher, NASA Administrator, said, "In an age when space research and technology was measured by its immediate value in the marketplace, studies of the planets often are considered little more than an exercise in curiosity. Nothing could be farther from the truth." When spacecraft were developed to meet the specialized needs of space travel and planetary exploration, technology was advanced. "NASA cannot lay claim to the development of the electronic computer, but it can take a large credit for spurring a revolution in the electronics industry." The need for lightweight components in spacecraft led to electronic microminiaturization. The Illiac 4 computer, developed at the Univ. of Illinois for DOD and now operated jointly by DOD and NASA, could perform 300 million calculations a second and store 1 trillion bits of information at a time, more than the entire surviving output of the Greek and Roman civilizations. Operational since 1973, Illiac's speed and refinement allowed scientists to simulate and study global climate changes, study long-range effects of pollution on the atmosphere, distinguish between earth tremors from natural sources and underground nuclear blasts, develop seismic pictures of the earth's outer shell, and enhance medical x-rays so that tumors could be discovered in earliest stage of development. Illiac 4 has helped solve problems that had eluded scientists for decades.

Other space technology was being used effectively to preserve and manage the earth's resources and enhance the quality of life for all its inhabitants. Communications satellites were making global communications a reality. Meteorological satellites had advanced the science of weather forecasting on land and sea. Sensors, surgical instruments, biological isolation garments, lenses, portable cardiac care systems, and eye switches were just a few examples of space technology used in the medical field. New alloys and materials developed to withstand the rigors of space were being used for longer-lasting automobile and aircraft engines; temperature-resistant lubricants and fireproof materials added to the safety of homes and buildings.

"The fast-growing list of things that are and *will be* as a result of our venture into the cosmos is almost beyond imagination. Everyone who lives will be affected by them. And because of them, the world a decade or so hence will be quite different from the world we now know." (LeRC Release 75–11; Fletcher speech, text)

*20–28 March:* U.S. and U.S.S.R. flight controllers and crews for the Apollo–Soyuz Test Project participated in simulations of phases of the joint mission to check out the interaction of control-center personnel under actual flight conditions, using voice, teletype, datafax, and TV communications. Beginning with rendezvous of Apollo and Soyuz, the simulations were conducted with Houston and Moscow control centers fully manned—including some specialists from the other nation—and U.S. and Soviet crew members were in simulators in their respective countries; events simulated were the launches, undocking, Apollo separation, 2nd and 3rd crew transfers, and joint activities. Further simulations were scheduled for May, June, and early July. (NASA Release 75–77; Tass, FBIS–Sov, 19 March 75, U1)

*21 March:* Dow Jones and Co.'s *Wall Street Journal* signed a 7-yr contract with American Satellite Corp. for the design and construction of earth stations and provision of high-speed communications that would permit a Florida-based *WSJ* printing plant to receive, via satellite, full printed pages for facsimile reproduction. Full pages would be transmitted, at an average rate of 3 min a page, from *WSJ*'s Massachusetts printing plant to a receiving antenna in Florida. The antenna would connect to printing equipment in the Florida plant, which would transform the communications signal into printed pages again. (AmSatCorp Release, 25 March 75; AmSatCorp Marketing Services Mgr., interview, 30 Nov 76; Jones, *W Post*, 1 April 75)

*22–23 March:* The International Association of Machinists continued its strike against McDonnell Douglas Corp. after turning down a contract offer. Although the union said it would return to the bargaining table to try to resolve the impasse, NASA announced that all Delta launches in the near future—including Telesat–C, Marisat–A, Nimbus–F, and OSO–I—would be postponed. (NASA PAO audio news recording, 24 March 75)

*24 March:* Radio transmissions from *Mariner 10* ceased after attitude-control gas was depleted and flight controllers at Jet Propulsion

Laboratory sent the command for the spacecraft to stop operations at 8:20 am EDT. Before shutdown, spacecraft engineering tests were conducted to try to open the stuck door on the plasma science experiment, rotate the solar panels to various positions, stow the scan platform at 180°, operate the stuck tape recorder, and provide data for future spacecraft design.

During its more than 16 mo in heliocentric orbit, *Mariner 10* (launched 3 Nov. 1973) had traveled more than 1.6 billion km and made four planetary encounters, one of Venus and three of Mercury. It had passed within 5800 km of Venus 5 Feb. 1974, and transmitted to earth the first closeup photographs ever taken of the cloud-shrouded planet.

*Mariner 10* had confirmed the high-speed circulation of Venus' upper atmosphere and revealed circulation patterns that swirled from the equator toward the poles. It had also revealed a long tail of charged particles trailing behind the planet and away from the sun; verified the presence of a bow shock wave created by the solar wind acting upon the dense atmosphere; detected hydrogen, helium, and argon in the Venusian atmosphere; and discovered temperature-inversion zones suggestive of deeper stratiform cloud layers.

First spacecraft to use the technique, *Mariner 10* used the gravity of Venus to speed it on toward Mercury. The probe flew by that planet three times, on 29 March and 21 Sept. 1974 and 16 March 1975. At Mercury, *Mariner 10* photographed a highly cratered lunar-like surface with large scarps hundreds of kilometers long and over 1.6 kilometers high. *Mariner 10* data also revealed the existence of a chemically differentiated light crust, lunar-like near the surface and earth-like at its interior, and an unexpected magnetic field 100 times smaller than earth's. A slight but measurable atmosphere was found, in addition to streams of high-energy electrons and protons on the planet's dark side. On its last encounter with Mercury [see 16 March] *Mariner 10* came within 327 km of Mercury to take the closest photographs ever made of another primary planet. (NASA MOR, 7 April 75; *NASA Activities*, April 75, 12–13; NASA Release 75–59; *A&A 1974*)

- The Saturn IB scheduled to launch the Apollo spacecraft for the Apollo-Soyuz Test Project in June left the Vehicle Assembly Building on its crawler to begin the 5-km journey to Launch Complex 39 Pad B. Witnessed by 6000 people, the rollout was completed at 3:15 pm, followed by a ceremony attended by the ASTP astronauts and KSC Deputy Director Miles Ross commemorating this as the last Saturn IB and Apollo rollout. Then the slow journey was resumed. (NASA PAO audio news service, 24 March 75; KSC Release 48–75)

- Assembly of the first main engine of the Space Shuttle had been completed by Rockwell International Corp.'s Rocketdyne Div. 1 mo ahead of schedule, NASA announced. After assembly the engine was automatically checked out—for the first time in history—by the engine's internal controller and integral digital computers for functional, electrical, and mechanical operations. Known as the Integrated Subsystems Test Bed, this engine was not built for flight but for static firing tests at the National Space Technology Laboratories.

The Space Shuttle would use 3 of the liquid-hydrogen/liquid-oxygen engines. Designed to perform at high internal pressures for greater specific impulse, each engine would produce more than 2.1 million newtons thrust and would be used for up to 55 missions before overhaul, with between-mission maintenance comparable to techniques used by commercial airlines. (NASA Release 75−83; Rocketdyne Release RD−7)

*25 March:* The computer aboard *Landsat 2* (launched 22 Jan.) was the most advanced of its type ever flown on an unmanned NASA satellite, NASA announced. Able to handle up to 55 ground-station commands controlling spacecraft operations for up to 24 hrs, the compact digital computer was a forerunner of the NASA Standard Space Computer that will be used in a wide variety of unmanned missions. Managed by Goddard Space Flight Center, the standard space computer program was part of an agency-wide program to develop standardized satellite components, both to reduce costs and to increase reliability. The program also covers other spacecraft hardware such as nickel−cadmium batteries, transponders, attitude-control thrusters, pyrotechnic initiators, silicon solar cells, and tape recorders. (NASA Release 75−85)

- Two solid rockets on the Space Shuttle acoustic model accidentally escaped from the test stand during a captive static firing. The model's liquid-propellant rocket engines had operated normally for a few seconds when the two solid rockets ignited on schedule. Both solid rockets then separated from the test stand. One was propelled north for 1.2 km, landing in a pine thicket; a brush fire caused by the rocket was quickly extinguished and no one was injured. A nine-member board was appointed to investigate the cause of the accident, determine the amount of property damage, investigate the adequacy of safeguards to prevent further incidents, recommend corrective measures, and assure containment of facilities and surrounding areas in case of recurrence.

  Although the test stand was damaged, the Orbiter and external tank of the model were not. The model would be returned to operation with the replacement of some wiring and instrumentation. (*Marshall Star,* 2 April 75, 2)

- The new Soviet 100-seat jet airliner, Yak 42, was being air-tested in the Soviet Union, Tass reported. The aircraft was expected to replace outdated airliners on Aeroflot's medium-distance routes. (Tass, FBIS−Sov, 26 March 75, U1)

*26 March:* Langley Research Center was flight-testing a new-generation light twin-engine Piper PA−34 Seneca aircraft—called ATLIT (Advanced Technology Light Twin)—fitted with several improvements using NASA advanced aerodynamic technology. One important new component, the GAW−1 (General-Aviation-Whitcomb) airfoil, was 27% smaller than the original wing but was expected to be safer and more efficient, and to reduce fuel consumption by as much as 10%. (NASA Release 75−84)

- The Japanese National Space Development Agency announced the award of a contract to the Mitsubishi Electric Corp. to design two experimental stationary communications satellites for launch in FY 1978. Design was expected to be complete by 20 Sept. Each 130-kg satellite, to be launched by a Japanese N rocket, would be used to study technology for launching stationary satellites, satellite-tracking technology, and attitude-control techniques. (Kyodo, FBIS-Japan, 3 April 75, C2)

*27 March:* Soviet scientists told U.S. scientists and journalists touring Soviet scientific facilities that they had successfully used a rocket engine and gigantic magnets to produce large amounts of less expensive electricity with less pollution than other methods. Called magnetohydrodynamics and first proved feasible in the U.S. in the early 1960s, the new system was one of the cooperative programs between the two countries.

A recent test that produced 12.4 mw of power for 1.5 hr was the first time the system had achieved that power and duration. Based on the knowledge that electricity could be generated by passing a charged gas through a magnetic field, the Soviet experiment burned 55 kg per sec of natural gas with oxygen-enriched air into which a potassium compound was injected to create a plasma of positive and negative particles; when passed between the poles of a great magnet, the charges separated onto electrodes and created large voltages. The *Washington Post* later reported that the Soviets were committed to building, by 1981, a 1000-mw magnetohydrodynamic plant with expected conversion efficiency of 50%. Conventional steam-turbine plants using oil or coal for fuel operate at 40% efficiency, those using nuclear fuel at 33%. (Toth, *LA Times* News Service, *W Post*, 28 March 75, A3)

- Canada's de Havilland Dash 7 four-turboprop short-takeoff-and-landing (STOL) transport made its first flight test near Ontario, Canada. The first of two preproduction prototypes of the 50-passenger aircraft got off the runway in 274 m. During the 2-hr 10-min flight, Dash 7 reached 2896 m altitude and a maximum speed of 204 km per hr. Purpose of the first flight was to perform power-on and power-off tests of longitudinal and latitudinal stability. (*Av Wk*, 7 April 75, 18)

- Langley Research Center announced the appointment of Robert E. Smylie as Director for Space, replacing Eugene S. Love, who was retiring. Smylie had been in NASA Hq as Deputy Associate Administrator for Space, Office of Aeronautics and Space Technology. In his new position he would direct Langley's work in space systems, space applications and technology, environmental quality programs, and the Scout launch vehicle. (LaRC Release 75-11)

- NASA Hq announced the disestablishment of its Office of Organization and Management. All personnel and responsibilities were assigned to other offices within NASA Hq. (NASA anno, 2 April 75)

*31 March:* Boeing Co. scientists were studying a way to cut aircraft fuel consumption as much as 40%, according to the *Baltimore Sun*. The technique involved cutting many tiny holes in the aircraft wings and applying suction from within. This would control the laminar flow of the airstream, preventing drag-producing turbulence from being created by the change in pressure as the air flows over the wing surface. In the 1960s Northrop Corp. had built the Air Force a plane incorporating the concept but technical problems had discouraged further development. Rising fuel costs and recent improvements in structural materials and manufacturing techniques had sparked new interest in the method. (AP, *B Sun*, 31 March 75, A7)

*During March:* NASA's *Ats 6* (Applications Technology Satellite, launched 30 May 1974) was the means for coordinating long-distance emergency operations of ships near the Azores Islands during search and rescue tests conducted by Goddard Space Flight Center. The tests evaluated the effectiveness of *Ats 6* functioning as a communications relay between a ship in distress and a coordination center thousands of kilometers away.

In the tests, a "stricken" ship dropped a buoy equipped with a position-indicating radio beacon into the ocean. *Ats 6* picked up the signals and transmitted the ship's identification and position to a NASA ground station, which relayed them to the Coast Guard's New York Rescue Coordination Center. The rescue center then sent the information to rescue aircraft and ships. As they neared the area, they could home in on the buoy's signals. Within 2 hr of the time the buoy was dropped, the ship and the buoy were located.

The search and rescue communications tests were part of continuing joint L-band experiments using *Ats 6* and conducted by GSFC and the Depts. of Commerce and Transportation, as well as by the European Space Agency and the Canadian Dept. of Communications and Ministry of Transport. (GSFC *Goddard News*, Oct 75, 3)

- Preparations continued at Kennedy Space Center for the launch of two Viking spacecraft to Mars. Lander 1 was mated with its Orbiter on 8 March to test compatibility. The Lander and Orbiter were mated with the Centaur launch vehicle shroud and moved to Launch Complex 41 on 27 March to undergo a flight-events demonstration, Orbiter and Lander precountdown tests, and terminal countdown demonstration with the complete vehicle.

    During a complete 18-hr plugs-out test of Lander 2 on 13 March, data were transmitted to Jet Propulsion Laboratory to provide training for the Viking flight-control team. (KSC *Spaceport News*, 6 March 75, 6; 20 March 75, 5)

# April 1975

*1 April:* NASA commemorated the 15th anniversary of the launch of *Tiros 1* (Television Infrared Observation Satellite), the world's first weather satellite. Since its launch 1 April 1960 at Cape Canaveral on a Thor-Delta booster, 30 experimental and operational weather satellites of increasing complexity had been launched to provide continuous information on earth's environment. Over the 15-yr span the weather satellites had taken more than 2.2 million photographs while traveling through 12.1 billion km of space. Not a single major hurricane or storm had gone undetected or untracked.

President Ford issued a statement in recognition of the *Tiros 1* anniversary: "More accurate daily weather forecasts. . .have had an immeasurable impact in making the lives of millions more pleasant, productive, and secure. No major storm anywhere goes undetected by NASA or NOAA [National Oceanic and Atmospheric Administration] satellites. We can be proud that people everywhere can benefit from this practical application of U.S. space science and technology." (NASA Release 75–73; *NASA Activities*, April 75)

- A prototype of a flexible tunnel to connect the Space Shuttle airlock with the forward end of the Spacelab was being tested at Marshall Space Flight Center. The tunnel would provide a pressurized passageway for crew members and scientists to move to and from the orbiting laboratory without spacesuits. The flexible circular tunnel, built by Goodyear Rubber Corp., would accordion-fold to 0.6 m and extend to a length of more than 4.3 m. Made rigid by steel rings, the tunnel was constructed of layers of aluminum foil, Capran film, and nylon cloth covered by a spongy meteoroid shield.

  MSFC engineers were testing the tunnel, which would be exposed to the space environment when the Orbiter bay doors were opened, for structural strength, airtight integrity, and materials compatibility. (MSFC Release 75–63)

- Marshall Space Flight Center had issued a single-source request to Sunstrand Corp. to submit a proposal for the design, development, manufacture, test, and delivery of the auxiliary power unit for the Space Shuttle Solid Rocket Booster's thrust-vector control subsystem. The APU would include a gas generator, turbine, gearbox, fuel pump, electrical controls, control valves, instrumentation, mounting system, and the mechanical and electrical connections required to interface with other SRB subsystems.

  Two APUs would be used during prelaunch and ascent phases of flight to provide hydraulic power to the thrust-vector control system of each booster. (MSFC Release 75–62)

- The Air Force was again studying the feasibility of a nuclear-powered aircraft twice the size of the C–5, the *Washington Star* reported. In an interview with the *Star*, Dr. Lawrence W. Noggle, program manager for nuclear aircraft-propulsion technology at Wright–

Patterson Air Force Base, said that nuclear power could be used as an alternative power source for large subsonic aircraft. Other potential fuels under study were liquid methane, liquid hydrogen, and coal. A nuclear-powered aircraft, which would require larger engines to carry a larger cargo, could use conventional fuel for takeoff and landing and then switch to nuclear power at cruise speed. The *Star* quoted Dr. Noggle as saying, "We could use modified jet engines. . . with the reactor totally isolated from the engine through containment." He predicted that nuclear-powered aircraft could fly for days without refueling. (*W Star*, 1 April 75, A4)

- NASA announced publication of the *Skylab Earth Resources Data Catalog* (JSC 09016), containing 35 000 photographs taken on 1973-74 Skylab Orbital Workshop missions. The volume described in detail the earth-resources equipment and techniques used aboard the station; explained how the photos could be used in disciplines related to land-resource management, marine resources and management, land surveys and mapping, and environmental applications; and supplied a detailed index for finding space photographs, as well as instructions on locating data available through other government agencies. (NASA Release 75-92)
- NASA executed a $3.65-million supplement to a cost-plus-award-fee contract with Serv-Air, Inc., to continue to provide aircraft for earth-observation and astronaut-proficiency training at Johnson Space Center. Contract provisions also called for maintenance, modification, and related ground support of the JSC aircraft; maintenance and ground support of transient aircraft; engineering, design, fabrication, and installation of electronic and mechanical systems, subsystems, components and equipment; and related logistic functions.

  This agreement provided for the third and final year of the contract which had an estimated total value of $7.57 million. (JSC Release 75-23)

*1-4 April:* In Montreal, the 89-nation International Telecommunications Satellite Organization held its third ordinary meeting of signatories in conjunction with the 25th anniversary of the Canadian signatory, the Canadian Overseas Telecommunications Corp.

Significant accomplishments of the meeting included agreement that domestic services provided for Chile by INTELSAT would be considered on the same basis as international services, thus increasing Chile's voting representation on the board of governors; that any signatory owning shares in INTELSAT totalling 1.25% or more would be entitled to representation on the board of governors; and that the capital ceiling of $500 million for payments and contractual commitments for development of Intelsat V satellites, planned for launch in the late 1970s, would have to be increased by from $200 to $400 million. (INTELSAT Release, April 75)

*3 April:* Robert G. Strom, Univ. of Arizona Lunar and Planetary Laboratory scientist, was quoted by the Baltimore *Sun* as saying in a 31 March interview that 75% of the 500 photographs taken during *Mariner 10*'s third and final pass of Mercury [see 16 March] "did not turn out." Because of a receiver malfunction at the Canberra, Australia, tracking station, *Mariner 10* could not be maneuvered quickly enough to photograph all the desired area. However, Strom emphasized that *Mariner 10*'s third pass was "actually a bonus" and the mission had been a complete success. (B *Sun*, 3 April 75, A8)

*4 April:* NASA had awarded a $1 388 498 firm-fixed-price contract to Goodyear Aerospace Corp. to produce a special-purpose processor to augment existing computing capability for NASA's Large Area Crop Inventory Experiment (LACIE), Johnson Space Center announced. The contract included design, fabrication, delivery, and installation of the processor and any associated system software. LACIE, a cooperative program shared by U.S. Dept. of Agriculture, National Oceanic and Atmospheric Administration, and NASA, aimed at improving wheat-production forecasts by use of satellite earth-resources data. (JSC Release 75–20)

- The testbed aircraft for the Air Force Airborne Warning and Control System (AWACS) was deployed to Europe for a series of exercises with North Atlantic Treaty Alliance (NATO) ground, air, and sea forces. The Air Force was developing AWACS as an airborne surveillance command and control center to detect and track airborne targets at any altitude, at long ranges, and over all types of terrain and water.

  During the test exercises the aircraft, topped by a 9-m-diameter rotating radome assembly, would work in the United Kingdom with British naval forces and with the U.S. Navy's sixth fleet, and in West Germany with the NATO and U.S. air defense ground environment system and with Hawk defense missile batteries. NATO Hq in Brussels would provide ground static displays for NATO officials.

  During all exercises AWACS would coordinate with ground, sea, and air forces through a time-division multiple-access data link.

  This was the second deployment of AWACS to Europe. Since 1973, when the system had been demonstrated there, it had been equipped with additional major elements including communications, data-processing, navigation, display, and identification instrumentation. (AFSC Release OIP 99.75)

- NASA announced the appointment of Kenneth L. Woodfin as Assistant Administrator for Procurement, replacing George J. Vecchietti who had retired. (NASA Release 75–94)

*5 April:* A Soyuz spacecraft, launched by the U.S.S.R. from Baykonur Cosmodrome and carrying Cosmonauts Vasily Lazarev and Oleg Makarov, was returned to earth shortly after launch when the launch vehicle failed to perform normally. Tass reported, "On the third-stage stretch the parameters of the carrier rocket's movement deviated from the preset values and an automatic device produced the command to discontinue the flight under the program and detach the spacecraft for return to earth." Tass also reported that the purpose of the mission had been to continue experiments aboard the *Salyut 4* space station (launched 26 Dec. 1974). The Soyuz softlanded southwest of Gorno-Altaisk in Western Siberia. The search and rescue service brought the cosmonauts, both in good condition, back to the cosmodrome.

During a telephone conference on 8 April, Prof. Konstantin D. Bushuyev, U.S.S.R. technical director for the Apollo-Soyuz Test Project, told his U.S. counterpart, Glynn S. Lunney, that the launch vehicle that failed was not the version of the booster that would be used for the July ASTP launch. He promised to provide Lunney with additional details on the failure after they became available. (Tass, FBIS-Sov, 7 April 75, U1; NASA Release 75-97; *W Star*, 7 April 75, A4)

*7 April:* Thor-Delta launch operations at the Eastern and Western Test Ranges had resumed on a limited basis despite the continuing strike by employees of the launch vehicle's manufacturer, McDonnell Douglas Corp. [see 10 Feb.and 22-23 March], *Aviation Week and Space Technology* reported. With supervisory personnel at the company's plant completing the hardware in the place of striking workers, McDonnell Douglas hoped to finish 10 of the 12 launch vehicles scheduled for delivery to NASA launch sites by the end of 1975. NASA's GEOS-C Geodynamic Experimental Ocean Satellite and Canada's Telestar-C communications satellite, both originally scheduled for March, had been rescheduled for April and May launch on Thor-Delta vehicles delivered before the strike began. Other launch dates would depend on the buildup and checkout of booster hardware at the launch site, and on availability of hardware still at the plant.

Other scheduled launches affected by the strike included Nimbus-F experimental satellite, OSO-I Orbiting Solar Observatory, and Communications Satellite Corp.'s Marisat-A maritime communications satellite. (*Av Wk*, 7 April 75, 17)

- The European Space Research Organization (ESRO) announced the award of a $4.4 million [3.5 million accounting units] contract to a European consortium to provide software for ESRO's Meteosat meteorological satellite, scheduled for launch in April 1977. Under the terms of the contract the consortium, consisting of companies in France, United Kingdom, West Germany, Belgium, Italy, and Sweden, would provide data-acceptance capability, image processing, operational mission support, data archiving, analysis and dissemination of image data, and monitoring of the overall data-processing system.

  Meteosat, ESRO's first applications satellite, would record and transmit data for more accurate weather forecasting. (ESRO Release, 7 April 75)

- The relationship of the scientific community to any White House advisory panel should be the "same as for any other pressure group," Haywood Blum said in a letter to the editor of the *New York Times*. Based on past performance, "it seems to be unlikely for the scientific community to agree on any substantive issue; witness the ABM [antiballistic missile], SST [supersonic transport], National Cancer Institute, underground nuclear testing, use of DDT and breeder reactor development fights, to name only a few." Further, the wisdom and perspective of the scientific establishment "remain in doubt." One example was in scientific training in a time of the over-production of Ph.D.s: Dwindling funds were being used not to develop young scientists but to support "the Grant Swinger," who flew from place to place around the world "making contacts while recent graduates were squeezed out of their disciplines." The President needed counsel but, because the scientific community could not provide truly objective and balanced advice, "would it not be better for [this] advice to be exposed to public scrutiny?" (*NYT*, 7 April 75, 30)

*9 April:* The House of Representatives, by a vote of 318 to 72, passed H.R. 4700, the bill authorizing NASA $3 585 873 000 in funds for FY 1976 and $922 450 000 for the transition period 1 July through 30 Sept. 1976. The bill was passed as reported out of the House Committee on Science and Astronautics 14 March. (*CR*, 9 April 75, H2549–75)

- NASA would store $900 million worth of surplus Apollo-Saturn hardware instead of scrapping it as planned earlier, Dr. James C. Fletcher, NASA Administrator, said in a letter to Rep. Olin E. Teague, Chairman of the House of Representatives Committee on Science and Technology. The flight hardware would be stored in a manner to minimize costs and to permit restoration, if required, to flight condition. Dr. Fletcher also wrote that, as Space Shuttle launch-readiness schedule and program costs required, NASA would convert Launch Complex 39 and its supporting facilities to handle the Space Shuttle.

    As requested by the Committee during a 25 March meeting with Associate Administrator for Manned Space Flight John F. Yardley, NASA would supply the Committee with a summary of options for missions that could be flown using the remaining Apollo-Saturn flight hardware.

    Dr. Fletcher also stated that, because of the "considerable cost" to the government of storing the equipment, NASA would like to discuss the matter with the committee once again during the FY 1977 authorization hearings. (Text, letter Fletcher to Teague, 9 April 1975; UPI, *NYT*, 17 April 75, 7)

- The *Helios 1* (launched 10 Dec. 1974 by NASA for West Germany) mission objectives—to investigate the fundamental solar processes and solar terrestrial relationships by the study of the solar wind, magnetic and electric fields, cosmic rays, and cosmic dust—had been accomplished and the mission was adjudged successful. (NASA MOR, 11 April 75)

- NASA announced that Dr. Rocco A. Petrone, Associate Administrator, would leave NASA in May to join the National Center for Resource Recovery as president and chief executive officer. Dr. Petrone, who in 1960 was assigned on loan from the Army to Kennedy Space Center as Saturn project officer, later became Apollo program manager responsible for planning, developing, and activating all launch facilities for the Apollo program. Upon retirement from the Army he served at KSC as director of launch operations from 1966 to 1969, when he was appointed Apollo program director. In 1972 he was assigned additional responsibilities as program director of the NASA portion of the Apollo-Soyuz Test Project. In Dec. 1972 he was appointed Director of Marshall Space Flight Center, a position he held until his appointment in 1974 as Associate Administrator. (NASA Releases 69–124, 75–98)

- Sen. William Proxmire (D–Wisc.) said in a speech on the Senate floor that he had asked the Central Intelligence Agency to assess the safety of Soviet manned space technology. "The inlaunch failure of another Soviet manned satellite last Saturday [7 April] reinforces my deep concern that the upcoming joint Apollo-Soyuz experiment may be dangerous to American astronauts. . . . The history of the Soviet-manned program shows an appalling lack of consistency. As soon as one severe problem is solved another occurs." (*CR*, 9 April 75, S5527)

- Johnson Space Center announced the award to Martin Marietta Corp. of a 10-mo $373 000 study contract to investigate current technology and develop several conceptual designs of a manned maneuvering unit. When the most promising concept was selected, Martin would execute a detailed design and high-fidelity mockup of the unit and its support station. The contract also called for Martin to design and build a prototype hand controller for the unit.

    The maneuvering unit, a direct spinoff from the "Buck Rogers" unit that successfully demonstrated on 1973–74 Skylab missions the precision and control with which an astronaut could maneuver in zero-g, would allow Space Shuttle crew members to maneuver outside the Shuttle for maintenance, inspection, repair, retrieval, assembly, and photography. (JSC Release 75–24)

*9–22 April:* NASA launched *Geos 3* Geodynamic Experimental Ocean Satellite from Western Test Range at 4:50 pm PDT on a two-stage Thor-Delta 1410 launch vehicle. The satellite entered orbit with an 844-km apogee, 837-km perigee, 101.9-min period, and 115.0° inclination. Forty-one investigators would participate in the mission, whose primary objective was to perform an in-orbit radar altimeter experiment demonstrating the feasibility and utility of mapping the sea surface with a 1- to 2-m precision. *Geos 3* would also measure wave heights and contribute to the development of technology for future satellite radar altimeter systems. Other objectives included support of the calibration of NASA and other agencies' ground C-band radar systems; demonstration of a satellite-to-satellite tracking experiment with *Ats 6* (launched 30 May 1974); comparison of new and established geodetic and geophysical measuring systems; support investigations in solid-earth dynamic phenomena such as polar motion, earth rotation, earth tides, and continental drift; and further definition of orbit determination techniques using precision tracking systems such as lasers and doppler ground stations.

All spacecraft subsystems were checked out 9–10 April with telemetry indicating normal performance. Gravity-gradient stabilization maneuvers were also begun with the firing of the squibs that released the end mass and extended the boom to 0.7 m. Additional boom extension and retraction maneuvers were carried out 11 April with the maximum extension of 6 m occurring that day.

The momentum wheel was energized, providing additional three-axis stabilization; the desired pitch damping of ± 1° was achieved by 20 April.

By 22 April ground controllers activated doppler and C-band transponders and began tracking. Solid acquisition of return-pulse data was occurring on all passes, and the radar altimeter had operated in the high-intensity mode twice over the Atlantic Ocean. Engineering tests of the *Ats 6–Geos 3* satellite-to-satellite tracking experiment were completed. Calibration of radar altimetry began on 21 April and would continue until 20 May.

First of its kind, the satellite-to-satellite tracking experiment by *Geos 3* and *Ats 6* was expected to provide orbit information more precise than that obtainable by ground observations. From its geosynchronous orbit *Ats 6* tracked *Geos 3* for more than half its orbit while *Geos 3* radioed ranging signals through its S-band transponder to *Ats 6*, which relayed the signals to ground stations.

*Geos 3* carried an array of quartz reflectors that permitted laser ranging measurements accurate to 10 cm. A laser beam sent to the satellite would be reflected, and then received back at the originating ground site, one of a network of laser ranging stations which would track *Geos 3* during its mission lifetime.

Powered by 16 solar cell panels, the 340-kg, 8-sided spacecraft was the third in a series of GEOS spacecraft designed to gain knowledge of the earth's shape and dynamic behavior as part of the National Geodetic Satellite Program (NGSP). The NASA-managed program, begun in 1964 with the cooperation of the Depts. of Defense and Commerce and several universities and international organizations, was initiated to generate a unified world-survey network and to develop a more precise mathematical model of the earth's gravitational field.

*Geos 1* (*Explorer 29*, launched 6 Nov. 1965) and *Geos 2* (*Explorer 36*, launched 11 Jan. 1968) had demonstrated that the U.S. could make most observations using one satellite and that errors in any system could be discovered and corrected by reference to other systems. Other missions in the NASA geodesy program included *Explorer 22* (launched 10 Oct. 1964), *Explorer 27* (launched 29 April 1965), *Pageos 1* (launched 1 July 1966), and the Earth Resources Experiment Package (EREP, launched 15 May 1973 aboard *Skylab 1*).

The *Geos 3* program was managed by Wallops Flight Center under the direction of the Office of Applications at NASA Headquarters. Mission operations and the Thor-Delta launch vehicle were managed by Goddard Space Flight Center. The spacecraft, designed and built by the Applied Physics Laboratory of Johns Hopkins Univ., cost about $12.5 million. (NASA MORs, 8 April 75, 22 April 75; *Wkly SSR*, 3–9 April 75; NASA Release 75–88)

*10 April*: The first ground-test hardware for the Space Shuttle, a hydraulic actuator for the main engine, had arrived at Marshall Space Flight Center for functional checkout, MSFC announced. Each of the Orbiter's three main engines would use two of the actuators to gimbal the engine for steering control. After checkout the actuator—one of 8 to be tested—would be sent to the National Space Technology Laboratories where the main engines would be tested. (MSFC Release 75–68)

• Dust storms had been observed on the moon every sunrise and sunset by a three-axis microparticle detector left on the moon by *Apollo 17* astronauts during their December 1972 mission, a *Nature* magazine article reported. The article, based on a report by scientists at Goddard Space Flight Center, said the GSFC scientists suspected that the dust was being lifted from the lunar surface and then driven away from the rising or setting sun by electrostatic fields associated with the transition from day to night, or vice versa.

The finding supported a theory advanced 20 yr ago by Dr. Thomas Gold of Cornell Univ., who had suggested that electrostatic forces transported dust from lunar highlands to lowlands, accounting for the smooth appearance of the lunar seas. Although the Apollo landings had shown that those seas had been formed by lava flooding rather than deep dust accumulations, their surface was covered with very fine dust. Also, because neither wind nor rain existed on the airless moon, some other process must have been responsible for the evident erosion. Electrostatic levitation was being suggested as one possible

explanation. Because the lunar vacuum acted as an insulator, the dust cloud hugged the surface, reaching only between 20 and 30 cm high, explaining why it had been detected from a low angle by instruments on the lunar surface but never by orbiting astronauts or earth-based equipment. (Hughes, *Nature*, 10 April 75, 481–2)
- The Air Force's new B–1 bomber broke the sound barrier for the first time while making its sixth flight test, from Vandenberg Air Force Base. During the 6-hr 23-min mission, the B–1 also took on fuel from a KC–135 tanker for the first time and opened its weapons-bay doors. (AP, *W Star*, 11 April 75, 4)

*11 April:* President Ford, in a White House ceremony, presented the National Space Club's Dr. Robert H. Goddard Memorial Trophy to *Skylab 4* Commander Gerald P. Carr on behalf of the nine astronauts who inhabited the Skylab Orbital Workshop (launched 14 May 1973) during periods in 1973 and 1974. The astronauts were cited for their "exceptional accomplishments in the Skylab program . . . . When the entire mission was jeopardized the first crew, through uncommon ingenuity, saved the mission. The scientific content of the program is of extraordinary significance, in all disciplines. The demonstration that men can live and work in space for extended periods sets the stage for future exploration of space."

At the Space Club's 18th annual Goddard memorial dinner, the following awards were presented:

| Award | Recipient (Affiliation) | Citation |
|---|---|---|
| Dr. Robt. H. Goddard Memorial Trophy | NASA Skylab Astronauts Charles P. Conrad Joseph P. Kerwin Paul J. Weitz Alan L. Bean Owen K. Garriott Jack R. Lousma Gerald P. Carr Edward G. Gibson William R. Pogue | For their exceptional accomplishments in Skylab program |
| Astronautics Engineer Award | John M. Thole (NASA) | For direction of the OSO and ATS programs |
| Nelson P. Jackson Aerospace Award | NASA-Fairchild Industries *Ats 6* team | For demonstration of a significant advance in space communications |
| Hugh L. Dryden Memorial Fellowship | Dr. Bruce F. Smith (NASA) | For demonstrated abilities in addressing, theoretically and experimentally, a broad range of astrophysical problems |
| Dr. Robt. H. Goddard Scholarship | Alice K. Babcock (Georgetown Univ.) | To continue higher scientific education |
| Dr. Robt. H. Goddard Historical Essay | James E. Oberg (USAF) | For essay, "Russia Meant to Win the Moon Race?" |
| Nat'l Space Club Press Award | Roy Neal (NBC) | For resourcefulness in reporting the U.S. space effort |

(*PD*, 14 April 75, 372; program)
- NASA's Apollo-Soyuz Test Project Director Chester M. Lee said at a NASA Headquarters press briefing that a number of modifications had been made on the Apollo command and service modules to accommodate the special requirements of the July U.S.–U.S.S.R. ASTP mission.

Changes to the command module included a modified umbilical to accommodate docking-module functions, an additional TV camera and recorder for coverage of joint activities, modified controls and displays for new equipment and experiments, additional equipment for communications and TV coverage via *Ats 6* (Applications Technology Satellite, launched 30 May 1974), modified stowage, added intervehicular intercom, and additional VHF-FM communications capability at the U.S.S.R. frequency.

Changes in the service module included added heaters and insulation of propellant systems for solar inertial attitude, added propellant storage module for increased attitude-control and backup deorbit capability, and deletion of the unused main propellant tank.

The docking module would be carried in the same position in which the lunar module had been during lunar landing missions. It would be extracted in the same manner, by separating the CSM from the booster once in orbit, turning around, and pulling out the docking module with the nose of the CSM.

In response to questioning, Lee said that the U.S. cost of the ASTP mission would fall within the $245-million congressional authorization. Although the Soviets had not quoted a specific amount, Lee believed that the launch of the two unmanned Soyuz spacecraft to check out modifications made for ASTP, the launch of a manned mission, and the preparations for two complete spacecraft and crews would indicate that the Soviets were investing "at least the equivalent" in resources. (Transcript)

*14 April*: Flight Research Center pilots began flight tests as part of a cooperative industry-government program to measure noise levels of five business-class jet aircraft during various landing approach maneuvers. Object of the tests was to define noise characteristics for this class of aircraft and the effectiveness of alternate landing-approach procedures in reducing community noise levels. Using a Rockwell Sabreliner, Grumman Gulfstream II, Gates Learjet, Beech-Hawker 125 provided by the National Business Aircraft Association, and Lockheed Jet Star provided by NASA, FRC pilots flew four landing approaches over a microphone array. A normal 3° approach provided baseline information; a 3° decelerating approach with engines at reduced power, a 4° approach with normal power, and a two-segment approach starting with 6° angle and shallowing into a 3° approach provided additional data. (NASA Release 75-103)

*14-30 April:* U.S. Apollo-Soyuz Test Project flight crews, accompanied by 32 members of Working Group 1, met in the U.S.S.R. with their Soviet counterparts for the final training session in Soviet Russia before the scheduled July launch. U.S. astronauts included prime crew members Thomas P. Stafford, Vance D. Brand, and Donald K. Slayton; backup crewmen Alan L. Bean, Ronald E. Evans, and Jack R. Lousma; and support crewmen Karol J. Bobko, Robert L. Crippen, and Eugene A. Cernan.

During practice sessions similar to those held in the U.S. 7 Feb.-1 March, the participants practiced communications skills, transfer procedures, joint activities, and contingency situations, using the Soyuz simulator at the Gagarin Cosmonaut Center in Star City.

On 19 April the crews visited the mission-control center at Kaliningrad and, on 28 April, became the second group of westerners to visit the Baykonur cosmodrome (French President Charles de Gaulle had been the first western visitor when he witnessed the launch of *Cosmos 122* on 25 June 1966). At Baykonur the U.S. crewmen saw the actual flight hardware as well as the primary launch pad 2 km away, and the prime and backup crews each spent an hour in the Soyuz spacecraft. The astronauts later reported to the press that they had found no significant differences between the actual spacecraft and the simulators used during training. (Ezell *et al.*, *The Partnership: A History of the Apollo-Soyuz Test Project*, 278—280; JSC Releases 75—22, 75—25; *W Post*, 30 April 75, A21; *SBD*, 18 April 75, 273; *A&A 66*, 223)

*15 April:* The Council of the European Space Research Organization met and approved appointment of directors proposed by ESRO's Committee for the soon-to-be-established European Space Agency. When ESA became operational at the end of May, the new directorate would include Roy Gibson (United Kingdom) as Director General, André Lebeau (France) as Director of Planning and Future Programs, Dr. Ernst Trendelenburg (West Germany) as Director of Scientific and Meteorological Programs, and George Van Reeth (Belgium) as Director of Administration.

Directors already in office who would continue after the establishment of ESA were Bernard Deloffre (France), Director of the Spacelab Program; Prof. Gianni Formica (Italy), Director of the European Space Operations Center; Ove Hammarström (Sweden), Director of the European Space Research and Technology Center; Dr. Walter Luksch (West Germany), Director of Communications Satellite Programs; and Prof. Massimo Trella (Italy), Technical Inspector. (ESRO Release, 21 April 75)

- Flight Research Center announced the award of a 30-mo cost-plus-fixed-fee contract, valued at $13 million, to Rockwell International Corp. to design and build two subscale model aircraft for NASA's Highly Maneuverable Aircraft Technology Program. The two-phase contract called for an initial 60-day effort for detailed program planning; upon satisfactory completion of phase 1, NASA would approve the final design and fabrication.

  The two vehicles were scheduled for delivery to FRC late in 1977 for flight-testing, using the Remotely Piloted Research Vehicle test technique [see 3 March] developed by FRC. Air-launched from a carrier aircraft, the vehicles would be flown through maneuvers by a ground-based pilot using TV, telemetry, and radar. (FRC Release 7—75)

- The Air Force Systems Command announced the award of a $100 000 contract to Aero Co. to study the feasibility of using parafoils half the size of a C—5 aircraft's wing to recover remotely piloted vehicles (RPVs). RPVs were currently retrieved by recovery helicopters, a reliable but expensive system affected by weather conditions. The parafoil, part of an AF effort to find cheaper ways to land RPVs, was a high glider which, when deployed like a parachute, would take on an airfoil shape and descend at a rate of 1 m forward for each 0.3-m loss in altitude. Under the terms of the contract Aero would design and build two 12- by 24-m parafoils for testing scheduled to begin in May. (AFSC Release OIP 81.75)

*16 April:* The U.S.S.R. was planning to flight-test the SLX−14, a launch vehicle larger than the Saturn 5, from Tyuratam during the summer of 1975, Kenneth W. Gatland reported in the *Christian Science Monitor*. Launch preparations, observed by the Air Force's Big Bird reconnaissance satellites, had been under way since 1974. Gatland reported that the project had been delayed for more than 5 yr when a previous prototype caught fire and exploded during a 1969 fueling test. Two others had failed in flight, in 1971 and 1972. After considering cancellation of the project, Soviet officials ordered a drastic design review and the program continued with increased emphasis on systems reliability and ground testing. (Gatland, *CSM*, 16 April 75, 1)

*18 April:* NASA announced the appointment of David R. Scott as Director of Flight Research Center to replace Lee R. Scherer, who had become Director of Kennedy Space Center. Scott had been an astronaut on *Gemini 8* (16 March 1966), on *Apollo 9* (3−13 March 1969), and on lunar landing mission *Apollo 15* (26 July−7 Aug. 1971). After leaving the astronaut corps in 1972, Scott served as technical assistant to the Apollo Program Manager at Johnson Space Center, and as special assistant for mission operations and government-furnished equipment in the Apollo Spacecraft Program Office. (NASA anno, 18 April 75)

*19 April:* The U.S.S.R. launched *Aryabhata*, India's first scientific satellite, from Kapustin Yar in the Soviet Union. The satellite entered orbit with a 611-km apogee, 568-km perigee, 96.5-min period, and 50.7° inclination. The Delhi Domestic News Service said that the 360-kg satellite, named for a fifth-century Indian astronomer and mathematician, carried instruments to measure x-rays from celestial sources, look for neutrons and gamma radiation from the sun, and measure ionospheric parameters. Ground tracking stations in India and the Soviet Union reported the instrumentation was functioning normally.

*Aryabhata*, designed and built by scientists at India's Space Research Organization with technical assistance from Soviet scientists and specialists, was completed under an agreement for joint Indian-Soviet collaboration signed on 10 May 1972. The press reported the estimated cost of the project to be $6 million. (Delhi Dom News Ser, FBIS−India, 22−24 April 75; Weintraub, *NYT*, 21 April 75, 49; GSFC *Wkly SSR*, 17−23 April 75; educ'l attaché, Indian Embassy, Wash DC, interview, 18 Nov 76)

*21 April:* West Germany was studying the possibility of purchasing a Spacelab pressurized module for use on the U.S. Space Shuttle, *Aviation Week and Space Technology* reported. The German Spacelab would be configured for research in space processing, a field of considerable interest to Germany. Its missions on Spacelab, launched by NASA aboard the Space Shuttle on a cost-reimbursable basis, would be separate from missions conducted by the European Space Research Organization. (*Av Wk*, 21 April 75, 27)

- The Air Force Aero Propulsion Laboratory and NASA Lewis Research Center had begun a joint research program to determine the feasibility of using synthetic fuels to power military and commercial jet aircraft, the Air Force Systems Command announced. As part of the 10-yr $8-million study, Exxon Corp. had been awarded a contract to study the refining processes like hydrogenation required to produce fuels from coal and oil-shale synthetic crudes, and to compare the chemical and physical properties of these fuels with those of petroleum derivatives. Costs of processing the synthetics would depend on definition of engine and fuel-system needs.

The decision to begin research on synthetic fuels was part of an Air Force conservation effort. U.S. aircraft, military and commercial, accounted for 8% of U.S. petroleum consumption; fuel costs for military jets alone had more than tripled in the previous 18 mo. The cost of fueling a C−5 cargo aircraft had risen from $6780 in April 1973 to $22 400 in April 1975. (AFSC Release OIP 070.75)

- A NASA-sponsored research program at Massachusetts Institute of Technology had demonstrated that ultraviolet light could convert water into a hydrogen fuel, NASA announced. MIT researchers placed a titanium dioxide crystal and a piece of platinum connected by a wire into a beaker of water in which inert ionic substances were dissolved to allow an electric current to flow. The titanium dioxide and platinum acted as electrodes, establishing electrical contact with the water. When ultraviolet light from a laser illuminated the titanium dioxide, application of electrical potentials as low as 0.2 volts produced hydrogen and oxygen gases.

  The process thus far was inefficient, only 1% of the light energy being converted into chemical energy. Researchers were experimenting with different electrodes and reaction temperatures to find methods that would decompose water using low-energy visible light instead of high-energy ultraviolet light, which was only a small fraction of the light available from the sun. (NASA Release 75−106)

*21–24 April*: The National Academy of Sciences held its 112th annual meeting and the National Academy of Engineering held its 11th annual meeting, in Washington, D.C. The state of the sciences and their potential contribution to the solution of national problems were discussed. During one session Dr. J. Herbert Holloman, Massachusetts Institute of Technology professor and former Assistant Secretary of Commerce for Science and Technology, said that the U.S. "no longer dominated the world's technological process. We have now become the strongest nation among many." Dr. Holloman said that the growth of productivity in the U.S. during the last 15 yrs had been "the lowest of any industrial nation in the world with the exception of the United Kingdom."

Dr. Michael Boretsky, Dept. of Commerce analyst, agreed, saying that the productivity decline could be attributed in part to the "relative decline in the overall rate of the United States technological advance." Since the mid-1960s, output per man-hour had been growing at half the average rate of the previous 100 yr, while productivity gains had been speeding up in other countries. Funds that could correct this situation by supporting the development of new technology had instead been channeled into social commitments.

Dr. Ralph Landau, chemical engineer and president of Halcon, Inc., said that, because labor costs in the U.S. and leading foreign nations were virtually equal, it was no longer advantageous for U.S. companies to sell licenses to foreign enterprises to use American patents. Instead pressures existed to keep the technology at home, creating jobs in the U.S., and to sell U.S.-made goods abroad.

During the business meeting on 22 April, 84 new members were elected to NAS "in recognition of their distinguished and continuing achievements in original research." NAE elected 86 U.S. engineers to its membership, including Dr. Kurt H. Debus, former Kennedy Space Center Director; Dr. Rocco A. Petrone, NASA Associate Administrator; and Dean R. Chapman, Ames Research Center scientist. (NAS Release, 22 April 75; NAE Release, 1 April 75; joint NAS-NAE Release, 9 April 75; McElheny, *NYT*, 27 April 75, 15)

*23 April:* NASA marked the 60th anniversary of the first meeting of NASA's predecessor institution, the National Advisory Committee for Aeronautics (NACA) Established by a law signed by President Woodrow Wilson 3 March 1915, NACA had 12 presidentially appointed members from the military and scientific communities "to supervise and direct the scientific study of the problems of flight, with a view to their practical solution" and "to direct and conduct research and experiments in aeronautics."

During its nearly half-century of operation, NACA was responsible for the technical proficiency of U.S. aviation. Early NACA wind-tunnel research yielded the unique NACA cowling and wing shapes that would dominate aeronautics into the 1940s. NACA's fundamental research and its direct application to industrial, military, and civil aviation helped create the world's greatest commercial air transportation network before World War II and the world's strongest air forces during that conflict. In 1947, it was the NACA-Air Force X−1 aircraft that broke the sound barrier.

The agency had remained small in size and budget through 1939 when 523 persons worked with $4.5 million in funds. By the mid-1940s the staff had grown to 6800 and its budget to an annual $40 million. In 1958 the National Space Act dissolved NACA, transferring its 8000 workers and "all functions, powers, duties and obligations" to the new National Aeronautics and Space Administration. (NASA Release 75−86; Anderson, *Orders of Magnitude*, 1−3)

- NASA announced the award of a $11-million cost-plus-fixed-fee contract to Cutler-Hammer, Inc., AIL Div., for acquisition of the Microwave Scanning Beam Landing System ground stations for the Space Shuttle Orbiter. AIL would perform the research and development necessary to build the ground stations for the Shuttle landing sites. Ground-station hardware would be able to transmit localizer and glide slope signals to the Orbiter and respond to distance-measurement interrogations from the Orbiter. (JSC Release 75−32)

*24 April:* NASA announced the award of a $2 376 400 contract to Reinhold Construction Co. for phase 2 of the Space Shuttle landing facility, including the landing aids control building, Orbiter landing instrumentation facilities, and related communications and utility systems. Phase 2 was scheduled for completion by the end of 1976.

Phase 1, which began in April 1974 as a $21 812 737 contract with Morrison Knudsen Co., included construction of the 4570-m-long, 91-m-wide runway, towway apron, and towway. Site preparation was nearing completion and the paving with nearly 200 000 cu m of cement would begin in July. Phase 1 was scheduled for completion by the summer of 1976. (KSC Release 62−75)

*25 April:* NASA announced the selection of Sperry Rand Corp. for the negotiation of a fixed-price $8-million contract for a Space Shuttle mission-simulator computer complex at Johnson Space Center. Sperry would design, develop, fabricate, test, and deliver a computing complex and all associated hardware for the Shuttle mission simulator. (JSC Release 75−37)

*28 April*: NASA announced the beginning of a broad program of stratospheric research to determine the effects of natural and manmade pollutants on the earth's ozone shield. In a cooperative effort with other Federal agencies, universities, and private industry to assess the threat posed by Freon, a trade name for gases used as refrigerants and in aerosol spray cans, NASA would use high-altitude U−2 aircraft, balloons, sounding rockets, and satellites to measure amounts of these compounds and their decomposition products already in the upper atmosphere. These data combined with data obtained in laboratories would help to determine the validity of predictions of Freon-induced ozone depletion in the stratospheric layer. Some scientists feared that a serious reduction of the protective ozone and a resulting increase in the amount of ultraviolet light striking the earth could increase the incidence of skin cancer and could change the average temperatures of the earth's atmosphere.

The NASA program, managed by the Office of Space Science, emphasized basic research needed to understand the physical and chemical processes occurring in the atmosphere; this understanding was necessary to the assessment of changes caused by human activity. (NASA Release 75−124)

*29 April*: NASA was inviting scientists to propose experiments for a 1979 mission that would fly by Jupiter in 1981 and then travel to a mid-1985 encounter with Uranus. Although the mission was not yet approved, NASA officials wanted to select flight investigations early to initiate necessary science planning and development. (NASA Release 75−127)

*30 April*: The first set of wings for NASA's Space Shuttle Orbiter began an 18-day land and water journey from Grumman Aerospace Corp.'s facility in New York to Palmdale, Calif., for final assembly with the Orbiter. The two wing panels, each 9 m long from tip to fuselage and 18 m wide at the fuselage edge, would travel aboard transporters specially designed to be pulled by a truck—at 16 km per hr—or loaded onto an ocean-going container ship. From New York the wings would make a 13-day trip through the Panama Canal to Long Beach, Calif., and then travel overland to Palmdale.

The double-delta aluminum wings had been designed, manufactured, and tested by Grumman under a $40 million subcontract awarded by the prime Space Shuttle contractor, Rockwell International Corp., in March 1973. (JSC Release 75−38)

- NASA announced the appointment of Dr. John E. Naugle, Deputy Associate Administrator, to be Acting Associate Administrator until a replacement was named for Dr. Rocco A. Petrone. (NASA anno, 30 April 75)

*During April*: Soviet spending for research and development had risen 50% between 1968 and 1973, representing an increase of from 2.5% to 3.8% of the Soviet gross national product, the *Air Forces Journal International* quoted a Dept. of Commerce report as saying. During the same period, outlays for U.S. R&D had fallen from 3% to 2.4% of the GNP. The number of Soviet scientists and engineers working in R&D functions nearly doubled in the period, from 594 000 to 1 069 000; by contrast, the U.S. technological pool had risen from 454 000 to 531 000. The growth stemmed from a Soviet decision, announced in 1968, to increase Soviet expenditures for science and technology by more than 60%. The report estimated that 90% of the Soviet technical workforce was in the military and space fields. (*AF Journ Int*, April 75, 20)

- NASA's flying laboratory *Galileo II*, an instrumented Convair 990, participated in a month-long cooperative investigation of the Arctic seas. Researchers from NASA, the U.S. Geological Survey, U.S. Army, U.S. Navy, National Oceanic and Atmospheric Administration, and several Canadian agencies hoped to find methods of measuring critical changes in the Arctic environment that would help environmentalists find out how best to manage Arctic resources.

  A major objective of the program was an experiment—called the AIDJEX, the Arctic Ice Dynamics Joint Experiment—to investigate the relationship among ice motion, weather, and ocean conditions. Combined data from aircraft- and ground-based investigations would be used to interpret satellite data obtained at the same time.

  BESMEX, the Bering Sea Marine Mammal Experiment, would study the migration patterns of marine mammals and the relationship between mammal population, migration, and sea-ice conditions.

  In a third experiment, scientists would measure microwave emission characteristics of polar ice, hoping to find a key to global monitoring of ice distribution by climate and size. This information would aid in predicting the survival rate of offshore drilling rigs. (ARC Release 75−11)

- Goddard Space Flight Center awarded a $20 000 research grant to the University of Florida to establish the feasibility of the electromagnetic wave energy-conversion (EWEC) concept. The EWEC converter used electromagnetic antennas to absorb the sun's electromagnetic radiation into an array of insulated components and then convert it directly to electricity. The absorber operated similarly to insect antennae and the retina of the eye: Insects communicated by the reception of infrared and other electromagnetic radiation or energy, using their antennae tuned to the proper frequency by nature's design. Once received, the energy was converted into electrical impulses. (NASA Release 75−102)

*During April−May*: NASA conducted Atmospheric Variability Experiment IV and Atmospheric Variability Severe Storms Experiments I and II in cooperation with the U.S. Air Force, Army, National Weather Service, and several universities. Coordinating data collected over the eastern two-thirds of the U.S. by satellites, aircraft, radar, and weather balloons, participants hoped to develop new techniques for predicting where and when tornadoes might strike. The data collected, including information on thunderstorms, updrafts, and wind and cloud data, would be compared with ground-truth observations to determine the relationship between severe storms and their environment. Participating in the experiments were 41 radiosonde stations, 18 radar stations, 7 instrumented aircraft, and 5 meteorological satellites. (MSFC Release 75−102)

## May 1975

*1 May:* A decision had been made to designate Viking Lander 2 and Orbiter 2 as Viking Spacecraft—A, Kennedy Space Center's *Spaceport News* reported. Lander 1 and Orbiter 1 would become Viking Spacecraft—B. The change had been made to permit early sterilization of Lander 2 to meet the 11 Aug. launch date. (*Spaceport News,* 1 May 75, 2)

- Dr. Noel W. Hinners, NASA Associate Administrator for Space Science, and Major General Soehardjono, Director General of Posts and Telecommunications for the Republic of Indonesia, signed an agreement for the July 1976 launch of Indonesia's first communications satellite on a NASA Thor-Delta rocket. The new comsat, in synchronous equatorial orbit over Indonesia where its signal could be transmitted to 40 ground receiving stations, would permit Indonesia to begin development of a national communications network. (NASA Release 75–134)

- A telescope aboard NASA's *Oao 3* Orbiting Astronomical Observatory (named *Copernicus* and launched 21 Aug. 1972) was observing three nearby stars—epsilon Eridani, tau Ceti, and epsilon Indi—for signs of other civilizations that might be trying to contact earth with laser beams.

  The telescope had scanned the first star in November 1974 for 14 orbits while the spectrometer scanned the ultraviolet spectrum for potential laser signals to earth. The data were being analyzed and scientists were planning to scan tau Ceti and epsilon Indi again during the summer and fall of 1975. (NASA Release 75–130)

- Under an agreement with the Air Force, Flight Research Center pilots would fly the YF–17 aircraft—contender for selection as the Air Force combat fighter—in a research program to acquire flight data of a high-performance aircraft for comparisons with wind-tunnel tests and other analytical methods, FRC announced. The comparisons would be used to update prediction techniques required for the design of future highly maneuverable aircraft. During the 25 flight hrs, pilots would measure performance and stability at high angles of attack at both high and low altitudes, and perform maneuverability tests and pilot physiological studies during sustained high acceleration levels. (FRC Release 11–75)

- An overactive sun and excessively bright moon had caused brief shutdowns of instruments aboard Intelsat (International Telecommunications Satellite) communications satellites, the *New York Times* reported. Showers of particles during solar flares on 11 March and 13 Oct. 1974 had built up excess electrical charges on satellite surfaces that had not been electrically grounded. Noise from the resulting electrical discharges confused readings of which way onboard antennas were pointing, causing the antennas to point away from the earth and temporarily cut off communications.

A similar problem occurred 27 March when a spinning comsat interpreted infrared light from an exceptionally bright moon as coming from earth. Because of this, the satellite's controls were receiving an earth indication twice as often as usual, thus commanding the pointing system to compensate by despinning at twice its normal rate. Communications were blacked out because the antennas were spinning relative to the earth rather than remaining stationary and pointing toward earth.

The *NYT* quoted Communications Satellite Corp. engineers as saying that these occurrences pointed out the need for more knowledge about the exact behavior of Intelsat satellites in space so that changes could be made in the design of future satellites. The solar-particle problem had led to more thorough electrical grounding aboard the newest Intelsat satellites. (McElheny, *NYT*, 1 May 75, 83)

*2 May*: Dr. John F. Clark, Director of Goddard Space Flight Center, and Daniel J. Fink, Vice President of General Electric Corp. Space Div., were awarded the National Aeronautic Association's Robert J. Collier Trophy for individual accomplishment in making the Landsat—formerly the Earth Resources Technology Satellite—program the outstanding aerospace event of 1974. The presentation was made at a dinner in Washington, D.C., jointly sponsored by the NAA and the National Aviation Club. Choice of the two officials as representative of the NASA-industry team "which abundantly proved during 1974 the value of remote sensing to space" was the unanimous decision of the selection committee of 26 distinguished leaders and authorities. The committee also paid particular tribute to Hughes Aircraft Co.'s Aerospace Group and RCA's Government and Commercial Systems Group for their roles in the earth-resources and environmental surveys by Landsat. (NAA Release, 6 March 75; *NASA Activities*, April 75; *Goddard News*, June 75)

• President Ford presented the Harmon Aviation Awards, given for outstanding feats of individual pilot skill, worthy of international recognition, and contributing to the art and science of flight.

The 1974 Astronauts' Trophy was awarded to the *Skylab 2* crew who, during their 25 May to 23 June 1973 mission, became the first astronauts to successfully accomplish major repairs on a spacecraft in space. *Skylab 2* commander Charles Conrad, Jr., accepted the award on behalf of himself and crewmembers Paul J. Weitz and Joseph P. Kerwin.

President Ford presented the 1974 Aeronauts' Trophy to the son of Malcolm S. Forbes for the latter's series of 21 flights in a hot-air balloon that took him 4000 km across the U.S. from Coos Bay, Ore., to Gwynn Island, Va.

The 1974 Aviators' Trophy went to Col. Edward J. Nash, Deputy Chief of Staff for Operations of the 21st Air Force, for his 21 000-km round-trip airlift mission in a C−5 aircraft from the U.S. to Israel, and for his direction of continuous C−5 and C−141 operations from the U.S. to the Azores and on to Israel between 14 Oct. and 14 Nov. 1973.

The 1973 Aviators' Trophy was awarded to L/C Edgar L. Allison for piloting an HC−130H aircraft of the Air Force 57th Aerospace Rescue and Recovery Squadron nonstop on 20 Feb. 1972 from Taiwan to Scott Air Force Base, Ill., a distance of 14 053 km, without refueling. This established a record for the longest straight-line nonstop flight in a turboprop aircraft. (*PD*, 5 May 75, 480−481)

*5 May:* The Senate Committee on Aeronautical and Space Sciences favorably reported out H.R. 4700 authorizing FY 1976 appropriations to NASA. Changes recommended by the committee would raise the NASA authorization by $5.7 million to $3.545 billion.

### Recommended Changes in the FY 1976 NASA Authorization by the Senate Committee

| Program | Difference from Submission (millions) | Recommended Authorization (millions) | Purpose |
|---|---|---|---|
| *Research and Development:* | | | |
| Space shuttle | 0 | 1206.0 | Accepted as presented |
| Space flight operations | −4.0 | 203.1 | Allocation of reduction left to discretion of NASA |
| Advanced missions | −1.5 | 0 | Recommended deletion of this item; activities to be integrated with similar and related functions in other programs |
| Physics and astronomy | +7.0 | 162.8 | To increase knowledge of upper atmosphere, identifying problems and developing new technology for dealing with them |
| Lunar and planetary exploration | 0 | 259.9 | Accepted as presented |
| Launch vehicle procurement | 0 | 166.9 | Accepted as presented |
| Space applications | +8.9 | 183.9 | To increase emphasis on weather research, LACIE program, advanced communications research, and Space Shuttle payload definition |
| Aeronautical research and technology | 0 | 175.4 | Accepted as presented |
| Space & nuclear research & technology | 0 | 74.9 | Accepted as presented |
| Energy technology applications | 0 | 5.9 | Accepted as presented |
| Tracking & data acquisition | −2.2 | 240.8 | To seek economies in the program |
| Technology utilization | 0 | 7.0 | Accepted as presented |
| R&D Totals: | +8.2 | 2686.6 | |
| *Research and Program Management:* | | | |
| Basic submission | 0 | 776.0 | Accepted as presented |
| R&PM Totals: | 0 | 776.0 | |
| *Construction of Facilities:* | | | |
| Modification of 11-by-11-ft transonic wind tunnel at ARC | 0 | 2.7 | Accepted as presented |
| Addition to Lunar Sample Facility at JSC | −2.5 | 0 | Deleted from funds |
| Addition for composite model and metal finishing shops at LaRC | 0 | 1.9 | Accepted as presented |

### Recommended Changes in the FY 1976 NASA Authorization by the Senate Committee (continued)

| Program | Difference from Submission (millions) | Recommended Authorization (millions) | Purpose |
|---|---|---|---|
| *Construction of Facilities: (cont.)* | | | |
| Space Shuttle facilities | 0 | 47.2 | Accepted as presented |
| Rehabilitation of facilities at various locations | 0 | 16.0 | Accepted as presented |
| Minor construction of new facilities or additions to old facilities | 0 | 5.0 | Accepted as presented |
| Facility planning and design | 0 | 9.3 | Accepted as presented |
| C of F Totals: | −2.5 | 82.1 | |
| Grand Totals* | +5.7 | 3544.7 | |

*Total does not include a supplemental authorization for increased pay costs introduced and considered as H.R. 13172.

For the transition period 1 July to 30 Sept. 1976, the committee recommended decreasing NASA's total budget request of $958 900 000 to $929 900 000. The requested $730 600 000 for research and development was reduced to $704 600 000, leaving NASA management to apportion the $26 million reduction. The committee reduced funds requested for construction of facilities from $14 500 000 to $11 500 000 in a series of minor economies. (Sen Com on Aero & Space Sci Rpt 94−103)

- The Apollo-Soyuz Test Project was "one of the few tangible survivors of that champagne-bubble atmosphere of détente along with the badly tarnished SALT [Strategic Arms Limitations Talks] agreements," *Aviation Week and Space Technology* said in an editorial. From the Soviet viewpoint the goals were obvious: to tap into the mainstream of U.S. technology, to blur the international image of U.S. superiority with the appearance of parity by flying a joint manned mission, and to develop the technical base for international space rescue capability. U.S. goals remained obscure. The mission would provide the answers to some experimental questions: Could the U.S. and U.S.S.R. work together in a joint space program? Could an international rescue capability be developed? Could the ASTP experience be expanded to future joint space ventures?

That ASTP had survived to the final countdown was a tribute to the flexibility, stamina, and ingenuity of the technical working groups of both countries. A formidable series of technical, philosophical, linguistic, and operational problems had been solved and the investment by both countries had been substantial.

ASTP had been "partially worth the effort... to have a sizable group of NASA and industry technicians get some practical hard-headed experience in working with their Soviet counterparts and acquiring a realistic view of their technical capabilities and philosophy. We think the rest of the value of the mission will come from determining just how serious the Soviets are about future joint ventures in space and how much they really want to give and how much they want to take." (Hotz, *Av Wk*, 5 May 75, 7)

*5–22 May*: Representatives from the five Apollo-Soyuz Test Project Working Groups, including 55 NASA employees, met in Moscow to conclude preparations for the scheduled 15 July launch date. Part of the U.S. contingent left for the Soviet Union on 5 May and the final group, led by NASA Deputy Administrator George M. Low, departed 16 May.

U.S. communications equipment checkout and docking target alignment tests were completed at the Baykonur cosmodrome by mid-May, in time for Dr. Low and a group of NASA officials to take an inspection and orientation tour of the cosmodrome. The group visited the launch pad from which the ASTP Soyuz spacecraft would be launched and the Soviet equivalent of Kennedy Space Center's Vehicle Assembly Building (VAB) where the group inspected the prime launch vehicle and prime backup Soyuz spacecraft.

At the 22 May flight readiness review (FRR) chaired by Dr. Low and Vladimir A. Kotelnikov, acting president of the Soviet Academy of Sciences replacing the ailing Mstislav V. Keldysh, Soviet and U.S. space officials signed the report indicating that the "Apollo-Soyuz Test Project was proceeding in accordance with the agreed schedule and ready to proceed toward the launching planned for July 15, 1975."

Before the signing, U.S. ASTP Technical Director Glynn S. Lunney and his Soviet counterpart Konstantin D. Bushuyev gave a 5-hr review of the technical histories of their respective spacecraft. Reporting that no major problems remained, Lunney and Bushuyev said that 133 documents had been negotiated and signed and only the postflight review remained to be prepared.

Working Group chairmen made their final presentations indicating that all major tasks had been completed. (NASA Release 75–156; Ezell *et al.*, *The Partnership: A History of the Apollo-Soyuz Test Project*, 285–301; Reuters, *NYT*, 23 May 75, 33)

*6 May*: President Ford announced the recipients of the 1974 Presidential Management Improvement Awards, given annually as the highest recognition for outstanding contributions in improving the effectiveness and economy of government operations. Among the recipients was NASA's Mariner–Venus 1973 management team at NASA Headquarters and Jet Propulsion Laboratory. (*PD*, 12 May 75, 504)

• Détente, while relaxing tension between the U.S. and Soviet Union, had produced "a technology drain from the United States to the Soviet Union that has cost the United States millions of dollars, thousands of jobs and some valuable scientific secrets," Jack Anderson and Les Whitten reported in a *Washington Post* article. Soviet technology, which lagged years behind the U.S. in fields including computers and space science, had little to offer and much to gain from the technological exchange. With official government encouragement U.S. firms had exported or were planning to export millions of dollars worth of technical information and products including aircraft and advanced computers.

Anderson and Whitten reported that the Soviets often put finished products to military use. They were using détente and some advanced U.S. computers to catch up in their 10-yr lag behind the U.S. in computer science, a key to the U.S.'s overall technological superiority. One U.S. firm, Control Data Corp., had signed preliminary agreements to help the Soviet Union build plants to manufacture computers and peripheral equipment.

The Soviets were also eager to catch up with the U.S. in the production of wide-bodied jets. Holding out possible contracts, the U.S.S.R. had obtained detailed plans for wide-bodied jets from Boeing Co., Lockheed Aircraft Corp., and McDonnell Douglas Corp. (Anderson et al., *W Post*, 6 May 75, B13)

- Astronomers in the People's Republic of China said they had discovered two minor planets between Mars and Pluto, reported a British press delegation visiting Nanking Observatory. The press quoted the Chinese as saying that the planets, observed in January, were rarely visible because of their position in the solar system. The Chinese had named the planets Purple Mountain One and Purple Mountain Two, for the name of the mountain on which the Nanking Observatory was located. (Reuter, *W Post*, 7 May 75)

- Rockwell International Corp., holder of the prime contract with NASA to build the Space Shuttle Orbiter, announced that more than $109 million in subcontracts for the Space Shuttle main engine had been awarded by the Rocketdyne Div. since work on the engine began 3 yr ago. Figures to date showed that firms in 40 states were participating in the Space Shuttle program.

  Rockwell reported that Space Shuttle development would funnel an estimated $5 billion into the U.S. economy by 1978 and create employment for approximately 50 000 workers. (Rockwell Int'l Release RO-9)

- Canada would develop the remote manipulator system (RMS) for NASA's Space Shuttle to permit astronauts inside the Shuttle Orbiter to deploy or retrieve payloads in space, NASA announced. Canada would fund the $30 million project, providing the first flight unit to NASA without charge in 1979. Canada would also supply flight units for follow-on Orbiters. Costs for these units would not include charges for Canada's research and development. (NASA Release 75-135)

*7 May*: An Italian launch crew launched *Explorer 53* (SAS-C) Small Astronomy Satellite for NASA from San Marco launch platform, located in the Indian Ocean off the coast of Kenya, at 6:45 pm EDT (1:45 am local time 8 May) on a Scout launch vehicle. The spacecraft entered near-equatorial orbit with a 523.35-km apogee, 502.18-km perigee, 94.88-min period, and 2.995° inclination.

Primary objectives of the *Explorer 53* mission were to measure the x-ray emission of discrete extragalactic sources, to monitor the intensity and spectra of galactic x-ray sources from 1.0 kev to 50 kev, and to monitor the x-ray intensity of the star Sco X-1.

During a preliminary in-orbit checkout, controllers noticed that a stuck nutation damper was causing the spacecraft nutation, or coning, angle to be 4° rather than the planned 0.2°. Energizing the z-axis coil reduced the coning angle to 0.5°. Although the damper could not be freed despite repeated attempts, the nutation could be kept within reasonable limits so that scientific objectives could be achieved. A committee would evaluate the problem.

Other spacecraft anomalies included intermittent operation of the spinning digital solar-aspect detector, and irregular and noisy operation of the spacecraft star sensor. Neither problem was expected to seriously degrade the mission.

The 195-kg satellite was designed to extend and complement the capabilities of *Explorer 42* (*Uhuru*, launched 12 Dec. 1970 as the first satellite dedicated entirely to the study of x-ray sources in space) and *Explorer 48* (launched 16 Nov. 1970 to study galactic and extragalactic gamma radiation). *Explorer 53*, with its broad spectral range, provided higher energy resolution than previous satellites and permitted a study of physical conditions near an x-ray source. The spacecraft's quasi-three-axis stabilization system allowed much longer continuous observation of sources. Data from *Explorer 53* would provide the basis for the more detailed x-ray studies to be made by High Energy Astronomy Observatories A and B scheduled for launch in 1977 and 1978.

The experiment aboard *Explorer 53* was designed and built by the Center for Space Research of the Massachusetts Institute of Technology. The spacecraft control section was designed by the Applied Physics Laboratory of the Johns Hopkins Univ. Goddard Space Flight Center managed the program for the Office of Space Science. Langley Research Center was responsible for the Scout launch vehicle. Italy was responsible for the assembly, checkout, and launch of the Scout on a cost-reimbursable basis. (NASA MORs 6 May 75, 15 May 75; NASA Releases 75−101, 75−121)

- Canada's *Anik 3* (Telesat−3) domestic communications satellite was launched by NASA at 7:35 pm EDT from Eastern Test Range on a three-stage thrust-augmented Thor-Delta launch vehicle. Originally scheduled for March, the launch had been postponed because of a strike by McDonnell Douglas Corp. workers [see 10 Feb.]. The satellite was placed in the planned highly elliptical transfer orbit with a 35 945-km apogee, 231.4-km perigee, 634.3-min period, and 24.75° inclination. Ground control fired the spacecraft's apogee kick motor at 1:30 pm EDT on 10 May, circularizing the geosynchronous orbit at 38 000 km; the spacecraft was drifting toward its final operational location at approximately 104° west longitude.

*Anik* (Eskimo word for brother) 3 was third in a series of Canadian domestic comsats; *Anik 1* (launched 9 Nov. 1972) and *Anik 2* (launched 20 April 1973) were operating at 114° and 109° west longitude, providing TV and telephone service to the most remote areas of Canada. Built by Hughes Aircraft Co., *Anik 3* was 1.8 m in diameter and 3.3 m high, weighing 544 kg at launch and 272 kg in orbit. With power from 23 000 solar cells and enough battery capacity to maintain full service during a solar eclipse. Anik had 13 channels that could transmit up to 10 color TV transmissions or 9600 telephone circuits or analog or digital data.

The NASA−Telesat Canada contract defined the NASA mission objective as the placement of the satellite in a synchronous transfer orbit of sufficient accuracy to allow onboard propulsion systems to place the spacecraft into a stationary orbit with enough residual stationkeeping propulsion to meet mission-lifetime requirements. This objective was met and the mission was adjudged successful on 20 May.

NASA would be reimbursed for launch hardware, services, and DOD range support. The project was managed by Goddard Space Flight Center under the direction of the NASA Office of Space Science. (NASA MORs, 25 April 75, 20 May 75; GSFC *Wkly SSR*, 1−7 May 75; NASA Releases 75−113, 75−131)

- NASA, in cooperation with the Dept. of Transportation, was investigating aircraft wake vortices—turbulent air trailing from an aircraft's wingtips and swirling at velocities up to half the flight speed of the aircraft. Safety required an aircraft to travel far enough behind a preceding airplane to permit the vortices to dissipate, adding to congestion at already busy airports. NASA's objective was to develop methods to reduce the intensity of the vortices and render them harmless 3 km behind the aircraft. This would decrease the amount of fuel wasted and the pollution produced when aircraft were subjected to holding patterns and waits to taxi to a runway, and would permit aircraft to land and take off at a faster rate.

    Engineers and scientists at Ames and Flight Research Centers were investigating the vortices by flying an instrumented Learjet and a T-37 jet trainer through wakes created by larger aircraft. Data collected during these flights, checked against wind-tunnel data, suggested several possible ways of alleviating the wake-vortex problem, including tailoring the lift across the aircraft wings, modifying engine placement, and adding wing spoilers and drag devices. (NASA Release 75-136)

*8 May*: A new antireflection-coated metal oxide semiconductor (AMOS) solar cell developed by Jet Propulsion Laboratory scientists might have potential use in the nation's solar energy program, JPL scientist Richard J. Stern reported at a Photovoltaic Specialists Conference in Phoenix. The solar cells, made from oxidized gallium arsenide with an extremely thin nearly transparent film of gold on the surface, had demonstrated an efficiency of about 15% in terrestrial sunlight, better than the average silicon solar cell now in standard use. The new cell also had superior resistance to radiation, promising a longer lifetime in space. Dr. Stern said that the technology was potentially adaptable to very low-cost polycrystalline thin films, with only a modest reduction in efficiency. (NASA Release 75-133)

- The International Telecommunications Satellite Organization announced the award of a $99 894 15-mo contract to the Research Foundation of the State Univ. of New York for a study of radiation-induced defects in solar cells. The Foundation would develop an information base for predicting electron damage to silicon solar cells over a wide range of parameters for the materials from which the cells were made. (INTELSAT Release 75-32)

*11 May*: Employees of McDonnell Douglas Corp. ratified a new contract by a 4609 to 2796 vote, ending a 13-wk strike [see 10 Feb.]. The contract agreement called for a 31% across-the-board wage increase during the first year and 3% boosts during the second and third years. The strike of 19 000 McDonnell aerospace workers, including 200 at Kennedy Space Center, had delayed all NASA's scheduled Thor-Delta launches. (*W Post*, 12 May 75, A16)

*12 May*: The Senate passed by voice vote H.R. 4700, the NASA Authorization Bill for FY 1976 and the transition quarter. Reported out of the Senate Committee on Aeronautical and Space Sciences 5 May, the Senate version of the bill authorized $3.545 billion for NASA in FY 1976, an increase of $5.7 million over the original request, and $930 million for the transition period 1 July 1976 to 30 Sept. 1976, a decrease of $29 000 million from the request. (*CR*, 12 May 75, S7890-96)

- Soviet officials gave western newsmen their first glimpse of a Soviet space center during a tour of the main Soviet flight-control center 24 km from Moscow. The *Washington Post* reported that the group led by the center's technical director, Dr. Alberty V. Militsin, toured a 5-yr-old six-story brick and sandstone building acknowledged by Dr. Militsin to be one of several space facilities near Moscow. The operations room, which was similar to the control room at Johnson Space Center, contained five rows of consoles, each with telephones and closed-circuit TV screens. Mission Control Director Aleksey S. Yeliseyev told the newsmen that the consoles were staffed during missions by from 15 to 24 officials and technicians. An illuminated world map on the wall showed tracking stations and the orbital paths of spacecraft. Electronic screens flashed flight information. Other support consoles, communications systems, and computers were located elsewhere in the building.

    Dr. Militsin said that the center had contact with Soyuz spacecraft in each orbit for 5 to 25 min, depending on the orbit.

    The newsmen also visited the three-room suite near the main operations room assigned to nine U.S. technicians for the July Apollo-Soyuz Test Project. (Osnos, *W Post*, 13 May 75, A10)
- Rockwell International Corp. Space Div. announced the award of an estimated $1 million subcontract to the United Technologies Corp. Hamilton Standard Div. to build the flash evaporator subsystem, a part of the Space Shuttle Orbiter's environmental control and life-support system. The evaporator would use excess heat generated by Orbiter subsystems to turn water from the spacecraft's three power-producing fuel cells into steam vented into space, handling as much as 60 kg of water per hr. (United Aircraft became UTC in May.)

    Rockwell held the prime contract with NASA to build the Orbiter and integrate the complete Space Shuttle system. The evaporator subsystem contract was the third of three contracts totaling more than $27 million awarded to Hamilton. Under previously awarded subcontracts, Hamilton was also developing the Orbiter's atmospheric revitalization subsystem, Freon coolant loop, water boiler hydraulic thermal-control unit, and ground-support equipment hydraulic cart. (Rockwell Int'l Space Div Release SP-17)

*13 May*: The Air Force Systems Command reported that researchers at the Air Force Avionics Laboratory were testing fiber optics—thin strands of glass that transmit light—to transmit avionic signals from one point to another in an aircraft. Fiber-optics data transfer began with an electronic signal that was converted into light by a transducer; the light traveled through glass strands and was reconverted at the other end into electrical signals, with minimal loss of energy. AFSC quoted Avionics Lab researchers as saying that, unlike the usual copper wire, fiber optics were unaffected by static from nearby electrical equipment and were immune to lightning, nuclear radiation, electrical sparking, fires, and grounding problems. Because extensive lab testing was nearly complete, the next step would be a low-cost flight-test program, perhaps on a remotely piloted vehicle (RPV).

The Avionics Lab also had awarded contracts to develop passive optical couplers, a fiber-optics data bus, and a wideband fiber-optics system that would handle large data rates. The proposed Air Force Advanced Development Program could put a total fiber-optics system on manned and unmanned aerospace vehicles by 1980. (AFSC Release OIP 087.75)

- Communications Satellite Corp. Chairman Joseph H. McConnell told a stockholders' meeting that ComSatCorp was finalizing a contract with Exxon Corp. for five shipboard terminals for use with ComSatCorp's Marisat maritime communications satellite scheduled for launch in summer 1975. McConnell also said that ComSatCorp had agreed to lease one shipboard terminal to Seagap, an oil exploration consortium led by Phillips Petroleum Co. Marisat would make available, for the first time, reliable voice, data, teletype, and facsimile communications to ships at sea and to offshore drilling rigs. McConnell said that the success of the Marisat program would depend on the Navy's use of the system and on ComSatCorp's ability to develop a substantial market in the commercial shipping industry. (Snider, *W Star*, 14 May 75, F1)

*13−14 May:* More than 500 representatives from government, industry, and universities attended a 2-day conference at Lewis Research Center to review progress in aerospace propulsion. *Aviation Week and Space Technology* reported that high-speed performance was no longer the top priority among the propulsion researchers, emphasis having switched to quiet, clean, and fuel-conserving engines. Presentations and panel discussions included aircraft noise-reduction technology, fuel-conservative engine technology, upper-atmosphere pollution-measurement programs, and engine systems and component technology.

LeRC's John B. Whitlow reported that new technology generated by NASA's supersonic cruise airplane research (SCAR) would develop an engine far different from the afterburning supersonic turbojet that was to have powered the first U.S. supersonic transport. A potential new engine might be a duct-burning turbofan with features such as a variable-geometry fan that allowed the engine to move as much as possible between straight turbojet and straight turbofan operational cycles. Other objectives for future NASA-developed SST propulsion systems included a 7400-km range with outstanding economy at speeds from mach 2.2 to mach 2.7, good subsonic cruise characteristics, good hold endurance and economy, and low takeoff noise.

Whitlow stressed that, because of noise limitations, future SSTs would not be straight turbojets, and SCAR researchers were concentrating on low-bypass-ratio turbofans. As bypass ratio also affects engine weight, fuel consumption, drag, and cost-effective performance, no single bypass ratio of a conventional engine could meet all desired goals simultaneously. Thus SCAR researchers were developing variable-cycle engine technology that could tailor the bypass ratio and perhaps the fan/pressure ratio for optimum performance. Candidate engines included a mini-bypass turbojet, a variable-bypass supersonic turbofan, a duct-burning turbofan, and a rear-valve variable-cycle engine. (LeRC Release 75−19; Yaffee, *Av Wk*, 14 July 75, 46−49)

*13−20 May:* The second in a series of flight simulations between flight controllers and Apollo-Soyuz Test Project crew members in Houston and Moscow was completed in preparation for the July mission. The first of the three simulations scheduled for the session began at 6:20 am CDT 13 May, 1 hr before Soyuz launch time, and continued for

25.5 hr into the simulated mission. With both Soviet and U.S. crewmen participating in simulators in their respective countries and control centers fully manned, the Apollo and Soyuz spacecraft were "launched" and put through various maneuvers. Communications between the two control centers included voice, teletype, datafax, and TV. During the simulation, a technical problem prevented observers in Houston from hearing the two Soviet cosmonauts; NASA officials said later that that problem should not happen during a real flight. No other major problems were reported.

A second simulation begun at 6:30 am CDT 15 May and continuing for 56 hr rehearsed Apollo-Soyuz rendezvous, docking, crew transfers, and undocking and final separation. A third simulation, begun at 6:30 am CDT 19 May, was cancelled when the command module simulator in Houston did not work properly. The simulation, a 9-hr rehearsal of rendezvous and docking, was successfully completed 20 May.

Final simulations by Houston and Moscow control centers and crewmen were scheduled for 30 June and 1 July. (NASA Release 75–141; Tass, FBIS–Sov, 16 May 75, U1; Ezell *et al.*, *The Partnership: A History of the Apollo-Soyuz Test Project*, 285; UPI, *NYT*, 14 May 75, 69)

*14 May:* The U.S.S.R.'s Baykonur cosmodrome, located in the city of Leninsk near Tyuratam, was three to four times as big as Kennedy Space Center, U.S. Apollo-Soyuz Test Project commander Thomas P. Stafford said at an ASTP press briefing at Johnson Space Center. ASTP astronaut Vance D. Brand said that 50 000 persons inhabited the launch complex. The briefing followed the crew's return from the Soviet Union where they had participated in the final training session before the scheduled July launch [see 14–30 April].

The astronauts said that at the Soviet spaceport they had been shown "everything we had a requirement to see," the launch pad, spacecraft, test and checkout facilities, automatic checkout equipment (ACE) rooms, and crew quarters. The one "drastic difference" between the tour received by the Soviet cosmonauts in the U.S. and that of the American astronauts in the U.S.S.R., said Stafford, was "we didn't go to Disneyland over there."

Astronaut Donald K. Slayton described the launch pad as "relatively simple." The spacecraft, assembled on top of a railroad car, was hauled to the pad about 4 days before launch. The stacked vehicle was jacked up and set on the pad; after checkout, the vehicle was rotated to the proper launch azimuth. The booster was held to the pad by four support arms which were popped out of the way by counterweights once the vehicle started to move.

Stafford said that Leninsk was a new city built in a desert. The continued construction activity indicated that the Soviets "are not slowing down one bit in their space program." The Soviets had told their U.S. counterparts that they would continue to have several manned flights a year operating in low earth orbit. (Transcript)

- Marshall Space Flight Center had issued a request for proposals for a 10-mo study of a space-based power-conversion and power-relay system, MSFC announced. The proposed study would examine alternate means for generating electrical power in space, with emphasis on identification and delineation of problem areas and new technology required for these power systems. (MSFC Release 75–90)

- Fifteen years ago the launch of two U.S. satellites on the same day would have "blasted the top off of Page One," Ernest B. Furguson said in a *Washington Star* article. But when it happened 7 May—*Explorer 53* and *Anik 3*—it "drew less coverage than two swallows returning to Capistrano." At a total cost of $44 million, to be shared by NASA and foreign friends, *Explorer 53* Small Astronomy Satellite would study x-ray sources within and beyond the Milky Way, and *Anik 3* would transmit communications back and forth across Canada.

   The NASA space budget for FY 1975 was nearly $3 billion. "Considering inflation, that seems a bargain." The current space budget was "not quite 12 times what the President has asked to handle the Vietnamese refugee effort for each of two years. You are free to say for yourself where the two programs stand on the relative scale of things." (Furguson, *W Star*, 14 May 75, A17)

*15 May:* The National Science Foundation marked the 25th anniversary of its establishment at a dinner at the Dept. of State in Washington, D.C. Dr. Norman Hackerman, chairman of the National Science Board, presided at the dinner hosted by the 25-member board, the policy-making body of NSF. Among those present were many former board members, including the six charter members appointed by President Truman in 1950, and representatives of the scientific community, industry, and the executive and legislative branches of government.

   President Ford sent a message saluting NSF: "In the last quarter century the National Science Foundation has become the Nation's principal agency for the support of basic research and education in all fields of scientific endeavor. The creation of the National Science Foundation represented a culmination of our national understanding of the role of science in our society. In the last two and a half decades, the Foundation has built a prestigious reputation for sound scientific exploration to improve the quality of life for all."

   Dr. H. Guyford Stever, NSF Director, announced plans to establish an Alan T. Waterman Award for research and advanced study in the mathematical, physical, medical, biological, engineering, social, and other sciences. Dr. Waterman had been the first NSF Director. (NSF Release PR75–47)

- NASA had awarded a definitive $140 577 924 contract to Thiokol Corp. Wasatch Div. for solid rocket motors for the Space Shuttle, MSFC announced. A previous letter contract issued 26 June 1974 had given Thiokol authority to proceed with the work which included design, development, test, and evaluation of the solid rocket motor. The contracts also called for Thiokol to provide support equipment, tooling and support parts, systems-integration support and special studies, and data and documentation. (MSFC Release 75–91)

- The National Science Foundation and the Univ. of Wyoming announced plans for construction of a 213-cm infrared telescope near Laramie, Wyo. To cost $1.6 million and be operational in late 1977, it would be the largest of its kind in the continental U.S. and one of the largest in the world.

   To be used for study of objects in our galaxy and beyond and for study of the more distant stars, the instrument would be a Cassegrain telescope mounted on a ring 3 m high and housed in a building with a 14-m dome capable of operating automatically in subzero weather. The facility would also contain equipment for infrared detection systems and a laboratory-dormitory for observers.

The state of Wyoming had committed $975 000 for construction and NSF was committing $625 000. Operating costs would be shared by the state and users. (NSF Release PR75—42).

- Kennedy Space Center announced the selection of Modular Computer Systems, Inc., for final negotiations on a $6.7-million fixed-price contract for equipment associated with the Space Shuttle launch-processing system. The company would provide minicomputers and their associated equipment, software, maintenance, spares, and engineering support. The launch-processing system now under development at KSC would provide automatic test, checkout, launch control, and operational management for NASA's Space Shuttle program. (KSC Release 69—75)
- Kennedy Space Center had issued $7.5 million in awards and contracts to small businesses during the first 9 mo of FY 1975, bringing the total awarded to small businesses since FY 1970 to $61 million. The KSC set-aside program required that all procurement requests exceeding $2500 be reviewed by the center's small-business specialist to determine if they could be set aside exclusively for small businesses. In addition, KSC had successfully participated in a program aimed at awarding contracts to minority-owned firms, with $7 million in awards made since the program began. (KSC Release 70-75)

*15—16 May:* U.S. and European scientists met at a Goddard Space Flight Center symposium to discuss the importance of a never-before-flown space mission out of the ecliptic plane, the pancake-like plane in which all planets orbit the sun. Symposium chairman Dr. Leonard A. Fisk said, "A view of the solar cavity [the region dominated by the sun through the flow of the solar wind] from outside the ecliptic plane is expected to expand greatly man's knowledge of interplanetary physics, solar physics, and plasma processes in the universe. To place a spacecraft in such an orbit brings expectations that we will witness new and exciting phenomena that we could neither observe nor predict from our present vantage point in the ecliptic plane."

The scientists agreed that the missions under consideration could be undertaken jointly by NASA and the European Space Research Organization: ESRO would have primary responsibility for the spacecraft, and NASA would provide the launch vehicle. (*NASA Activities*, June 75)

*16 May:* Marshall Space Flight Center had issued a request for proposals for an x-ray source system for the x-ray telescope facility being constructed at MSFC for use in the High Energy Astronomy Observatory (HEAO) project, MSFC announced. The proposed contract would include design, fabrication, assembly, preliminary in-plant checkout, installation, and verification. The contractor would also have the option of building a cover for the system. Under the terms of the contract, the x-ray source system would be installed on a 40.6-cm-by-39.3-cm steel flange by 20 Feb. (MSFC Release 75—93)

- Flat conductor cable developed by NASA for rocket and spacecraft wiring was being installed in Marshall Space Flight Center's experimental solar house, MSFC announced. The surface-mounted cable was being considered for commercial and residential applications, to provide a wiring system that was lower in cost, easier to install, adaptable to all types of wall construction, and as safe as or safer than conventional roundwire electrical systems. Flat conductor cable could be routed under all paneling, tile, and carpet, and the low-voltage flat cable for lights and receptacles could be routed under wallpaper or simply painted. The cable had already been installed successfully in an MSFC conference room and in six apartments in Yonkers, N.Y., as a joint effort between NASA and the New York State Urban Development Corp. (MSFC Release 75—94)

*17 May:* France launched two satellites, *Castor* and *Pollux*, on a single Diamant BP—4 booster from Kourou, French Guiana. It was the first time France had orbited two satellites on a single booster. *Castor* entered orbit with a 1269-km apogee, 273-km perigee, 100.3-min period; and 30° inclination; *Pollux* entered orbit with a 1283-km apogee, 270-km perigee, 100.4-min period, and 30° inclination. The press reported that *Castor* carried an accelerometer to measure changes in the earth's atmosphere for a 5-to-6-yr period. *Pollux* carried a micrometer for experiments in satellite propulsion. (GSFC *Wkly SSR*, 15—21 May 75; *SF*, Nov 75, 406; Agence France-Presse, *Today*, 18 May 75)

*18 May:* Equipment aboard U.S. reconnaissance satellites could photograph a pack of cigarettes or monitor a radio conversation between foreign pilots from an orbit of 130 km, the *Philadelphia Inquirer* reported. Satellites could also instantly detect the launch of an intercontinental ballistic missile and determine where it was going.

Early reconnaissance satellites were of two types: SAMOS (satellite and missile observation system) satellites carried wide-area cameras to determine ground targets of interest; film was processed on board and imagery radioed to ground. Discoverer satellites from lower orbits used high-resolution cameras for closeups; the satellite ejected its film, which parachuted down for midair recovery by Air Force aircraft.

These satellites were replaced, in the 1970s, by the 11 340-kg "Big Bird" spacecraft. The *Inquirer* reported that technical experts believed that Big Birds carried TV cameras which allowed human monitors to maintain continuous watch on ground; if they saw something of interest, they could order immediate photographs by the satellite's high-resolution cameras. The experts also believed that Big Bird carried side-looking radar to penetrate cloudcover and equipment sensitive enough to distinguish between old and new grass. Resolution was no longer measured in meters but in centimeters.

The *Inquirer* also reported that other satellites carrying infrared and other detectors were used to monitor missile launches, detect hidden heat sources such as underground powerplants, and spot the warmth of buried missile silos. (Coughlin, *P Inq*, 18 May 75, 5E)

*19 May:* The unemployment rate for scientists and engineers, which had been 1.1% in 1970, rose to 1.9% in 1972 and then fell to 1.1% again in 1974, the National Science Foundation reported. Of the employed engineers and scientists, only 0.5% were not working in their field because of a lack of available positions. Business and industry was by far the largest employer, with 60% of all the employed engineers and scientists throughout the 1970—74 period. Educational institutions employed about 13%; Federal and state and local governments employed about 10.5 and 8%, respectively. Between 1972 and 1974 the proportion of engineers in the group had increased from 62% to 64%. Physical scientists constituted 12% of the science and engineering workforce. Environmental scientists were the smallest group, making up 2.5% of the science and engineering workforce. (NSF *Highlights*, 19 May 75)

*20 May:* NASA's *Ats 6* Applications Technology Satellite (launched 30 May 1974 into geosynchronous orbit) began a 6-wk move from its position 35 900 km over the Pacific Ocean west of the Galapagos Islands to a new location above Lake Victoria in eastern Africa. The satellite, which was expected to reach its new location by 1 July, would be checked out in time to participate in the Apollo-Soyuz Test Project in mid-July. During that mission *Ats 6* would track the Apollo and Soyuz spacecraft and relay TV and data from the two spacecraft to earth, marking the first time a satellite had been used to relay TV from a manned spacecraft.

The new location also would bring *Ats 6* within range of India for the 1-yr Satellite Instructional Television Experiment (SITE) scheduled to begin 1 Aug. During the experiment—a cooperative effort by NASA and the Indian Space Research Organization—the Indian government would relay daily educational TV programs to 5000 villages and cities throughout India. After the experiment, the satellite would be moved back to a position within range of the U.S., where it would be used for further experimentation. (NASA Release 75—153)

- Two U.S. Air Force communications satellites, launched from Eastern Test Range, failed to reach orbit when the third stage of the Titan IIIC launch vehicle malfunctioned and tumbled out of control. United Press International quoted an Air Force spokesman as saying the two satellites, valued at $57 million, were to have been placed in geostationary orbits, one over the Pacific Ocean and the other over the Indian Ocean. The 544-kg comsats—designed to handle as many as 13 000 telephone calls or to relay televised or printed messages without danger of jamming or eavesdropping—would have completed the worldwide military communications network begun in 1973. Officials believed that a power problem was responsible for the failure. (UPI, *NYT*, 21 May 75, 40)

- Frank Briscoe Co., Inc., and Santa Fe Engineers, Inc., were the apparent successful bidders on two major Space Shuttle contracts which would be awarded by Kennedy Space Center, KSC announced.

  Frank Briscoe Co. had bid $8 733 300 on a contract for the construction of the Orbiter processing facility, Orbiter towway, and associated site work and utilities at KSC. It was in the processing facility that the Orbiters would be safed, serviced, and maintained after returning from flights into space and then prepared for their next mission. Phase 1 of the construction—scheduled for March 1977 completion—included a 5000-sq-m building with a low bay area 88 m long, 30 m wide, and 8 m high and two high bay areas 61 m long, 46 m wide, and 29 m high.

  Santa Fe Engineers had bid $3 567 567 on a contract to construct the Shuttle approach and landing test facilities at Flight Research

Center. The contract would include site preparation, construction of a maintenance hangar and shops, Shuttle Orbiter mating structure area, hypergolic fuel and oxidizer area, fire protection pump station, concrete towway and service aprons, parking area and road paving, and necessary utilities. The project was scheduled for completion in September 1976. (KSC Release 73–75)

- Air Force Systems Command Aerospace Medical Research Laboratory scientists were studying changes in human body weights and sizes as part of a NASA-requested anthropometric study to help engineers design cockpits and crew stations for future use. NASA engineers who wanted to lay out the Space Shuttle work station were concerned that body sizes of the pilots might change before the station was used. The scientists performing the study, using both current data and data that went back as far as the Civil War, estimated that in 1985 a pilot would average nearly 2 kg heavier and 1.5 cm taller than in 1967, when he had weighed 78.5 kg and was 177 cm tall.

  In addition to statistics, NASA requested drawing board manikins for designing the Shuttle's work areas. The study had produced two 2-dimensional manikins like cardboard dolls with movable joints; each manikin was articulated in nine places to represent typical ranges of body movement. A 95% manikin represented a person 189.0 cm tall; a 5% manikin, representing a 168.3-cm astronaut or pilot, was designed so that only 5% of the 1985 pilot-astronaut population would be smaller. An engineer using both manikins could design a crew station or cockpit that would accommodate 90% of all astronauts and pilots. (AFSC Release OIP 072.75)

*21 May:* NASA's Airborne Infrared Observatory, a four-engine C–141 jet transport equipped with a 91.5-cm infrared telescope, was dedicated to the memory of Dr. Gerard P. Kuiper, the founder and former director of the Univ. of Ariz. Lunar and Planetary Laboratory. The Kuiper observatory, based at Ames Research Center, would fly at altitudes up to 13 500 m with the telescope operating through a cavity in the aircraft's fuselage.

  Dr. Kuiper, who had died in 1973, had been one of the foremost authorities on lunar science. He had played an active role in NASA's Ranger and Surveyor programs and, as head of the Ranger scientific team, had directed the photographic analysis to select landing sites for the Apollo astronauts. He also had participated in the *Mariner 10* mission, launched 3 Nov. 1973 to Venus and Mercury, and the *Pioneer 10* mission, launched 2 March 1972 toward a December 1973 encounter with Jupiter. (NASA Release 75–149; *A&A 73,* 350)

- NASA announced the appointment of Robert E. Smylie as Deputy Associate Administrator for Aeronautics and Space Technology, effective immediately, and the return of J. Lloyd Jones, Jr., to Ames Research Center as Chief of the Planning and Analysis Office.

  Smylie had been Deputy Associate Administrator (Space), OAST, since July 1974. From October 1973 to July 1974 he had been Deputy Associate Administrator for Technology after serving as chief of the Crew Systems Div. at Johnson Space Center.

  Jones had come to Hq. in 1972 as director of the Aerodynamics and Vehicles Systems Div. in the Office of Aeronautics and Space Technology, a position he held until appointed Deputy Associate

Administrator (Aeronautics), OAST, in July 1974. (NASA anno, 21 May 75)
- The Energy Research and Development Administration had awarded a 45-mo $17 736 000 contract to General Electric Co. Space Div. to provide nuclear generators for NASA's two Mariner–Jupiter–Saturn missions scheduled for launch in late 1977. GE would build seven radioisotopic thermoelectric generators for use during the mission; three of the nuclear generators—each providing 150 watts—would be used for each spacecraft, and the seventh would serve as a spare. (ERDA Release 75–79)
- Marshall Space Flight Center had issued a request for proposals for a study of alternative means for generating electrical power in space and relaying the power to earth, with emphasis on problem areas and new technology, the *Marshall Star* reported. Previous studies had defined concepts for the use of large solar-cell panels to generate power in geosynchronous orbit and transmit it to earth via microwaves; also studied had been the transmission via an orbital microwave system of power generated at a remote location on earth and transmitted long distances to the user location. (*Marshall Star*, 21 May 75, 1)

*22 May:* NASA successfully launched *Intelsat IV F–1* for the Communications Satellite Corp., as agent for the International Telecommunications Satellite Organization, from Eastern Test Range at 6:04 pm EDT. The 1400-kg satellite entered the desired synchronous transfer orbit with a 35 870-km apogee, 534-km perigee, 10-hr 40-min period, and 26° inclination. ComSatCorp successfully fired the apogee motor at 6:45 pm EDT on 24 May, placing the satellite in synchronous orbit over the Indian Ocean (36 184-km apogee, 35 704-km perigee, 24-hr 4-min period, and 0.5° inclination) with a final operational position to be at 61° east longitude. When it became operational in June, it would replace *Intelsat IV F–5* as the primary operating satellite in the Indian Ocean area in the global 6-active-satellite Intelsat network; it would be able to transmit 3500 two-way telephone conversations or 12 TV channels.

The launch was the second recent attempt to place an Intelsat IV satellite over the Indian Ocean; on 20 Feb., the first attempt failed and Intelsat IV F-6 had to be destroyed when the Atlas-Centaur launch vehicle tumbled out of control.

INTELSAT launched its first comsat, *Intelsat I F–1 (Early Bird)*, on 6 April 1965. Between 1967 and 1970 12 Intelsat II and III satellites were launched before the Intelsat IV launches began in 1971. (NASA MORs, 16 May 75, 25 Sept 75; NASA Release 75–125; NASA Audio News, 23 May 75; GSFC *Wkly SSR*, 25–28 May 75; Salmanson, Atlas mgr., NASA Hq., interview, 25 May 77)
- A giant 283 200-cu-m balloon carrying a Marshall Space Flight Center gamma ray detector was launched from Palestine, Tex., by the National Center for Atmospheric Research. The balloon reached an altitude of 40 km and traveled 800 km before touching down near Ft. Stockton, Tex., 23 hrs later. The purpose of the flight was to detect the sources and determine the nature of sudden bursts of gamma rays coming from various directions in space. (MSFC Release 75–100)

- *Landsat 2* (launched 22 Jan.) was monitoring water resources in Miss. and the effects of pollution on the water system, Marshall Space Flight Center announced. As part of the NASA earth resources program, three data-collection buoys placed in the Pearl River were transmitting water-quality data to *Landsat 2*, which relayed the data to a tracking station at NASA's National Space Technology Laboratories (NSTL) for processing. Processed data on water temperature, dissolved oxygen, acidity/alkalinity, and conductivity were received in the Miss. Air and Water Pollution Control Commission, where state investigators could identify a pollution source within 15 min of a satellite pass over the buoys.

  A similar experiment was being conducted in Mobile Bay by NASA and the Marine Environmental Sciences Consortium. Data from the experiments would help evaluate the usefulness of the system and suggest improvements. (MSFC Release 75–97)
- President Ford told a group of eight congressmen at a White House meeting that he would work to reestablish the Office of Science and Technology as a permanent part of the White House organization, the *New York Times* reported. The office had been abolished by President Nixon 1 July 1973. The *Times* quoted Director of the Domestic Council James Cannon as saying that President Ford would propose legislation creating the post of science advisor with a staff of 10 to 15 persons and an annual budget of $1 million to $1.5 million. (*NYT*, 23 May 75, 14)
- Rockwell International Corp. announced that its Electronics Research Div., under a contract awarded by NASA on 18 Feb. 1974, had developed a single-chip bubble-domain memory element with a capacity of 102 400 bits, believed to be the largest capacity memory device ever built. The new data recorder—a major milestone in NASA's continuing program to develop an all-solid-state data recorder for future U.S. spacecraft—was one-third the size and one-half the weight of current equipment. It was expected to use half the power and have 10 times the life expectancy of present recorders. Objective of the effort was to demonstrate the feasibility of fabricating high-capacity memory elements with reasonable yield and operating margins. (Rockwell Int'l Release ERD–1; LaRC proj off, interview, 18 Jan 77)

*22 May–7 June:* The U.S. and Peru launched more than 40 sounding rockets and balloons in a joint project, called Antarqui for the Inca god of flight, to study the earth's atmosphere and ionosphere from Chilca Launch Range in Peru, a location on the magnetic equator where the earth's magnetic field is horizontal. Instruments aboard the rockets and balloons measured the composition of the neutral and ionized atmosphere, density and temperature, wind shear, and turbulence. The measurements would be compared with data from *Explorer 51* (Atmosphere Explorer–C, launched 15 Dec. 1973) and the ground-based Jicamarca and Huancayo Observatories. The overall mission included the launch of 4 Nike-Tomahawks, 8 Nike-Apaches, 7 Super Arcas, and 10 Super Lokis, as well as two balloons measuring 36 m in diameter and 12 ozonesonde balloons.

Project Antarqui was a continuation of a NASA sounding-rocket program conducted during the 1964–65 International Quiet Sun Year when 77 sounding rockets had been launched during the first 4 mo of 1965 from a sea-going launch platform, the USNS *Croatan*.

Wallops Flight Center managed the program for NASA; the Instituto Geofisico del Peru arranged for the use of the range, supplied range operation and logistics support in Peru, correlated ground-based measurements from Jicamarca and Huancayo, assisted in recovery operations, and performed some preliminary work at the range.

In addition to WFC, Goddard Space Flight Center, Dudley Observatory, Univ. of Pittsburgh, Univ. of Ill., Penn. State Univ., Univ. of Denver, and GCA Corp. supplied scientific payloads for the sounding rockets and balloons. (WFC Release 75–6; *NASA Activities*, June 75, 14; WFC proj mgr, interviews, 11–12 Jan 75)

*24 May–26 July:* The U.S.S.R. launched *Soyuz 18*, carrying cosmonauts Pyotr Klimuk and Vitaly Sevastyanov, from Baykonur cosmodrome at 7:58 pm local time (10:58 am EDT). According to the Tass news agency, the spacecraft entered orbit with a 247-km apogee, 193-km perigee, 88.6-min period, and 51.6° inclination. Tass announced that the purpose of the mission was "to conduct further experiments with the orbital research station 'Salyut–4,' started on January 12, during the joint flight of the delivery spaceship 'Soyuz–17' and the station 'Salyut–4' as well as to test individual elements and systems of the spaceship in various modes of flight."

During the first day in orbit Klimuk and Sevastyanov checked out essential spacecraft and life-support systems and prepared for orbital maneuvering. The crew switched on the main engines twice on 25 May to raise the spacecraft's orbit in preparation for rendezvous and docking with *Salyut 4*. On 26 May *Soyuz 18* approached the station by an automatic control system. At a distance of 100 km the crew took over control and manually docked the spacecraft with the *Salyut 4* station in pitch darkness. After docking, the crew switched on the station's lights and began activation and checkout of the onboard systems. The crew transferred equipment and boarded the station where they found a "Welcome to our common home" sign left by the station's previous tenants, *Soyuz 17* cosmonauts Aleksey Gubarev and Georgy Grechko. The cosmonauts reported that all systems were operating normally and that they themselves were in good health. By 28 May the station and spacecraft were in an orbit with a 349-km apogee, 338-km perigee, 91.4-min period, and 51.6° inclination.

During their 63 days in orbit, the cosmonauts studied the sun, planets, and stars in various bands of the spectrum of electromagnetic radiation using several x-ray telescopes, including one called "Filin"; investigated geological-morphological objects on the earth's surface, photographing the U.S.S.R. in the medium and southern latitudes; studied physical processes in the earth's atmosphere and in space; performed medical and biological research with particular emphasis on the impact of weightlessness on the human organism; and tested the station's design and onboard systems. The cosmonauts also raised a garden of onions and peas as a forerunner of providing self-sufficient food supplies in space. Tass reported that, for the first time during a space flight, the polar lights were investigated using integral photography and spectrography methods.

Twice during the mission Klimuk and Sevastyanov spoke with cosmonauts Aleksey A. Leonov and Valery N. Kubasov who were also in orbit participating in the joint U.S.–U.S.S.R. Apollo-Soyuz Test Project begun 15 July.

By 23 July the cosmonauts were preparing the space station for automatic operation. On 26 July at 3:56 pm Baykonur time (6:56 am EDT), *Soyuz 18* and its crew undocked from *Salyut 4*. They softlanded 56 km northeast of Arkalyk in Kazakhstan at 7:18 pm Baykonur time (10:18 am EDT).

Tass reported that by 27 July the cosmonauts' slight pulse and blood pressure fluctuations noted immediately after landing had returned to normal. Decreases in erythrocytes and increases in leucocytes disappeared after a few days, and in 1 or 2 wk orthostatic stability and resistance to functional tests had almost reached preflight baselines. Both Klimuk and Sevastyanov had lost some weight— 3.8 kilograms and 1.9 kilograms respectively—during the mission.

The *Soyuz 18* crew's 62-day 23-hr 40-min mission broke the previous 29-day 13-hr 20-min Soviet record for time in space set by the *Soyuz 17*. The current world's record for time in space was held by the *Skylab 4* astronauts for their 84-day 1-hr 16-min mission, 16 Nov. 1973 to 8 Feb. 1974.

*Salyut 4*, which had been visited by the two crews of *Soyuz 17* and *18*, was the fourth space station orbited by the U.S.S.R. *Salyut 1* had been launched 19 April 1971. A three-man crew, launched on *Soyuz 10*, docked with the station on 24 April but returned to earth without entering. The *Soyuz 11* crew, launched 6 June 1971, boarded the station and conducted experiments for 24 days but were killed during a reentry accident. *Salyut 2* was launched 3 April 1973 but an explosion of wildly firing thrusters sent the station out of control, tearing off the solar panels and making the station uninhabitable. *Salyut 3* was orbited 25 June 1974 and a two-man crew, carried by *Soyuz 14*, lived and worked aboard the station from 3–19 July 1974. A second crew, launched 26 Aug. 1974, returned to earth in 48 hr after an unsuccessful attempt to dock with the station, which itself reentered the earth's atmosphere 26 Sept. 1974. (Tass, FBIS–Sov, 24 May–28 July 75; GSFC *Wkly SSR*, 22–28 May 75; AP, B *Sun*, 27 July 75, A3; Lib of Congress, "S&T News Alert," No 2600, 19 Sept 75)

*27–28 May:* The Saturn IB launch vehicle scheduled to launch three U.S. astronauts into orbit in July as part of the joint U.S.-U.S.S.R. Apollo-Soyuz Test Project underwent flight-readiness tests at Kennedy Space Center. During abort tests on 27 May, automatic checkout equipment was connected and the mobile service structure placed around the vehicle. Saturn IB heavy ordnance and the spacecraft's launch escape system were installed but not connected. Two abort-procedure rehearsals were held. After a 28 May simulated countdown sequence, including propellant loading, umbilical ejection, ignition, holddown arm release, and liftoff, a 6-hr simulated mission was held. The launch team and a crew of astronauts conducted post-orbital and insertion checks, launch-vehicle guidance tests, computer updates, simulated burns of the spacecraft's engine, deorbit, reentry, and splashdown. (KSC Release 74–75; *Spaceport News*, 30 May 75, 1)

*27 May–14 June:* A one-third-scale model of the Space Shuttle Orbiter was tested in Ames Research Center's 40- by 80-ft wind tunnel, to gather low-speed flight data in support of approach and landing tests of the first full-scale Orbiter at Flight Research Center in 1977. In addition, data were gathered to calibrate the vehicle's air data probes. The model, built by Shuttle prime contractor Rockwell International

Corp., was 13.1 m long and weighed 20 400 kg. Additional wind-tunnel tests were scheduled for November 1975 and February 1976. (*Astrogram*, 5 June 75, 1; Young, JSC aerodynamic subsystems mgr, interview, 21 April 77)

*28–29 May:* More than 200 persons from 30 countries participated in an observance of "European Space Days," a demonstration of Europe's space capability held at the European Space Research and Technology Center (ESTEC) in Noordwijk, Netherlands, by the European Space Research Organization [scheduled to become the European Space Agency on 30 May].

In a speech, Secretary General of the International Telecommunications Union M. Mili said ITU encouraged Europe's intention to play an important role in the peaceful uses of outer space; the programs of ESRO–ESA offered a particularly well chosen range of satellites for research and applications.

Y. Demerliac, Secretary General of Eurospace—Groupement Industriel Européen D'Etudes Spatiales—reminded the audience that since 1962 Europe had developed 43 satellites, including four communications satellites. In addition, European industry had built 36 of the 104 ground stations or antennas in the network of the International Telecommunications Satellite Organization (INTELSAT). More than two-thirds of these had been sold to non-European countries.

Director General of ESRO Roy Gibson (United Kingdom) said the creation of ESA was of more significance than a mere change in name: the convention of the new agency gave it a specific mandate to work toward a rationalization of Europe's space program which its predecessors ESRO and ELDO (European Launcher Development Organization) did not have. He hoped that the 3-day presentation of ESRO–ESA and other European programs would give the visitors an impression of the total competence and capability of Europe's up-to-date, rather dispersed space effort. (ESRO Release 28 May 75)

*29 May:* Lewis Research Center was testing a version of an ion-thruster engine with reduced propellant requirements that would permit payload increases of up to 20% on future spacecraft, NASA announced. The ion-thruster propulsion system used solar cells to convert solar energy into electrical power, tailored to the current and voltage necessary for the ion thruster. The propellant—mercury, cesium, or relatively stable gas—was ionized in the engine and electrically exhausted to produce thrust.

During the tests the NASA version of the engine had become the first electric propulsion system to exceed 10 000 hr of operation. LeRC was continuing the testing toward an endurance goal of 15 000 hr. The engine was the second of its kind to be tested by NASA. The first had experienced internal sputtering and erosion which had degraded performance. Testing had been stopped at 9715 hr for modifications. (NASA Release 75–150)

- Kennedy Space Center announced the selection of General Electric Co.'s Space Div. and Martin Marietta Corp. for competitive negotiation of a contract to check out, control, and monitor subsystem hardware for the Space Shuttle launch processing system. The contractor would design, develop, build, test, and install the hardware. The launch processing system would use modern industrial automation techniques, applied to a modular concept, to meet maintenance and refurbishment requirements for turn-around and launch of the Space Shuttle. The other candidates for the contract had been Grumman Aerospace Corp. and Aeronutronic Ford Corp. (KSC Release 94-75)

*30 May:* Both Viking Landers were 1 wk behind schedule because of an electrical interface problem between two onboard power control units, Kennedy Space Center's *Spaceport News* reported. (*Spaceport News*, 30 May 75, 2)

- A team of Federal and university scientists were using three instrumented aircraft and an elaborate ground network to construct a three-dimensional map of the atmosphere's electrical fields around Kennedy Space Center and to determine the conditions under which a launch penetrating these fields would trigger a lightning strike. By July the team hoped to be able to predict what the electrical conditions would be at the launch pad 10 to 20 min before liftoff. The project, directed by the Dept. of Commerce's Environmental Research Laboratories and funded by NASA, included scientists from NASA, National Oceanic and Atmospheric Administration, Naval Research Laboratory, N. Mex. Inst. of Mining and Technology, and Univ. of Ariz. The experimenters would conduct a month of aircraft flights under and around various convective clouds, measuring electrical fields and making penetrations of frozen cloud anvils—cirrus clouds swept off the tops of thunderstorms by upper level winds. In addition, data from the mapping ground network—24 electrical-field-measuring stations installed around KSC by NOAA in 1973—were fed into a computer whose video component displayed a map of the area and ground stations, total electrical activity in nearby storms, weather radar contours of precipitation, electrical-field contours, and the positions of lightning strikes.

  The project was a continuation of one begun for NASA by NOAA after the *Apollo 12* launch on 12 Nov. 1969 when lightning struck the spacecraft, shutting off the spacecraft's electrical power and setting off numerous alarms. The spacecraft automatically switched to backup battery power while the crew restored the primary power system. (NOAA Release 75-91; *A&A 69*, 372-378)

- Scientists believed that light flashes observed by *Skylab 4* crew members during their 16 Nov. 1973 to 8 Feb. 1974 mission as they flew through the South Atlantic Anomaly—a slight asymmetry in the earth's magnetic field where the Van Allen radiation belt dipped unusually low over Brazil—were a previously unsuspected form of radiation, *Science* magazine reported. The radiation seemed to consist of atomic nuclei heavier than hydrogen which, when stripped of electrons, would carry multiple positive electric charges.

  The astronauts had seen the flashes when the spacecraft cabin was dark and they were preparing for sleep. Scientists suspected that the flashes were stimulated within the retina of the eye by the high-energy cosmic-ray particles. Normally earth was shielded from cosmic rays by the magnetic field, but strong penetrating cosmic rays could reach as far into the field as the Skylab orbit.

  Because the magnetic-field shielding of the earth weakened toward the geomagnetic poles, it was possible to predict variations in cosmic-ray exposure as Skylab moved away from the poles. During two test periods, light-flash observations of Astronaut William R. Pogue showed a close correlation with exposure variations. (Pinsky *et al.*, *Science*, 30 May 75, 928-930; Sullivan, *NYT*, 29 May 75, 10)

*30–31 May*: The Convention of the new European Space Agency was formally adopted and opened for signing at a conference of plenipotentiaries held 30 May in Paris and presided over by French Minister of Industry and Research Michel d'Ornano. Under the terms of the Convention, ESA came into de facto operation 31 May. The convention was signed by the 10 member-countries of the former European Space Research Organization—Belgium, Denmark, France, West Germany, Italy, The Netherlands, Spain, Sweden, Switzerland, and United Kingdom—and would remain open for signature by other countries until 31 Dec. 1975. Ireland, which had had observer status in ESRO, had indicated its intention to sign.

The new space agency's symbol was a rounded "e," for European, on a globe representing earth, with fine lines suggesting both orbit and axis and a white dot representing a satellite in space. (ESRO–ESA Releases 20 May 75, 23 May 75, 2 June 75)

*30 May–8 June*: The 31st Annual Paris Air Show was held at Le Bourget Airport in France. New restrictions on flight demonstrations by high-performance aircraft, imposed by the French government after the 1973 Air Show crash of the Soviet Tu–144 supersonic transport, had placed the show's emphasis on displays of the latest aerospace hardware and technology from throughout the world. *Aviation Week and Space Technology* reported that the general turndown in the economy throughout the western world had caused a decline of 20% in exhibitors from the 1973 show.

France emphasized the cooperative nature of its major programs, displaying the French–British Concorde SST, the French–German Alpha Jet trainer-attack aircraft, the French–British Jaguar close-air support aircraft, and the multinational A–300B twin-engine wide-body transport.

The U.S. pavilion reflected the changing environment at the show, reserving 66% of its floor space for company exhibits restricted to industry visitors; the remaining 33% contained exhibits open to the general public. In 1973 60% was exhibitor space and 40% for the general public.

Boeing Co. and McDonnell Douglas Corp. displayed aircraft that were potential replacements for the Boeing 707 and McDonnell DC–8 on commercial airlines. The Boeing aircraft, a three-engine member of the 7X7 family, incorporated the NASA-developed supercritical wing.

"Blue Planet" was the theme of the public section of the U.S. pavilion which emphasized technological fallout and practical uses on earth of space technology. Included were full-scale models of the Viking spacecraft scheduled for launch to Mars later in the year, and of Martin Marietta Corp.'s large space telescope.

The 80 participating British companies emphasized their country's capability in avionics and equipment and aerospace technology. Among the new aircraft exhibited by the British were the BN2 Islander, a floating aircraft for service between islands; the Lockspieser LDA–1, for cargo and agricultural uses; and a new version of the BAC 111 fighter aircraft.

West Germany's display stressed its participation in international consortiums as a major hope for continuing growth. Messerschmitt–Boelkow–Blohm GmbH pointed out in its display that 70% of its revenues in 1974 had been derived from multinational projects. (*Av Wk*, 2 June 75, 14–20; 9 June 75, 12–17)

*31 May*: President Ford's announcement that he intended to restore the White House Office of Science and Technology was discussed in a *Washington Post* editorial: most scientists were heartened by President Ford's announcement. They had unanimously deplored President Nixon's decision to abolish the office, believing that it was vital to the nation's well-being for the President to get the most objective possible scientific assessment of new weaponry, as well as of medical, economic, and environmental policies. However, scientists also felt that the establishment of the office and the possible appointment of National Science Foundation Director Dr. H. Guyford Stever as its head would not solve the problem. A report drafted by a committee of 13 leading scientists had spelled out what was needed: establishment of a Council for Science and Technology that would carry the same weight as the Domestic and Economic Councils and could hold its own with the Office of Management and Budget and DOD.

Any serious candidate for the top science job would want to consider whether the President "really intends to make constructive use of a science adviser. Past experience tells us that the best mechanism for advice from the scientific community is of no avail if the President isn't listening." (*W Post*, 31 May 75, A12)

*During May*: Preparations continued for late summer launches of the two Viking spacecraft to Mars. A Lander 1 plugs-out test was completed 8 May, 2 days later than planned, because of a grounding problem. By 26 May all flight equipment, except for the gas chromatograph mass spectrometer, had been installed on Lander 1. Lander 2 was mated with its protective aeroshell on 28 May.

Orbiter 1 was undergoing final mechanical assembly as the flight subsystems became available after individual reworking. By 31 May all temperature-control louvers, protective covers, and thermal blankets had been installed. Initial Orbiter 2 system testing was completed 5 May, and Orbiter block validation tests were begun. All solar panels for both Orbiters had been assembled and swing-tested, and were awaiting buildup of the low-pressure gas system.

On 13 May a major hardware review of both landers was held at Kennedy Space Center with 100 NASA and industry participants. The review was the final chance to discuss modifications in the spacecraft before closure of the bottom plates and sterilization. Following the review, Langley Research Center's Viking Project Manager James S. Martin, Jr., said, "Everything is going just great and right on schedule." (Viking launch and mission operations status bulletins 4, 5, 6; *NASA Activities*, June 75)

- The vertical tail assembly for the Space Shuttle arrived at Rockwell International Corp., in Palmdale, Calif., after being shipped by Shuttle subcontractor Fairchild Republic Co. The tail unit, more than 8 m high, had a root chord of nearly 7 m. (FRC *X-Press*, 6 June 77, 2; *Spaceflight*, Jan 76, 25)
- NASA Life Sciences Director, Dr. David L. Winter, issued an invitation to scientists to help plan NASA's future life sciences program in space. The invitation, sent to 30 000 biological scientists in the U.S., said that NASA was considering a series of manned laboratories staffed by scientists, engineers, and technicians to be launched aboard the

Space Shuttle. One entire lab devoted to life-science experimentation would be launched every 6 mo. Other small highly automated life-science experiments could be performed during Spacelab and Space Shuttle missions dedicated to other disciplines; NASA estimated that as many as 200 life-science flight experiments could be conducted in space during the 1980s.

The invitation stated that "in order to take maximum advantage of future research opportunities, NASA intends to involve a large cross section of the Life Sciences community in shaping scientific program objectives, in selecting spacecraft laboratory equipment, and in the planning and execution of Flight Experiments."

NASA estimated that approximately 200 scientists could participate as principal investigators of future flight experiments and that 50 of those could perform their research in space as crew members.

As part of the first phase of the life-science program, NASA requested that interested scientists suggest general topics for future research and identify equipment needed to carry out life-science investigations. (Winter letter, text; NASA Release 75–140; UPI, *NYT*, 9 May 75, 7)

- The launch of *Skylab 1* on 14 May 1973 had torn a temporary "hole" in the ionosphere, *Physics Today* reported. The journal said scientists speculated that the hole, a depletion of the total electron count, was caused by molecular hydrogen and water vapor from the exhaust of the Saturn V engines which had continued to burn well into the lower region of the ionosphere. The exhaust caused the oxygen atoms in the ionosphere to recombine, losing one electron each; the electrons were removed faster than could be replenished by the sunlight, thus the hole. The sun did replace the electrons, patching up the hole within 3 hr after the spacecraft passed.

  The electron loss triggered by the Skylab launch was similar to ones caused, on a larger scale, by magnetic storms. Because the launch occurred on the second day of such a storm, scientists found it difficult to trace any radio interruptions—typical during decreases in the electron content of the ionosphere—to Skylab. (*Physics Today*, May 75, 17–18)

- The Air Force Systems Command *Newsreview* reported that the Air Force had awarded an estimated $17 649 042 contract to McDonnell Douglas Corp.'s Douglas Aircraft Co. for the development, manufacture, and testing of a wide-body cargo aircraft fuselage segment using adhesive bonding instead of conventional riveting of the primary structures. Either the YC–14 or the YC–15, advanced medium-short-takeoff-and-landing transport prototypes, would be used as the engineering baseline against which the technology could be developed and evaluated. The contract was AFSC's primary adhesively bonded structure (PABST) portion of the advanced metallic structures (AMS) program to improve structural integrity and durability of future aircraft while minimizing weight and costs of acquisition and maintenance. (AFSC *Newsreview*, May 75, 4)

- *Flying* magazine discussed "Relevant Research" at NASA: NASA had often been accused of neglecting aeronautics research in favor of space. But the $6.2 million for general aviation, of NASA's total $3.5 billion budget, was "a lot of money, even for a Government agency, and if used wisely. . .can make significant contributions in the many areas where technology advancements are sorely needed."

NASA was particularly well suited to investigate stalls and spins and weather prediction—two of the most prominent factors in general aviation accidents—as well as to develop simple autopilots and flight-control systems.

With the aircraft industry becoming increasingly competitive, NASA was "one place to seek help" to produce a technically superior product. NASA was responding to the need, allocating manpower and dollars to relevant general aviation research programs.

More important was the new attitude of NASA and industry toward communicating about capabilities and needs. In the past NASA did its research "in a vacuum, oblivious to the needs of the aviation community and to the unique conditions that constrained the industry's ability to apply NASA research results." But recently NASA initiated a series of workshops between key industry technical experts and their NASA counterparts, designed to coordinate industry technological problem areas and NASA research programs. Also, in 1974, NASA established the General Aviation Technology Advisory Panel and gave it equal stature with the committees and panels that constituted the influential NASA Research and Technology Advisory Council.

The atmosphere for effective communication and relevant research was improving. "While the threat of bureaucratic inefficiency and self-serving growth seems ever-present in Government endeavors, we see a conscientious effort on the part of NASA to apply their resources to the needs of general aviation." (*Flying* magazine, May 75)

# June 1975

*2 June:* The Netherlands, Denmark, and Norway had signed preliminary contracts and memorandums of understanding with the U.S. to purchase 348 General Dynamics Corp. F−16 fighter aircraft for a total of $2.1 billion, *Aviation Week and Space Technology* reported. The countries were part of a four-nation consortium—with Belgium—to evaluate the F−16, French Dassault-Breguet Mirage F1E, and Sweden's Saab Viggen as possible replacements for the Lockheed Aircraft Corp. F−104 operated by these countries. The preliminary contract was contingent on a full commitment to the F−16 by all four countries by 15 June.

*Aviation Week* reported that Belgium had requested a two-week delay in making its choice but then announced its selection of the F−16 on 7 June after the countries had received assurance that the consortium would be permitted to produce, in Europe, 10% of all parts for the F−16. In addition, the four countries would produce 15% of the aircraft for sales to third world countries as well as 40% of the components for the aircraft bought by the consortium. (*Av Wk*, 2 June 75, 21; 16 June 75, 17; AFSC Release 280−75)

- Private industry and government agencies in the U.S. and in Europe were investing money in designs and prototypes of giant dirigibles that could carry cargo and passengers more cheaply than conventional aircraft, *Newsweek* reported. In May a 9-m-wide helium-filled prototype dirigible named Skyship, shaped like a flying saucer, had been flight-tested at a Royal Air Force station in England; a flight model was expected to go into commercial operation within 3 yr. Shell Oil Co. had spent more than $1 million studying the possibility of developing dirigibles to transport natural gas, *Newsweek* said.

  The U.S. Air Force was investigating dirigibles as a means of ferrying missiles from silo to silo, and the Navy was studying their ability to track submarines.

  A reason for the resurgence of interest in lighter-than-air craft was the energy crisis. Airships require hardly any fuel to get aloft and use very little for propulsion, and they also produce less noise and less atmospheric pollution.

  New designs for dirigibles included the substitution of inert helium for highly flammable hydrogen as the lifting gas, and the replacement of the familiar cigar shape with other configurations to do away with the problem of instability near the ground. (*Newsweek*, 2 June 75)

- U.S. helicopter manufacturers were increasing production to capitalize on heavy worldwide demand for helicopters, *Aviation Week and Space Technology* reported. In 1974 U.S. manufacturers exported 420 helicopters at a value of $124 million, an increase of $39 million over 1973 sales and double 1971 sales. (*Av Wk*, 2 June 75, 114)

- *Luna 22*, launched 29 May 1974 by the U.S.S.R., had been in lunar orbit more than a year. Tass reported that, after 3296 orbits of the moon, the planned program for a comprehensive exploration of the moon and near-moon space had been fully carried out and additional exploration was continuing. All systems aboard the spacecraft were functioning normally. (Tass, FBIS-Sov, 18 June 75, U1)

*2-7 June*: Three NASA scientists and two Europeans conducted a Spacelab simulation mission aboard NASA's Galileo II airborne laboratory. The series of five night flights was designed by NASA and European Space Agency representatives to permit Spacelab mission planners to evaluate experiment techniques and operations. The simulations, part of a study program called the Airborne Science Spacelab Experiments System Simulation (ASSESS), would provide valuable information on the interaction between scientific investigators on the ground and experiment operators aboard Spacelab and would aid in estimating the training needed by Spacelab experiment operators.

Experiments flown in the simulation mission included infrared observations of the earth's upper atmosphere, Venus, stars, and other celestial features, and ultraviolet measurements of planetary atmopheres. (NASA Release 75-177; ESA Releases, 27 May 75, 18 June 75)

*3 June: Explorer 49*—Radio Astronomy Explorer B, launched 10 June 1973 into lunar orbit to measure galactic and solar radio noise at frequencies below ionospheric cutoffs and beyond terrestrial background interference by using the moon for occultation, focusing, or aperture blocking for increased resolution and discrimination—had completed nearly 2 yr of successful operation and was adjudged successful.

*Explorer 49*'s lunar orbit, away from the noisy environment of earth, had increased the effectiveness of the instrumentation. The spacecraft had been able to detect very weak solar-noise bursts up to 13 mhz and as low as 25 khz. Its predecessor, *Explorer 38*—Radio Astronomy Explorer A, launched 4 July 1968—had been able to observe relatively intense solar-noise bursts only in the 200 khz to 9 mhz range because of its orbit within the earth's magnetosphere. *Explorer 49* also was able to observe and follow solar bursts out of the ecliptic plane.

Data from *Explorer 49* had verified two major magnetospheric noise regions, one on the morning side of the earth and the other on the evening side. In addition, spacecraft data showed that intense noise occurred in the 8 to 12 pm time zone with less intense noise in midmorning. The intense late-evening noise was generated within 10 000 to 12 000 km of the earth's surface, and the dayside emission seemed to come from an altitude of less than 4000 km. (NASA MOR, 3 June 75)

*3-6 June:* A Kennedy Space Center team tested a parachute system designed to lower an instrumented probe through the dense atmosphere of Venus to its surface. The tests—made for Ames Research Center in preparation for a Pioneer mission to Venus in 1978—were conducted in KSC's Vehicle Assembly Building, an ideal wind-free testing facility. Full-scale 3.7-m parachutes with simulated pressure vessels weighing 204 kg were dropped 13 times from heights of up to 137 m to the floor of the VAB with impact cushioned by a honeycomb cardboard impact arrester. The performance of the parachute system was monitored photographically during the drop to determine the aerodynamic trim characteristics of the parachute.

The 1978 Pioneer–Venus mission would include the launch of two separate spacecraft toward Venus. The first would orbit the planet collecting data for about 250 days; the second would carry one large probe, which would use the parachute to descend to the surface, and three small probes decelerated by aerodynamic drag only. The parachute on the largest probe would be used to separate an instrument-bearing pressure vessel and stabilize it for a series of critical scientific measurements as it passed through the Venusian cloud layer. None of the probes was designed to communicate with earth after impact. (KSC Release 98–75; Lockyer, Public Affairs Office, interview, 6 Oct 75)

## FY 1976
## NASA Budget Authorization
## Congressional Adjustments

(millions)

| Program | Budget Request | House Changes | Senate Changes | Conference Changes | Total Conference Recommendation |
|---|---|---|---|---|---|
| *Research & Development* | | | | | |
| Space Shuttle | 1206.0 | 0 | 0 | 0 | 1206.0 |
| Space Flight Operations | 207.1 | −4.0 | −4.0 | −4.0 | 203.1 |
| Advanced Missions | 1.5 | +1.5 | −1.5 | +0.5 | 2.0 |
| Physics & Astronomy | 155.8 | +1.0 | +7.0 | +7.0 | 162.8 |
| Lunar & Planetary Exploration | 259.9 | −1.0 | 0 | 0 | 259.9 |
| Launch Vehicle Procurement | 166.9 | 0 | 0 | 0 | 166.9 |
| Space Applications | 175.0 | +6.5 | +8.9 | +6.5 | 181.5 |
| Aeronautical R&D | 175.4 | 0 | 0 | 0 | 175.4 |
| Space & Nuclear R&D | 74.9 | +2.0 | 0 | 0 | 74.9 |
| Energy Technology Applications | 5.9 | 0 | 0 | 0 | 5.9 |
| Tracking & Data Acquisition | 243.0 | −2.2 | −2.2 | −2.2 | 240.8 |
| Technology Utilization | 7.0 | +2.0 | 0 | +1.0 | 8.0 |
| R&D Totals | 2678.4 | +5.8 | +8.2 | +8.8 | 2687.2 |
| *Research and Program Management* | | | | | |
| Basic Submission* | 776.0 | 0 | 0 | 0 | 776.0 |
| R&PM Totals* | 776.0 | 0 | 0 | 0 | 776.0 |
| *Construction of Facilities* | | | | | |
| Modification of 11- by 11-ft Transonic Tunnel (ARC) | 2.7 | 0 | 0 | 0 | 2.7 |
| Modification of 40- by 80-ft Transonic Tunnel (ARC) | 0 | +12.5 | 0 | +12.5 | 12.5 |
| Addition to Lunar Sample Facility (JSC) | 2.5 | −2.5 | −2.5 | −2.5 | 0 |
| Addition to Composite Model & Metal Shop (LaRC) | 1.9 | 0 | 0 | 0 | 1.9 |
| Construction of Transonic Tunnel (LaRC) | 0 | +27.5 | 0 | 0 | 0 |
| Space Shuttle Facilities | 47.2 | −0.9 | 0 | 0 | 47.2 |
| Rehabilitation of Facilities | 16.0 | 0 | 0 | 0 | 16.0 |
| Minor Construction | 5.0 | 0 | 0 | 0 | 5.0 |
| Facility Planning & Design | 9.3 | +4.5 | 0 | +4.5 | 13.8 |
| C of F Totals | 84.6 | +41.1 | −2.5 | +14.5 | 99.1 |
| Grand Totals* | 3539.0 | +46.9 | +5.7 | +23.3 | 3562.3 |

*Does not include a supplemental request for pay increases introduced as H.R. 13172 and signed into P.L. 94–303 on 1 June 1976.

## Transitional Period
## NASA Budget Authorization
## Congressional Adjustments

(millions)

| Program | Budget Request | House Changes | Senate Changes | Conference Changes | Conference Recommendation |
|---|---|---|---|---|---|
| Research & Development | 730.6 | −30.0 | −26.0 | −30.0 | 700.6 |
| Research & Program Management | 213.8 | 0 | 0 | 0 | 213.8 |
| Construction of Facilities | 14.5 | −6.5 | −3.0 | −3.7 | 10.8 |
| Totals | 958.9 | −36.5 | −29.0 | −33.7 | 925.2 |

*4 June:* A House of Representatives and Senate conference committee reported out H.R. 4700, the FY 1976 and transitional-period NASA authorization bill, after resolving the disagreeing votes of the two Houses.

In addition, for the transition period 1 July through 30 Sept. 1976, the conference committee recommended a total authorization of $925 150 000. (H.R. Comm. Rpt 94−259; *CR*, 4 June 75, H4874−77)

- The Central Intelligence Agency reported to the Senate Appropriations Committee on its secret monitoring of the Soviet space program and on the safety risks that might exist during the Apollo-Soyuz Test Project mission, the *Washington Post* reported. Details of the report, which was requested by Sen. William Proxmire (D−Wisc.), would be made public by the committee within a few weeks. (AP, *W Post*, 5 June 75, A3)

- A new medical diagnostic system known as time-delay spectrometry, using high-frequency sound waves, had been developed by scientists at Jet Propulsion Laboratory. The new technique passed continuous, varying high-frequency sound waves through the body; the waves were received and displayed as pictures on a cathode-ray-tube screen. Existing ultrasound systems had operated less effectively, sending into the body a pulse of sound waves and timing the echo after the waves were reflected from the body's various internal organs.

  A picture from the JPL system looked like an x-ray and took shape in 2 to 4 min. Bones, muscles, organs, and differences in soft tissue could be clearly seen. The varying pitch of the sound waves and the differences in frequency response and absorption properties of various body tissues would aid the diagnostician in discriminating between, for example, a cyst and a tumor.

  JPL researchers were planning to evaluate the system's usefulness in hospital applications, first by demonstrating its ability to detect and identify tumors in the female breast. (NASA Release 75−162)

- Marshall Space Flight Center announced the selection of Bendix Corp. Guidance Systems Div. for negotiations leading to a contract for integrated electronic assemblies for the Space Shuttle solid rocket booster. The cost-plus-incentive-fee contract would cover design, development, test and evaluation, and fabrication of assemblies and assorted test equipment for the first 6 Shuttle flights. The initial contract would call for 33 units, including flight articles, spares, and development and test versions, with delivery to begin in 1976.

Each solid rocket booster would have two assembly units, one forward and one aft. Ignition commands would route from the Orbiter through the aft assembly to the forward assembly. During launch, the aft assembly would route commands from the Orbiter to the thrust-vector control system. The forward assembly would release the nose cap and frustum, jettison the solid rocket nozzle, detach the parachutes from the solid rocket booster, and turn on recovery aids. (MSFC Release 75−106)

- The *Christian Science Monitor* quoted Dr. H. Guyford Stever, National Science Foundation Director, on the future role of science: Dr. Stever said that scientists should stop looking back to days when research money flowed freely and science was valued for its own sake. Scientists must face the fact that " 'the social environment in which science is performed has changed.' " Scientists were bewildered as they watched research support tightening and jobs dwindling, and public questioning of the values of basic research grants. Yet science would become involved, as never before, in the economic success or failure of this country and the world. Society wanted research effort devoted to solving mankind's problems of food, energy, and environment. Dr. Stever noted that the era was long gone when the scientific community could conduct its affairs as " 'a pure search for truth apart from serious considerations of its human consequences.' " From now on " 'the drive to understand nature. . .must be seen to be carried on in the public interest and not merely to satisfy the private curiosity of an intellectual elite.' " (Cowen, *CSM*, 4 June 75, 25)

*5 June:* The U.S.S.R. launched comsat *Molniya I−30* and France's *Sret 2* on a single booster from Plesetsk. *Molniya I−30* entered elliptical orbit with a 39 876-km apogee, 448-km perigee, 712.2-min period, and 62.8° inclination. *Sret 2* entered orbit with a 40 824-km apogee, 512-km perigee, 737.8-min period, and 62.8° inclination.

Tass reported that *Molniya I−30* would ensure long-range telephone, telegraph, and radio communications as well as transmit TV programs to Orbita network stations in extreme northern Siberia, the Far East, and central Asia.

*Sret 2* (Satellite de Récherches et d'Environment Technique), according to Tass, would study the efficiency of a radiation system of thermal protection in outer space. It was the second of three spacecraft in a program of Soviet−French cooperation. *Sret 1*, launched 4 April 1972, had studied the characteristics of solar batteries for space operations and the degeneration of solar cells from cosmic-ray exposure in the Van Allen Belt. (GSFC *Wkly SSR*, 5−11 June 75; Tass, FBIS−Sov, 17 June 75, U1; *A&A 1972*, 128)

*6 June*: The establishment of a science advisory office in the White House, when it has to be urged on the President rather than conceived by him or his close associates, "is virtually irrelevant to the workings of the White House," Daniel S. Greenberg said in a *Washington Post* article. Greenberg pointed out that the President already had National Science Foundation Director Dr. H. Guyford Stever, whose office was assigned a pared-down version of the presidential advisory function when President Nixon dispensed with in-house science advice in 1973. Although Dr. Stever possessed all the right qualifications and was eager to serve, "Mr. Ford has had more conversations with golf pros over the years than he has had with Mr. Stever."

The science community had agitated for the return of a science adviser by appealing to Congress and appointing committees of distinguished scientists who duly certified that the President suffered from the absence of a science adviser close at hand.

Although President Ford decided in favor of the appointment, still unresolved was whether a science office would be more useful than the NSF-based operation. Past Presidents had not often used their science advice. "The bookshelves of the Executive Office Building bear innumerable aging reports of early warnings by scientists concerning problems of energy, food, resources, pesticides, pollution, drug safety, and so forth, but few [Presidents] chose to listen."

Science advice was a valuable resource for a President who chose to employ it, "for as grating as they often are with their contentions of superior wisdom, scientists do possess experience and talent that can be put to good political use." President Ford's move had little significance as far as the Ford administration was concerned, and it would "add up to nothing more than a cosmetic device" if the new science office had no jurisdiction over military research and development, a function sliced away from the advisory role under President Nixon's reorganization.

"But if a science office is in place at the White House, the next administration can get good service from it if it chooses to do so. The sad history of the matter is that, with rare exceptions, Presidents have not made that choice." (Greenberg, *W Post*, 6 June 75, A28)

- During recovery operations of the U.S. Apollo-Soyuz Test Project astronauts, the Apollo spacecraft would not be retrieved by helicopter as usual on Apollo missions, Dr. Donald E. Stullken, chief of Johnson Space Center's Mission Support Branch, said during a press briefing. Instead, the recovery ship would steam up alongside the command module and haul it aboard using the ship's boat and aircraft crane. Navy personnel would open the hatch and the crew would step out on the ship's deck.

Arnold D. Aldrich, deputy manager of the Apollo Spacecraft Program Office at JSC, said that the most extensive modification required for the ASTP mission was redesigning the Soyuz spacecraft to accommodate the newly added docking system. The front end of the Soyuz had been modified with respect to structural configuration, thermal design, and the requirement to fit all the new system under the launch shroud that covered the entire Soyuz during the boost phase.

John E. McLeaish, JSC Public Affairs Officer, said that NASA would provide U.S. and Soviet air-to-ground communications, air-to-air communications, and U.S.- and U.S.S.R.-originated commentary for newsmen during the mission. All Russian would be translated simultaneously into English, using a "voice-over" technique; a few communication lines would transmit raw Russian with no translation. The U.S.S.R. would provide an interpreter to assist newsmen in the U.S. with interpretation or explanation, if necessary.

Questioned about the political significance of the ASTP mission, U.S. crew member Donald K. Slayton replied, "I think we'd be naive to assume that this program is going to end all conflicts between our two societies, certainly. I think . . . it's a step in that direction, but there's little doubt in my mind that we're going to continue to have conflicts in many different areas. However, if we can kind of break the ground and get at least one area where we continue to work con-

structively on something, we're certainly not going to end up in any major confrontation." (Transcripts)
- Marshall Space Flight Center issued a request for quotations for a proposed study contract to design, build, test, and deliver a full-scale solar-array wing for solar electric propulsion and Space Shuttle payload applications. The selected contractor would establish design requirements and develop techniques for low-cost fabrication and testing incorporating all previously developed technology. (MSFC Release 75-109)

*7 June*: Rockwell International Corp. Rocketdyne Div. successfully conducted the first ignition test of the Space Shuttle's Integrated Subsystem Test Bed engine at the National Space Technology Laboratories. The test, conducted for Marshall Space Flight Center, lasted for 0.8 sec. It was first of a series of 8 ignition tests leading up to the firing of the Space Shuttle main engine. Initial evaluation of data indicated that all the test parameters were satisfactory and that the test objective was achieved. All ignition tests would be of short duration, none reaching full engine thrust. (*Roundup*, 20 June 75, 4; *Marshall Star*, 11 June 75, 4)

*8 June*: If NASA had unlimited funding, "I would move fairly quickly on a manned station because that will become an important outpost for anything that we do in space," Dr. George M. Low, NASA Deputy Administrator, said in an interview with the *San Diego Union*. Dr. Low added that he would move quicker in launching a space telescope and bringing back soil samples from Mars, "both enormously important from a scientific point of view."

Dr. Low expressed concern that "we really aren't being inventive enough. We haven't discovered a new propulsion system." All NASA's rockets so far, including the Space Shuttle, had used conventional means of propulsion, and with it man was limited. Dr. Low said that the moon landing "was enormously difficult because we are tied to chemical propulsion systems." NASA was experimenting with solar electric propulsion but to date no one had invented a fission or antigravity system. (*San Diego Union*, 8 June 75)

*8–17 June:* The U.S.S.R. launched *Venera 9* and *10* to continue Soviet exploration of the planet Venus. *Venera 9* was launched on 8 June at 7:37 am local time (10:37 pm EDT 7 June) from Baykonur Cosmodrome, near Tyuratam, on course for a rendezvous with Venus in October. Tass announced the purpose of the mission was "to carry on scientific research on the planet of Venus and the surrounding space, which is carried out by means of an automatic craft." Tass also said that, on the earth–Venus flight path, the probe would study the physical characteristics of interplanetary space including magnetic fields, solar winds, and ultraviolet radiation. Tass said that *Venera 9* was the heaviest in the U.S.S.R.'s Venera series and had "certain distinguishing design features."

On 14 June at 9:30 am local time (12:30 am EDT) *Venera 10* was successfully launched toward Venus from Baykonur Cosmodrome. Tass reported that "in design and purpose it is analogous to the '*Venera-9*' station launched on June 8" and that all spacecraft systems were functioning normally. *Venera 10* was also expected to reach Venus in October.

In an interview 17 June with the *Christian Science Monitor*, Boris N. Petrov, chairman of the U.S.S.R. Council of International Cooperation in the Study and Exploration of Outer Space, revealed that the Venera mission called for a landing rather than just a flyby. (GSFC *SSR*, 30 June 75; FBIS—Sov, 8—17 June 75; Pond, *CSM*, 17 June 75, 1; *W Post*, 8—17 June 75; *NYT*, 8—17 June 75)

*9 June:* The House of Representatives, by voice vote, agreed to the conference report on H.R. 4700, the NASA authorization bill for FY 1976 and the transitional period 1 July through 30 Sept. 1976. [See also 4 June.] The conference committee authorized a total of $3 562 310 000 for FY 1976, an increase of $23 310 000 over the budget request, and $925 150 000 for the transition period, a decrease of $33 750 000 from the budget request. (*CR*, 9 June 75, H5106—08)

- President Ford transmitted to both Houses of Congress proposed legislation to create, within the Executive Office of the President, an Office of Science and Technology Policy headed by a director who would also serve as presidential science and technology advisor.

  In his letters of transmittal to Speaker of the House of Representatives Carl Albert and President of the Senate Vice President Nelson A. Rockefeller, President Ford said that the new director would "identify new opportunities for using science and technology to improve our understanding of national problems and to contribute to their solution. He will also chair the Federal Council on Science and Technology, and I expect him to provide advice on the scientific and technological considerations in Federal policies, programs, and budgets."

  The President said that the director and deputy director would be presidential appointees. The office would draw extensively on the nation's scientific and engineering community for advice and assistance, and the director and staff would also call upon Federal agencies for assistance in carrying out their responsibilities. (*PD*, 16 June 75, 610)

- President Ford transmitted to Congress the *Aeronautics and Space Report of the President: 1974 Activities.* The President said that the "Nation's activities in aeronautics and space continued to produce significant benefits, to experiment with and develop new applications, to increase scientific knowledge, and to advance technology." He cited the continued expansion of international communications satellites; testing of earth-observation satellites for crop surveys, water resource management, and pollution management; completion of the successful Skylab manned space-station missions; continued planetary exploration, with successful missions to Jupiter, Venus, and Mercury; continued development of new aeronautical technology including the reduction of energy requirements, noise, and pollution, and modernization of the traffic-control system; and milestones in military aircraft development, including the rollout of the B—1 bomber and delivery of the F—15 fighter aircraft. (*CR*, 9 June 75, H5106)

- The Paris Air Show [30 May—8 June] reflected the international trends in aerospace with a depth and clarity far beyond what the most intensive study by individual nations could yield, an *Aviation Week and Space Technology* editorial said. Among trends evident at the 1975 show at LeBourget Airport was the "tremendous resurgence of the helicopter market, particularly in the commercial field where sales were now approaching parity with the traditional military market and promise to outstrip them during the next few years." Helicopters had been found particularly useful in the exploration, development, and production of oil and minerals.

Other expanding markets included the executive jet and short rugged-field transports for both military and civilian duty.

Within the military market, fighter aircraft had moved into a new era of technology with a new emphasis on weapons for both air-to-air and air-to-ground operations. "It is evident that they [fighter aircraft] must serve as platforms for new generations of weapons and advanced delivery systems that can produce a degree of military efficiency commensurate with their high cost."

The American presence was strong at LeBourget and the "U.S., through the performance of its new stable of fighters—F−14, F−15, and F−16. . .—and its current generation of advanced technology transports is demonstrating what a truly tough competitor it can be." The Europeans were going to have to streamline their management, better organize their resources, and cut through national bureaucracy if they were to have any hope of keeping pace. "Europe cannot continue at the same glacial development pace of the Concorde and MRCA [multirole combat aircraft] and expect to produce products that are technically competitive in the global market." However, its success with helicopters, tactical missiles, small transports, and the Airbus was indicative of what could be accomplished.

Noteworthy at' LeBourget was the evidence of the "slow pace of Soviet civil aviation caused by a series of development difficulties across their whole spectrum of transports and helicopters." The most visible evidence was the Tu−144 supersonic transport, which appeared in the fifth version off the assembly line, only the third such aircraft produced by the Soviet Union in the previous 2 yr. *Av Wk* quoted an expert as saying there was an estimated 1-yr slippage in Aeroflot plans to begin regular supersonic service. (Hotz, *Av Wk*, 9 June 75, 7)

*10 June:* The Senate, by voice vote, agreed to the conference report on H.R. 4700 authorizing NASA $3 562 310 000 in funds for FY 1976. [See also 4 June.] The authorization was $23 310 000 more than the budget request. In addition, the conference committee authorized $925 150 000, a decrease of $33 750 000 from the request, for the transition period 1 July through 30 Sept. 1976. The House of Representatives had agreed to the conference report on 9 June. (*CR*, 10 June 75, S10157)

- The U.S.—represented by Dr. James R. Schlesinger, Secretary of Defense—signed a memorandum of understanding with Belgium, Denmark, the Netherlands, and Norway that finalized selection of the General Dynamics Corp. F−16 air combat fighter as the four countries' replacement for the F−104G. The U.S. and the four countries entered into a co-production agreement under which the Europeans planned to produce up to 348 aircraft—at a value of $2.1 billion—for their own use. The proposal permitted aircraft manufacturers in these countries to share in the manufacture of aircraft and parts for U.S. and third world sales, which could more than offset their initial investment [see 2 June]. The Europeans would share logistics, maintenance, and training facilities in Europe.

The agreement would help enable the North Atlantic Treaty Organization (NATO) and allied forces to standardize their weapon systems. The U.S. Air Force had announced earlier plans to procure at least 650 F−16s, with a large number of these deployed to European bases. (AFSC Release 280−75)

*11 June*: *Pioneer 10* (launched 2 March 1972), on its way out of the solar system 1.2 billion km from earth and halfway between the orbits of Jupiter and Saturn, was making unexplained orientation changes, Ames Research Center announced. The spacecraft's axis of rotation was changing by 1° every 2 wk. Project officials had said that the changes could be caused by a slight malfunction or by some interplanetary phenomenon, but evidence was contradictory. Further tests were planned to find the cause of the change. (ARC Release 75−30)

- Rep. Olin E. Teague (D−Tex.) and Rep. Charles A. Mosher (R−Ohio) introduced H.R. 7830, the President's bill [see 9 June] "to strengthen staff capabilities for providing advice and assistance to the President with respect to scientific and technological considerations affecting national policies and programs." The bill was referred to the Committee on Science and Technology.

  H.R. 7830 called for establishing an Office of Science and Technology with a director who would be the Presidential science adviser, a deputy director, and a 15-member staff. The office would have a budget of $1.5 million.

  H.R. 7830 was the second bill concerning Presidential science advice introduced by Reps. Teague and Mosher; H.R. 4461 had been introduced 6 March to establish a national science and technology policy. It also included provisions for a council of science advisers and a cabinet-level Dept. of Research and Technology. (*CR*, 11 June 75, H5346; 6 March 75, H1464)

- Marshall Space Flight Center announced the award of a $3.2-million contract by Space Shuttle prime contractor Martin Marietta Aerospace Co. to Avco Corp. Aerostructures Div. to manufacture the intertank section of the shuttle's external tank. The intertank would provide support between the liquid-oxygen tank and the larger liquid-hydrogen tank. The disposable external tank, 47 m long and 8.4 m in diameter, would carry liquid propellant for the Space Shuttle's three main engines. (MSFC Release 75−116)

*12 June*: NASA launched *Nimbus 6* on a thrust-augmented Thor-Delta booster from the Western Test Range at 1:12 am PDT after a 6-min delay to clear a ship from the jettison trajectory of the Delta's solid rockets. The spacecraft entered sun-synchronous polar orbit with a 1103-km apogee, 1094-km perigee, 107-min period, and 99.96° inclination. Separation of the spacecraft from the launch vehicle was normal, at 59 min into the flight. The solar panels were deployed and the attitude-control system quickly acquired earth. A wheel control provided three-axis stabilization. Checkout of the power system, telemetry, command links, and tape recorder was completed on the first day of operation.

At 7 hr after launch, a cryogenic vent valve associated with the limb-radiance inversion radiometer was pyrotechnically actuated as planned. This caused unexpected high torques and excessive control-gas usage. *Nimbus 6* was put into a safe wideband gravity-gradient mode to conserve gas, and an investigation was begun to find the cause of the problem and determine possible remedies. By 24 June the 9 experiments and all spacecraft systems had been checked out and all instrumentation, except for 2 tape recorders, was working normally.

The primary objective of the *Nimbus 6* mission was to contribute to the Global Atmospheric Research Program (GARP) by refining and extending the capability of vertically sounding the temperature and moisture of the atmosphere, with particular regard to altitude resolution and interference effects of clouds. GARP, a cooperative worldwide weather program, was established to improve understanding of meteorological atmospheric processes in the tropics and to improve observation and computing systems for weather prediction and analysis. *Nimbus 6* would also monitor environmental conditions such as sea ice and rainfall and would measure the earth's radiation budget with a precision never before attained.

Secondary objective of the mission was to demonstrate the ability to track and relay data from a low-altitude polar-orbiting spacecraft using a geosynchronous satellite as a communications relay. *Ats 6*, the Applications Technology Satellite launched on 30 May 1974, would be the relay comsat.

A data-collection, processing, and relay-to-earth system was also on board to collect data from approximately 300 balloon-borne platforms, designed to float at an altitude of 20 km, and a large number of earth- and sea-based data platforms. *Nimbus 6*, unlike previous spacecraft, could receive and process up to eight incoming signals at a time.

The satellite carried nine new instruments to extend measurements of atmospheric parameters to significantly higher altitudes than had previously been possible: the tracking and data-relay (T&DR) experiment; a high-resolution temperature sounder (HIRS) and a scanning microwave spectrometer (SCAMS) to take temperature soundings in the atmosphere outside the tropics; a tropical wind-energy conversion and reference level experiment (TWERLE) to provide data on tropical areas; a limb-radiance inversion radiometer (LRIR) and a pressure-modulated radiometer (PMR) to provide high-altitude temperature soundings; an earth radiation budget (ERB) experiment to establish the earth's atmospheric radiation balance; an electrically scanning microwave radiometer (ESMR) to map the liquid-water content of the clouds, ocean rainfall, and distribution and variation of sea-ice cover; and a temperature/humidity infrared radiometer (THIR) to provide data on cloud cover, ground temperatures, and water-vapor distribution.

*Nimbus 6* was next to last in a planned series of seven experimental meteorological satellites. *Nimbus 1* (launched 28 Aug. 1964) and *Nimbus 2* (launched 15 May 1966) achieved all their objectives. Launch of the third Nimbus, on 19 May 1968, failed when the launch vehicle malfunctioned; an identical spacecraft launched 14 April 1969 successfully functioned as *Nimbus 3*. *Nimbus 4* and *5* (launched 8 April 1970 and 10 Dec. 1972) had both completed all mission objectives and were still operating.

General Electric Co. was the prime contractor responsible for *Nimbus 6* integration and test, stabilization control, control-subsystem integration, and spacecraft structures and antennas. Seven U.S. and two British companies, in addition to the Jet Propulsion Laboratory, provided the nine instruments. Goddard Space Flight Center managed the Nimbus program under the direction of the NASA Office of Applications and was also responsible for the Thor-Delta launch vehicle. (NASA MORs 21 May 74, 12 June 75, 24 June 75; NASA Release 75-145)

- An Apollo-Soyuz Test Project flight readiness review at Kennedy Space Center was attended by Dr. George M. Low, NASA Deputy Administrator; ASTP Program Director Chester M. Lee; and top management from NASA Headquarters and Marshall, Johnson, and Kennedy Space Centers.

  During the meeting Lee said that "ASTP is on schedule and 'go' for launch on July 15." Top management had reviewed and closed out all aspects of flight preparations. ASTP Technical Director Glynn S. Lunney reported that no hardware problems remained that might interfere with an on-time launch.

  MSFC Saturn Program Manager Ellery B. May reported that the 8-yr-old Saturn IB launch vehicle had been carefully checked out and all components were being monitored by periodic inspections; during one of these inspections corrosion cracks had been discovered in the fins, requiring them to be replaced [see 19 Feb.-19 March]. May reported that the vehicle was ready to fly.

  William H. Rock, manager of the Sciences and Applications Project Office at JSC, said that modifications, including a larger, more effective lightning rod on the launch tower, had been made to the launch pad to minimize the hazards from lightning strikes. Jesse R. Gulick, KSC meteorologist, said that thunderstorm probability for launch day was 23% and that the probability of a tropical storm or hurricane winds was less than 3%. Based on previous years' data, chances were good that, even if bad weather postponed the primary launch date, the Apollo spacecraft could be launched within 4 days. (MSFC Release 75-119; KSC Release 104-75; *Marshall Star*, 18 June 75, 1; Ezell et al., *The Partnership: A History of the Apollo-Soyuz Test Project*, 310-12)

- Viking Lander 1 would undergo a critical milestone in its development, terminal sterilization, beginning 13 June, Robert S. Kramer, NASA Director of Planetary Programs, said at a Viking news conference in Washington, D.C. The Lander would remain in the oven for 2.5 days and then be removed and checked out. Past missions had failed because "maybe we knew how to sterilize a spacecraft but we didn't know how to make one work after it was sterilized," Kramer said. However, "we think we know a lot more now than we did 15 years ago. We believe we have developed the techniques to be fully reliable, but when we complete that test along about next Tuesday morning, then we are going to find out for sure."

  A. Thomas Young, Viking Mission Operations Manager at Langley Research Center, said that plans called for launch of Viking 1 on 11 Aug. and launch of Viking 2 on 21 Aug. Mission planners had deliberately arranged for the first Viking to arrive at Mars on 18 June 1976, after 10 mos of travel, and Viking 2 to arrive on 7 Aug. 1976, about 1 yr after it was launched. Both spacecraft would spend about 17 days in orbit examining landing sites "to assure ourselves that they continue to be safe." The sites, which had not been examined since 1971, must be checked again because Mars was a dynamic planet.

Lander 1 was scheduled to begin its descent 4 July and start black-and-white photography immediately upon landing. Color photography would begin 18 hr later. The first soil sample would be collected on the eighth day. Analysis of the sample would take 12 days to complete, and 20 days would be required to complete one cycle of all surface science. Young noted that, to avoid confusion during the mission, Martian days would be called "sols" because they were a half-hour longer than earth days.

During surface operations, the Orbiter passing overhead once a day would remain synchronized with the Lander; on the pass, the Orbiter, acting as a communications satellite, would relay information from the Lander to earth. When the Lander completed its work, the Orbiter would perform observations of its own, studying Martian surface features, including the proposed Lander B landing site, from orbit. (Transcript)

- A theory that the moon was actually a piece of the earth broken away early in the planet's formation 4.5 billion yrs ago was presented in a paper by Dr. John A. O'Keefe, a Goddard Space Flight Center scientist, and Professor Harold C. Urey, Nobel Prize winner from the Univ. of California. The paper, presented by Dr. O'Keefe at a meeting of the Royal Society in London, stated that although the moon had little, if any, metallic core, data from Apollo lunar-landing missions and earlier unmanned lunar flights had provided chemical evidence that lunar rocks were once a part of a mass that had included a considerable portion of molten iron.

  The higher proportion of molten metal in the earth supported the theory that, before the split, the iron in the earth had sunk to the center, drawing with it the gold, platinum, and other rare metals found in molten rock originally mixed with the iron. According to the theory, the moon was pushed to its present distance from the earth by interaction with the tides of the body of the earth.

  Other theories about the formation of the moon included the possibility that the moon had been captured by the earth's gravitational field when it passed near the earth, or that it had been formed simultaneously with the earth. Neither of those theories explained the moon's lack of a substantial metallic core. (NASA Release 75–171)

- S. Neil Hosenball, NASA's Deputy General Counsel, had been appointed General Counsel, effective immediately, replacing R. Tenney Johnson, who had been named General Counsel of the Energy Research and Development Administration in February.

  Hosenball, who had been Deputy General Counsel since 1967, had been Assistant General Counsel for Procurement from 1966 to 1967 and previously had served for 4 yr as Chief Counsel at Lewis Research Center. (NASA Release 75–173)

*14 June:* France would begin deactivating its space center at Kourou, French Guiana, in September, the *New York Times* reported. A space center spokesman was quoted as saying, "We'll have to put the space center to sleep for a while because of lack of funds."

The deactivation did not mean that the French were abandoning their ambition to achieve independence from the U.S. in space operations but that it would take longer than originally thought. The European space program at Kourou would resume with launches of several experimental rockets scheduled in 1979 and a satellite launch planned for the 1980s.

Since its establishment in April 1968, the Kourou center had launched more than 225 balloons, satellites, and sounding rockets. Its location 5° north of the equator made Kourou especially suitable for equatorial launches. The largest of its disappointments had been the failure of the Europa 2 launch vehicle in 1971, a program finally abandoned in 1973 in favor of the Ariane booster. (Howe, *NYT*, 14 June 75)

*15–17 June:* The first [VLC 2; see 1 May] of 2 Viking Landers scheduled for launch in August underwent more than 47 hr of sterilization at Kennedy Space Center. Preliminary electrical testing indicated that all subsystems survived the process, including 30 hrs at the maximum chamber temperature of 112° C. Additional tests would check to make sure that the onboard science experiments were in good working condition. (Daspit, LaRC Viking Proj Off, interview, 21 June 77; Viking Proj Off Rpt M75–155–0–1; Viking Status Bulletin 7, 23 June 75; NASA Release 75–181)

*16 June:* Western Union Telegraph Co. announced that after *Westar 1* and *2* (launched 13 April 1974 and 10 Oct. 1974) began video transmission, users would save up to 66% of the cost of land-based transmissions. Rates of the new system, filed with the Federal Communications Commission, were scheduled to go into effect 14 July. The multipoint reception capability of the Westar system would attract a variety of broadcast applications, including independent networks and links to cable systems.

Savings by users of the satellite system would depend on the type of service and the route selected: Westar offpeak video service between Los Angeles and New York would cost $715 for the first hour, a savings of 61% over the $1832 charged for a landline transmission. For each additional hour the Westar rate drops to $620, a reduction of 66% over landline transmissions.

Initially, cities with earth stations—New York, Los Angeles, Chicago, and Dallas—would receive the service, which could be extended beyond those cities via connecting carrier. (WU release, 16 June 75)

- Scientists at Itek Corp. had developed a new telescope optical system that would dramatically reduce atmospheric distortion, the Baltimore *Sun* reported. The new system, which could improve viewing resolution 10- or 20-fold, used mirrors and a computer to cancel out distortions so that observers could get a clearer image of celestial objects.

In the system, a shearing interferometer and a detector measured the amount of distortion in the light entering the lens; the computer calculated the amount of correction needed in the light to cancel out the distortion, and a special mirror that changed its shape when electric charges were applied corrected the light before the viewer saw it.

Itek was manufacturing a prototype that would be tested on a working telescope in 1976. (AP, B *Sun*, 16 June 75, A3)

*17 June:* Kennedy Space Center announced the award of a $1 612 948 extension to a $600 000 contract to Reynolds, Smith, and Hills Co. for Space Shuttle architectural and engineering services. The contract, which would run until June 1977, provided for modification to Pad A and Mobile Launcher 3 at Launch Complex 39, both being reconfigured for the Space Shuttle. (KSC Release 109–75)

*18 June:* Western Union Telegraph Co. and RCA Global Communications, Inc., had been selected for negotiations leading to phase-one contracts to provide detailed systems-design proposals for the Tracking and Data Relay Satellite System (TDRSS), NASA announced. TDRSS, to be developed and operated by industry to meet NASA service requirements, would consist of three or more specialized satellites in synchronous orbit to continuously relay tracking information and data and voice commands to and from NASA satellites.

Western Union and RCA would both provide a technical, cost, and business proposal for leased services to be provided in the phase-two operational period. The proposals would be evaluated and a single contract for phase two would be awarded. The selected contractor would provide tracking, command, and data-acquisition services for essentially all spacecraft orbiting below 5000 km, including the Space Shuttle and Spacelab, as well as automated spacecraft to be orbited by the Space Shuttle, for a 1- to 10-yr period beginning in 1979. TDRSS spacecraft and ground stations would be the property and responsibility of the selected contractor. (NASA Release 75–176)

- Scientist Herbert F. Wischnia was using *Oao 3* (*Copernicus*, Orbiting Astronomical Observatory launched 21 Aug. 1972) satellite and its onboard optical telescope to look for possible flashes of an ultraviolet laser from another world, the *Christian Science Monitor* reported. Although radio signals would be a better way to communicate once contact was established, Wischnia believed that an optical telescope could pick up a whole spectrum of frequencies at a glance. A radio telescope required years to cover a significant band of frequencies because of the complex data processing necessary; an optical signal, once detected, would probably indicate which radio frequency would provide further information. The biggest difficulty was making the initial contact.

    *CSM* reported that Wischnia was looking at three stars likely to have planets, all older than the sun and about 11 light yr distant. (Cowen, *CSM*, 18 June 75, 25)

- Marshall Space Flight Center announced award of two contracts for parallel studies on space-processing equipment for Space Shuttle and Spacelab missions. TRW Systems Group was awarded $299 981 and General Electric Co. was awarded $284 974 to provide preliminary designs for equipment that could be used to process various materials, such as metals and crystals, in space. (MSFC Release 75–125)

*19 June:* President Ford signed Public Law 94–39 (H.R. 4700, the NASA Authorization Act for FY 1976 and the transition period 1 July 1976 through 30 Sept. 1976). P.L. 94–39 authorized NASA $3 562 310 000 for FY 1976 and $925 150 000 for the transition period. (*PD*, 23 June 75, 664)

- The House of Representatives' Committee on Appropriations reported out H.R. 8070, the Dept. of Housing and Urban Development–Independent Agencies Appropriations Bill for FY 1976.

    The committee proposed the continuation of the Space Shuttle and Viking projects; various planetary, weather, and scientific satellites; and further development of a strong aeronautics program. No new starts were proposed. The committee also directed deferral of the Pioneer–Venus mission for 1 year to permit a budget-priority decision in 1977 between the Large Space Telescope and Pioneer–Venus. The committee recommended reducing the $57 600 000 requested for Pioneer–Venus to $9 200 000 to maintain a management capability during the 1-yr deferral.

## NASA FY 1976 Budget
## Comparison of Request, Authorization, and House-Recommended Appropriations

(millions)

| Item | Budget Request | Authorization (diff. from req.) | H.R. Appropriations Committee (diff. from req.) |
|---|---|---|---|
| Research & Development | 2678.4 | 2687.2 (+8.8) | 2629.0 (−49.4) |
| Construction of Facilities | 84.6 | 99.1 (+14.5) | 82.1 (−2.5) |
| Research & Program Management* | 776.0 | 776.0 (0) | 775.5 (−0.5) |
| Totals | 3539.0 | 3562.3 (+23.3) | 3486.6 (−52.4) |

*Does not include a supplemental request for pay increases, introduced as H.R. 13172 and signed into P.L. 94-303 on 1 June 1976.

The committee also recommended denial of $1 000 000 of the $5 000 000 request for continued studies of the LST, and reduction to $2 000 000 of the $3 000 000 requested for the transition period. The remaining amounts were judged sufficient to complete LST studies.

Within the Construction of Facilities category, the committee denied funds requested for construction of the Lunar Curatorial Facility. No funds were to be used for modifying the 40- by 80-ft wind tunnel at Ames Research Center until the committee had reviewed a formal budget request.

Reduction in the Research and Program Management request reflected a 10% reduction in the payment of General Services Administration space-rental charges. The committee also recommended language in the bill to permit the replacement of five older aircraft with more modern aircraft for greater efficiency and safety. This would reduce operating costs by $1 300 000 annually.

The committee recommended a total appropriation of $925 028 000 for the transition period 1 July through 30 Sept. 1976. This total was $33 872 000 under the total budget request and $122 111 less than the authorization.

## NASA Transition Budget
## Comparison of Request, Authorization, and House-Recommended Appropriations

(millions)

| Item | Budget Request | Authorization (diff. from req.) | H.R. Appropriations Committee (diff. from req.) |
|---|---|---|---|
| Research & Development | 730.6 | 700.6 (−30.0) | 700.6 (−30.0) |
| Construction of Facilities | 14.5 | 10.8 (−3.8) | 10.8 (−3.8) |
| Research & Program Management | 213.8 | 213.8 (0) | 213.7 (−0.1) |
| Total | 958.9 | 925.2 (−33.8) | 925.0 (−33.9) |

(H.R. Com Rpt No. 94–313; Budget Chron Hist FY 1976, NASA Ofc of Budget Operations, 16 June 75)

- The first radar probes of Jupiter's moon Ganymede had been made by Dr. Richard M. Goldstein of Jet Propulsion Laboratory, NASA announced. Ganymede, at the time some 600 million km distant from earth, was found to have a rougher surface than the inner planets Mercury, Mars, or Venus. Dr. Goldstein scanned Ganymede three or four times on each of six nights in August 1974, using the 64-m antenna at JPL's Goldstone station of the Deep Space Network; the radar employed a 400-kw microwave beam at a frequency of 12.6 cm.

Writing in *Science* magazine, Dr. Goldstein and co-investigator George A. Morris said the most likely surface of Ganymede was rocky metallic material imbedded in a matrix of ice. Such a surface could be smooth with a top layer of ice rubble but would appear rough to the radar because the ice would be transparent to the microwave beam.

The theory was particularly interesting in view of the finding by the *Pioneer 10* and *11* flybys (3 Dec. 1973 and 3 Dec. 1974) that Jupiter itself seemed totally gaseous, with no solid surface to produce a radar echo. The JPL scientists theorized that Ganymede was probably meteoritic in origin. (NASA Release 75–170; Goldstein et al., *Science*, 20 June 75, 1211–1212)

- Flight Research Center announced the award of a $146 000 cost-plus-fixed-fee contract to LTV Aerospace Corp.'s Vought Systems Div. to study the possibility of flight-testing the oblique wing on an F-8 jet aircraft. LTV would investigate technical and mechanical problems in modifying the test aircraft and would estimate the cost of modifications.

The oblique wing, developed by Dr. Robert T. Jones of Ames Research Center, was a conventional straight wing mounted on top of an aircraft so that it could be swiveled around a central pivot point. The wing could be fixed at zero degrees—perpendicular to the fuselage—for takeoff and landing, and moved to various sweep positions at different speeds for the best performance. Wind-tunnel tests had indicated the new design could operate with maximum efficiency over a wide range of flight conditions, offering either increased speed or significant fuel savings. (FRC Release 17–25)

*20 June:* Dr. Harold Brown, President of California Institute of Technology, announced the appointment of Dr. Bruce C. Murray as Director of Jet Propulsion Laboratory, succeeding Dr. William H. Pickering who was retiring after 21 yr in the position.

Dr. Murray, a Caltech professor of planetary sciences for 15 yr, had also been a researcher on JPL space missions for more than 10 yr. He was completing scientific descriptions of Venus and Mercury from *Mariner 10* data, gathered during a February 1974 encounter with the planet, and comparing the surface histories of Mercury, Venus, Mars, and the moon with the history of the earth.

Dr. Pickering had been with JPL since 1944 and had served as its Director since 1954. As Director, Dr. Pickering led JPL in designing and building the first orbiting U.S. satellite, *Explorer 1* (launched 31 Jan. 1958). Also under his leadership, JPL had designed the *Ranger*, *Surveyor*, and *Mariner* spacecraft that were launched to gather information on the moon and nearby planets. (NASA Release 75–179)

- Sen. Frank E. Moss (D–Utah) introduced, for himself and Sen. Barry Goldwater (R–Ariz.), S. 1987, the President's bill [see also 9 June] to "strengthen staff capabilities for providing advice and assistance to the President with respect to scientific and technological considerations affecting national policies and programs." The proposed bill would create an Office of Science and Technology Policy within the Executive Office of the President.

The bill was referred jointly to the Senate Committee on Aeronautical and Space Sciences, Committee on Commerce, and Committee on Labor and Public Welfare. S. 1987 was similar to H.R. 7830 introduced 11 June by Rep. Olin E. Teague (D–Tex.) in the House of Representatives. (*CR*, 20 June 75, S11117)

- Marshall Space Flight Center announced the formation of a solar heating and cooling task team to support requests made by the Energy Research and Development Administration (ERDA). The task team, under the leadership of Donald R. Bowden, would direct MSFC's role in ERDA's program to demonstrate heating and cooling applications of solar energy. The team would acquire examples of existing solar heating and cooling systems and would direct additional development of these systems to increase efficiency and lower costs.

  ERDA had authorized approximately $50 million to MSFC for work which would cover 5 yr and use 100 persons at MSFC. (MSFC Release 75–130).

- Indecision and lack of a national policy had prevented the U.S. from having its own domestic communications satellite system after 9 yr of planning, a *Washington Star* article quoted Communications Satellite Corp. President Joseph V. Charyk as saying. In less than 10 yr a global comsat system developed under U.S. leadership was carrying more than two-thirds of the world's transocean communications. Canada had had its own since 1972. The problem, Charyk said, was that, after the success of the global system, "everybody wanted in." Companies, including American Telephone & Telegraph Co., Western Union Telegraph Co., Hughes Aircraft Co., RCA Corp., and Fairchild Industries, Inc., expressed interest in developing a domestic system. The Federal Communications Commission, whose job it was to decide who would operate such a system, seemed only intent upon keeping ComSatCorp and AT&T from dominating the field. Quoting Charyk, the *Star* said the net effect of this lack of a positive policy by the White House, Congress, and the regulatory agencies was that all the U.S. had to date were two or three earth stations using reproductions of the Canadian system.

  Despite the problems, Charyk said that two of ComSatCorp's domestic programs were nearing operational stages: Marisat, a maritime comsat planned for a summer 1975 launch, and Comstar 1, scheduled for launch in 1976. ComSatCorp would also participate with several European countries and Canada in the Aerosat program to test the use of comsats for aircraft navigation.

  However, Charyk said that, under the "open skies" policies of the FCC, ComSatCorp had no guarantees of permanent service. "It is hard to get customers with these rules in effect." (Snider, *W Star*, 20 June 75, D7)

- A digital Stoland navigation, guidance, control, and display system in an augmentor wing aircraft (a modified C–8A Buffalo) guided the aircraft to an automatic landing at the short-takeoff-and-landing (STOL) test facility at Crows Landing, near Patterson, Calif. The automatic landing, first for a STOL aircraft, was a milestone in the flight acceptance of the Stoland system being developed by Sperry Flight Systems under contract to Ames Research Center. The system would be tested on other STOL aircraft. (ARC *Astrogram*, 17 July 75, 2)

- A *Science* magazine article detailed some early findings from the Atlantic Tropical Experiment (GATE) of the Global Atmospheric Research Program (GARP). GATE, a 72-nation research effort during the summer of 1974 to gather meteorological information on the tropics, found very few cumulus clouds. Clouds in weather disturbances tended to be grouped in patterns, usually in roughly linear bands. Prior to GATE the only visible indicators of these weather-disturbance patterns were cloud clusters seen on satellite photographs, but the details were unknown; GATE demonstrated that these clusters were high cirrus clouds that formed above the weather disturbance and might persist long after the rainclouds themselves had dissipated. The organization of active rainclouds then came under intensive study as the essential feature of atmospheric convection.

  Although scientists had previously thought that cloud bases formed at a given level in the atmosphere, clouds in the GATE area were found to form at many different altitudes; many did not reach the heights often associated with tropical cumulus clouds until the latter part of the summer. The strength of weather disturbances intensified during the experiment, indicating that the tropical atmosphere probably altered over the course of the summer as its moisture content and convection increased, causing the clouds to grow taller as higher and higher levels of the atmosphere were warmed.

  GATE scientists also investigated the air layer between the cloud bases and the sea surface, finding that the subcloud layer was profoundly altered by the passage of a cloud. Warm moist air taken up into the cloud to fuel the convection was replaced by cooler drier air from higher in the atmosphere. Scientists believed that the resulting mix might explain cloud spacing and lifetime, and might indicate a continually changing subcloud layer in marked contrast to conditions prevailing in fair weather. GATE experiments documented the existence of atmospheric waves in the tropics and showed that weather disturbances developed near the low-pressure portion of the wave.

  Scientists also discovered a peculiar rapid meandering of the equatorial undercurrent—an intense eastward flowing stream 100 m below the sea surface at the equator—that shifted north and south about 1° in latitude on either side of the equator with an apparent period of 16 days. (Hammond, *Science*, 20 June 75, 1195−98)
- Kennedy Space Center announced the award of a $6 473 074 fixed-price indefinite-quantity contract to Modular Computer Systems, Inc., for minicomputers and associated equipment and services for the Space Shuttle. The contract included $4 226 646 in hardware and services already ordered and $2 246 428 in additional hardware. Modular Computer Systems would supply minicomputers, peripheral equipment, special interface devices, associated software, engineering and maintenance support, spares, and documentation. (KSC Release 114−75)

*21 June*: NASA successfully launched *Oso 8* Orbiting Solar Observatory at 7:43 am EDT from Eastern Test Range on a thrust-augmented Thor-Delta booster, after a 1-day delay because of an electrical malfunction in the launch vehicle. The spacecraft entered a near-perfect circular orbit with a 560-km apogee, 544-km perigee, 95.74-min period, and 32.94° inclination. By 23 June *Oso 8* was stabilized and was accurately tracking the center of the sun as planned. All spacecraft systems were operating normally. By 27 June all eight experiments had been activated and were acquiring data.

Primary objective of the *Oso 8* mission was to investigate the sun's lower corona and its chromosphere, and their interface in the ultraviolet spectral region, to better understand the transport of energy from the photosphere into the corona. Secondary objective was to study solar x-rays and earth-sun relationships and to investigate the background component of cosmic x-rays.

The 1064-kg spacecraft had two sections: A rotating cylindrical bottom section—the wheel—accommodated experiments that did not require solar reference pointing or that scanned the celestial sphere. The nonspinning rectangular sail section mounted on top of the wheel accommodated experiments that required stable pointing at the sun.

The spacecraft carried eight experiments. The two on the sail section were a Univ. of Colo. high-resolution ultraviolet spectrometer to measure solar-ultraviolet line profiles in the range of 1050 to 2300 Å and their variation with time and position, and a spectrometer and a cassegrain telescope that would observe the solar chromospheric structure simultaneously in six lines from 1000 to 4000 Å originating from different levels in the sun's atmosphere. This experiment was provided by France's Laboratory for Stellar and Planetary Physics under a cooperative agreement between NASA and the French Centre National de la Récherche Scientifique.

Other experiments—all on the wheel portion of the spacecraft and supplied by U.S. universities, industry, and government agencies—included a high-sensitivity crystal spectrometer to monitor the sun's emission in the 2- to 8-kev range, obtain a complete spectrum of the sun every 10 sec during flares, and obtain high-resolution spectra of many celestial x-ray sources; a mapping x-ray heliometer to measure location, spectrum, and intensity of intermediate-energy x-rays from individual solar-active regions and to acquire data on extrasolar x-ray sources; an investigation of soft x-ray background radiation using proportional counters; a cosmic x-ray spectroscope to determine the spectra of sources and the diffuse cosmic x-ray background in the energy range of 2 to 60 kev, and to measure intensity variations and identify possible emission lines of discrete x-ray sources; a high-energy celestial x-ray experiment to measure the spectrum of all point x-ray sources observable in the energy range of 0.01 to 1 megaelectron volts, and to search for temporal variations in the intensity and spectrum of the point sources detected; and a study of extreme ultraviolet radiations from earth and space, to determine the behavior of atoms such as hydrogen and neutral and ionized helium in the earth's atmosphere by measuring the intensity and distribution of solar radiation resonantly scattered by these atoms.

*Oso 8* was the ninth, and last, in a series of spin-stabilized orbiting solar observatories designed to gather new knowledge of the sun, the earth's atmosphere, and sun-activated terrestrial phenomena over a broad range of the electromagnetic spectrum not detectable by ground-based equipment. Of the eight OSO launches that began with *Oso 1* on 7 March 1962, only one, OSO−C, was unsuccessful, because of a launch-vehicle failure. Discoveries made by OSO spacecraft included gamma-ray emissions from the galactic center; a previously unknown isothermal plateau in the chromosphere of the solar atmosphere; differences in the structure of the extreme ultraviolet corona above quiet and active regions on the solar disk; evidence that the corona above the solar poles was significantly cooler than nonactive coronal regions near the equator; and synoptic observations of "holes" in the corona that survived several solar rotations.

*Oso 8*—built by Hughes Aircraft Co.—was managed by Goddard Space Flight Center under the direction of NASA's Office of Space Science. GSFC was also responsible for the Thor-Delta launch vehicle. (NASA MORs, 4 June 75, 23 June 75, 27 June 75; NASA Release 75−158)

*23 June:* NASA and the Energy Research and Development Administration signed an interagency cooperative agreement to enhance the national energy effort. ERDA and NASA management would identify specific program tasks that could be undertaken by NASA centers in support of ERDA programs. ERDA would use NASA's basic and applied research capabilities in areas including solar heating and cooling systems, gas turbines, fuel cells, hydrogen technology, ground-propulsion technology, bearings, seals, combustion, automatic control systems, and materials and structures technology.

NASA would submit proposals and plans to ERDA for specific technology developments including testing, evaluation, and demonstration of projects or hardware; ERDA would also call upon NASA for specific short- and long-term technical and administrative expertise such as technical review boards, evaluation groups, and other assessment techniques.

The agreement would establish an ERDA−NASA program committee to annually review NASA research and development work in support of ERDA. (NASA Release 75−182; ERDA Release 75−99)

- NASA was conducting design studies and component testing for an experimental laser-energy system using a nuclear-fission reactor fueled with gaseous uranium or a gaseous uranium compound such as hexafluoride, Dr. Karl Thom said in an article released by NASA. Whereas conventional laser power used various processes of energy conversion that required costly equipment and resulted in net loss of energy, a nuclear system could produce large amounts of high-grade power very cheaply and efficiently.

Some scientists were using various nuclear-powered lasers for experiments, but most used systems that transferred energy from external reactors to the laser gas within the laser tube. NASA, with its interest in gaseous fuel reactors as high-performance space power and propulsion systems, was looking at a more efficient system in which the fission fragments that energize laser gas were energized within the laser tube itself creating, at fission, a powerful chain reaction.

Dr. Thom said that, in combination with the research on nuclear-pumped lasers, Los Alamos Scientific Laboratory scientists were testing gaseous fuel reactors using components, reflectors, and controls salvaged from the nuclear rocket program. If the system proved practical "a new era of nuclear energy utilization will begin, in space and on Earth." Space engineers could create a nuclear-powered laser station in earth orbit to beam power via lasers to various customers in space, to other space probes for propulsion, to lunar bases for their power needs, and to earth for clean and abundant energy. (NASA Release 75−169)

*24 June:* The House of Representatives, by a vote of 391 to 25, passed H.R. 8070, the Dept. of Housing and Urban Development and Independent Agencies Appropriations Bill for FY 1976 and the transitional period from 1 July 1976 through 30 Sept. 1976. The bill—appropriating for NASA $3.489 billion for FY 1976 and $925.0 million for the transition period—was passed as reported out by the House of Representatives Committee on Appropriations on 19 June. (*CR*, 24 June 75, H6003–12)

*24 June:* Apollo-Soyuz Test Project crew members Thomas P. Stafford, Vance D. Brand, and Donald K. Slayton began a 3-wk preflight medical isolation at 12 pm CDT. Called the flight health stabilization plan, the isolation limited the astronauts to specific working and training areas at Johnson Space Center. In offhours they were being housed in special mobile homes. Only previously screened personnel, wearing surgical face masks, could come into face-to-face contact with the astronauts. (JSC Release 75–59: Ezell *et al.*, *The Partnership: A History of the Apollo-Soyuz Test Project*, 313)

- Rockwell International Corp. Rocketdyne Div. successfully test-fired the main chamber of the Space Shuttle main engine for the first time, at NASA's National Space Technology Laboratories. Initial evaluation of ignition-test data indicated that all test objectives were achieved. During a second firing the engine's high-pressure fuel pump reached a maximum speed of almost 7000 rpm, 19% of full power. The test was seventh in a planned series of 10 to 15 short-duration ignition tests that began 11 June and would culminate in the operation of the engine at 20% of maximum thrust level of 2 090 664 newtons (470 000 lbs). Subsequent tests would attempt higher mainstage thrust levels to evaluate engine-starting characteristics and performance.

  The Space Shuttle main engine, the most advanced liquid-fueled rocket engine ever built, was the first to use an electronic digital computer for automatic control. Three of these engines, designed for reuse up to 55 times, would power the Space Shuttle Orbiter. (*Marshall Star*, 2 July 75, 1; Rocketdyne Release RD–15; MSFC Release 75–115)

- The Air Force System Command's Arnold Engineering Development Center was developing a technique to study, under flight conditions, the physical behavior of the spray-on insulating material proposed for the Space Shuttle, AFSC announced. Using procedures similar to those used in analyzing aerial photographs, a series of reference lines was projected onto the original surface and photographed before wind-tunnel testing. The same pattern was then projected onto the surface during the testing. The shift in the apparent position of the projected lines as the surface eroded from the position shown on the original photograph allowed analysis of the behavior of the materials. (AFSC Release OIP 157.75)

*24–25 June:* At its first meeting, the council of the European Space Agency (ESA) elected Wolfgang Finke (West Germany) as chairman for 1 yr. General Luis de Azcarraga (Spain) and Jan Stiernstedt (Sweden) were elected vice chairmen. (ESA Release, 27 June 75)

*25 June:* Marshall Space Flight Center announced the award of two 9-mo contracts for parallel design studies of a biological holding facility for Spacelab.

Lockheed Missiles and Space Co. received $242 640 and McDonnell Douglas Astronautics Co. received $177 360 to develop conceptual designs of habitats for live biological specimens supporting life-sciences space flight research programs. The habitats—consisting of the structure and the environmental-control, food and watering, waste-management, lighting, and instrumentation systems—were to be suitable for a wide variety of specimens with emphasis on the adult rhesus monkey and the adult laboratory rat. (MSFC Release 75−133)

*25 June−3 July*: The countdown demonstration test, a step-by-step dress rehearsal for the July Apollo-Soyuz Test Project, was successfully completed at Kennedy Space Center. The rehearsal culminated in actual fueling of the Saturn IB and a simulated launch of the Apollo at 3:50 pm EDT on 2 July. The cryogenic propellants loaded for the "wet" portion of the test were off-loaded and the final portion of the count repeated with crew members Thomas P. Stafford, Vance D. Brand, and Donald K. Slayton aboard the spacecraft. A second simulated ignition and liftoff occurred at 2 pm EDT on 3 July.

During the test, a leak detected in Brand's spacesuit was traced to one of the pressure-sealing slide fasteners. Modifications were made and the spacesuit was successfully retested. (*Spaceport News*, 26 June 75, 1; KSC Release 115−75; Ezell *et al.*, *The Partnership: A History of the Apollo-Soyuz Test Project*, 314)

*27 June*: The European Space Agency announced the appointment of Hans Hinterman (Switzerland) as director of the European Space Research and Technology Center (ESTEC), effective in September. Hinterman would replace Ove Hammarström who was leaving to accept a post in industry. Hinterman was director of research of the Laboratoire Suisse de Récherche Horlogères in Switzerland. (ESA Release, 27 June 75)

- Most of the 40 participants at a space-food exhibition banquet in Zvezdny Gorodok, U.S.S.R., found the meal tasty but not very filling, the *Washington Post* reported. Presiding over the affair, cosmonaut Aleksey A. Leonov, Soviet commander for the scheduled July Apollo-Soyuz Test Project, said that after the third day of spaceflight, when an astronaut's appetite improved, he could consume four meals a day. The banquet menu included fruit juices, coffee, borsch, cabbage soup, bite-sized bread, and a compote of cucumber, chicken necks, and kidney.

  During the dinner Leonov revealed that the *Soyuz 18* cosmonauts Pyotr Klimuk and Vitaly Sevastyanov, launched 24 May and in orbit aboard the *Salyut 4* space station, would continue in orbit during the ASTP mission and would probably remain in orbit for another month. (Piper, *W Post*, 27 June 75, A2)

- U.S. airlines carried more than 207 million passengers during 1974, an increase of 2.6% over 1973 and a new record, the Air Transport Association announced. In addition, U.S. lines carried 4.9 billion ton-miles of freight, an increase of 3.2% over 1973. Total operating revenues in 1974 were $14.7 billion, up from $12.4 billion for 1973. Net profit rose by 41.9%, from $226.7 million in 1973 to $321.6 million in 1974. (ATA Release, 27 June 75)

*28 June*: When commercial communications via satellite began with the 6 April 1965 launch of *Early Bird 1*, the U.S. was expected to be the prime user of the new global communications system, a *New York Times* article said. Experts assumed the system would be used chiefly for occasional live events such as sports, state affairs, space missions, etc. However, these uses had actually accounted for only a fifth of transglobal comsat TV volume. A number of poor and developing countries had come to depend heavily on the International Telecommunications Satellite Organization's seven Intelsat comsats. Although major events were still the subject of much global TV traffic, the most extensive and consistent use of commercial comsats was for daily news.

News packages were sent from one country to another. U.S. networks used almost daily satellite news reports from correspondents in various parts of the world. Spain and Mexico were linked full time by satellite. Madrid also transmitted 15 min of daily news to Argentina, Brazil, Chile, Colombia, and Venezuela. When events warranted, these countries transmitted news to Spanish TV. France transmitted daily news to Israel, Iran, Jordan, and Martinique as well as several North African countries. News from London was transmitted daily to Australia.

The *NYT* quoted officials of Communications Satellite Corp., U.S. manager of the Intelsat system, as saying that in 1965 the comsats carried about six TV transmissions a month for a total of less than 6 hours of programming; in 1975, the Intelsat system was handling 400 transmissions a month, more than 200 hrs of programming not including the 200 hrs a month carried on the Spain–Mexico channel. (Brown, *NYT*, 28 June 75, 43)

*29 June:* Final joint U.S.–U.S.S.R. simulations for the Apollo-Soyuz Test Project crews and flight controllers began at Johnson Space Center and at the Moscow Mission Control Center at 6:30 am CDT. The simulations picked up the count at 47 hrs 10 min into the mission and continued for 56 hrs, covering the joint portion of the mission including rendezvous, docking, crew transfers and joint activities, and undocking of the Apollo and Soyuz spacecraft. U.S. and Soviet crewmen participated in simulators in their respective countries while both flight-control centers were fully staffed. The U.S. crew would also participate in the countdown demonstration test at Kennedy Space Center on 3 July. (JSC Release 75–61; Ezell *et al.*, *The Partnership: A History of the Apollo-Soyuz Test Project*, 314)

*30 June:* NASA's permanent employment level decreased by 521, or 2.1% during FY 1975, NASA's Office of Personnel reported in "The In-House Work Force." Average age of NASA employees had risen 0.3 yr to age 43 and the average executive age fell 1.3 yr to age 49.6. Average grade of permanent employees remained at GS–11, constant since FY 1973.

New hires and separations were down by 4.4 and 6.5%, respectively. Marshall Space Flight Center had had the only NASA reduction-in-force, resulting in 76 separations. The 62 intercenter professional transfers were down from 72 in FY 1974, and the 789 retirements were 30% below the FY 1974 level, although disability retirements continued to rise, totaling 220 in FY 1975—a 147% increase over FY 1970. (Off of Personnel Rpt, "The In-House Work Force," 30 June 75)

- The Skylab program was closed out at Marshall Space Flight Center and personnel assigned to the program had been reassigned, the *Marshall Star* announced. Responsibility for the deactivated offices, including storage of hardware and phaseout of remaining contractual and documentary activities, was assigned to the Administration and Program Support Directorate's Logistics Office. (*Marshall Star*, 2 July 75, 4)

- "After spending some $1.5 billion over more than 10 years, the United States, through the loss of its will to go forward in space, gave up on this next generation of rockets even though the nuclear technology worked," Gary L. Bennett said in a letter to the editor of the *Washington Star*. Noting that U.S. basic rocket technology was close to 40 yrs old, Bennett wrote, "We have shown no inclination to develop new rocket concepts despite the fact that these concepts have definite terrestrial advantages in terms of energy and employment. It's sad to think that we came so close to realizing our destiny in space and then we gave it all up for a senseless enervating policy of introspection." (*W Star*, 30 June 75)
- NASA announced the award of a $287-million contract to Air Products and Chemicals, Inc., to supply liquid hydrogen for the government's East Coast requirements over a 12.5-yr period beginning 1 July 1975. The liquid hydrogen would be used primarily in Space Shuttle engine testing at the National Space Technology Laboratories and for Space Shuttle launches. (NASA Release 75–192)

*During June*: Activities continued at Kennedy Space Center in preparation for the Apollo-Soyuz Test Project launch in July. During the week of 2–6 June, KSC engineers and technicians checked the Apollo spacecraft for propellant leaks, filled the liquid-oxygen storage tank at the pad, loaded oxygen and nitrogen onto the docking module, and installed conax valves on the Saturn IB first stage.

Spacecraft ordnance was installed 9–13 June and the command module was checked for leaks and prepared for hypergolic loading. Also during the week the liquid-hydrogen tank was filled.

Prime crew members—Thomas P. Stafford, Vance D. Brand, and Donald K. Slayton—and backup crew members—Jack R. Lousma, Alan L. Bean, and Ronald E. Evans—arrived at KSC on 10 June for command and docking-module crew compartment fit and functional checks. The crew also familiarized themselves with stowage arrangements. (*Spaceport News*, 12 June 75, 5; 26 June 75, 3)

- Development of the Spacelab by the European Space Agency for NASA's Space Shuttle continued.

An annual Spacelab status review 4–5 June attended by Dr. James C. Fletcher, NASA Administrator, and Roy Gibson, ESA Director General, established schedule milestones, determined readiness for the preliminary design review, and discussed arrangements for the first Spacelab flight. They also discussed plans for the instrument-pointing system and the first Spacelab payload, follow-on Spacelab procurement by NASA, and terms and conditions for use of the Space Shuttle–Spacelab system.

On 6 June, the recommendations of 10 teams—which had met at Marshall Space Flight Center in May to discuss Spacelab systems, avionics, structures, environmental control and life support, software, test and integration, and payload operations and accommodations—were presented to the NASA Office of Manned Space Flight for concurrence. At the same time ESA was preparing similar recommendations at the European Space Research and Technology Center (ESTEC).

Beginning on 9 June, combined NASA–ESA teams met in Noordwijk to consider the 1772 review-item discrepancies prepared by both agencies and, after 2.5 days, had processed all of them. (*Spacelab Newsletter* 75–2, 3 July 75)

- Goddard Space Flight Center's *Goddard News* reported that the John C. Lindsay Memorial Award, given annually to a GSFC employee in recognition of an outstanding contribution to science and technology, had been presented to Dr. Norman F. Ness. Dr. Ness had received the award for his pioneering work in the investigation of magnetic fields of planetary bodies and the interaction of the solar wind with these magnetic fields. Dr. Ness, who had been a participating scientist on the Explorer Interplanetary Monitoring Platform (IMP) satellites and *Mariner 10*, and Pioneer satellites, spearheaded the first measurements of the earth's distant magnetic fields and magnetic tail. He was also the first to explore the moon's magnetic field. On the *Mariner 10* flybys of Jupiter, Dr. Ness made the surprising discovery of a small, but measurable, intrinsic magnetic field. (*Goddard News*, June 75, 1)
- The Air Force continued its development of the remotely piloted vehicle (RPV), the Air Force Systems Command's *Newsreview* reported. The Air Force awarded three 1-yr firm-fixed-price contracts for definition studies of an advanced remotely piloted vehicle and associated elements. Rockwell International Corp. received $699 684; Boeing Aerospace Co., $646 750; and Northrop Corp., $499 614, to produce designs of an RPV for use in the 1980s. The system was expected to provide an improved cost-effective capability of carrying out electronic warfare, reconnaissance, and strike missions. The three companies would define the vehicle and its avionics, options of launch and recovery, ground-support elements, and systems maintenance to permit rapid mission turnaround.

  The *Newsreview* reported the award of a $100 000 contract to Aero Co. to study the feasibility of using parafoils, half the size of a C−5 aircraft wing, to recover RPVs returning from missions. Aero Co. would design and build for flight testing two 12- by 24-m parafoils—high-glide parachutes that, when deployed like a parachute, take on airfoil characteristics, descending at a rate of 1 m forward for each 0.3-m loss in altitude. (AFSC *Newsreview*, June 75, 1)

## July 1975

*1 July:* Marshall Space Flight Center observed the 15th anniversary of the transfer of personnel, facilities, and responsibilities from the Department of Defense's U.S. Army Ballistic Missile Agency to NASA. The center had been formally dedicated at a ceremony 8 Sept. 1960 attended by President Dwight D. Eisenhower; Dr. T. Keith Glennan, NASA Administrator; Alabama Governor John Patterson; MSFC Director Dr. Wernher von Braun; and Mrs. George C. Marshall, widow of the statesman, soldier, and Nobel Peace Prize winner for whom the center was named.

With the transfer had gone responsibility for continued development of the Redstone and the Saturn I; soon to come would be the Saturn IV and V launch vehicles that would carry U.S. astronauts into earth orbit and, eventually, to the moon. More than 100 000 employees in 12 000 companies across the U.S. worked during the peak of developing the Saturn family and its 31 launches with 100% launch success rate. MSFC developed the Lunar Roving Vehicle that carried moon-based astronauts several km from the landing site across rugged lunar terrain.

MSFC was also responsible for the *Skylab 1* Orbital Workshop. Launched into earth orbit 14 May 1973, *Skylab 1* carried other MSFC developments including solar-observation instruments, the docking adapter, and many experiments. Three three-man crews occupied the space station for a total of 171 days.

To date, as the last Saturn booster was preparing to carry three U.S. astronauts to an earth-orbital rendezvous with two Soviet cosmonauts for the joint Apollo-Soyuz Test Project, MSFC was heavily involved in development of the Space Transportation System, including the Space Shuttle, Space Tug, Spacelab, and related payloads. MSFC employees were also working on the development of solar heating and cooling systems and other applications projects.

In a message to MSFC Director Dr. William P. Lucas, President Gerald R. Ford said that "From its inception, the Marshall center has continued in the forefront of this Nation's tremendous advancement in the exploration of space. We look forward to more vital contributions . . . from the Marshall center in the years ahead as the United States continues to lead the way in using space for the benefit of all mankind." (MSFC Releases 75–131, 75–140; *Historical Origins of The George C. Marshall Space Flight Center*, MSFC Historical Monograph No. 1, Dec 1960)

- Rockwell International Corp. Space Div. was studying the economic and technical feasibility of using a rotating flywheel for electrical storage in commercial, utility, and transportation applications. The 9-mo study for the Energy Research and Development Administration was based on technology developed for a prototype spacecraft energy-momentum flywheel that Rockwell had built for NASA's

Langley Research Center. The small prototype spinning at speeds up to 35 000 rpm could provide 2500 w of electrical power as well as spacecraft attitude control.

In the flywheel concept, electric motor generators would spin up specially constructed flywheels which kinetically stored energy by their rotation. When electrical power was needed, the rapidly spinning flywheels would drive generators. A significant application of the flywheel would be to store energy for electric utility companies; the flywheels, spun up during off-peak hours, would drive power-producing generators during peak-demand periods. (Rockwell Int'l Release SP-18)

- NASA announced award of a $46.8-million cost-plus-award-fee contract to International Business Machines Corp. for developing and testing the Space Shuttle avionics software system. IBM would design, develop, test, and maintain the avionics software for the data-processing system on the Shuttle Orbiter that would include electric and electronic systems for guidance, navigation, and control capability; communication; computation; displays and controls; instrumentation; and electrical power distribution and control for the orbiter, external tank, and solid-rocket boosters. (JSC Release 75-63)

*2 July:* Ats 6 Applications Technology Satellite arrived at its new operating station at 35° east longitude over equatorial East Africa where it would relay communications from the docked Apollo and Soyuz spacecraft during the July Apollo-Soyuz Test Project mission. Upon completion of ASTP, *Ats 6* would be used by India to transmit educational TV programs to several thousand remote villages.

The satellite had experienced a malfunction in a drive circuit of one of the three momentum wheels used to control spacecraft attitude. A group of hydrazine gas thrusters was being used as a backup system for stabilization while ground controllers at Goddard Space Flight Center were analyzing the problem and developing remedies. (NASA Release 75-194)

- Prof. Hyron Spinrad, Univ. of Calif. astronomer, had photographed a galaxy 5 to 10 times larger than the Milky Way. Prof. Spinrad photographed the galaxy, designated 3C123, using the 120-in telescope at Lick Observatory and a new device that filtered out backlighting. The huge galaxy was moving away from earth at a speed of about 4.3 trillion km per year. (AP, B *Sun*, 2 July 75, A3)

- Three high-speed F-104G aircraft formerly belonging to the German Air Force arrived at Flight Research Center to be used for research purposes. FRC had been flying F-104As, an early version of the same aircraft, but replacement parts were becoming scarce. Parts for the F-104G would be available through the 1980s. The three aircraft left Jever Air Force Base in West Germany on 27 June and made stops in Scotland, Iceland, Greenland, and Labrador before arriving at FRC. (*X-Press*, 18 July 75, 4)

*3 July:* The *Washington Post* reported that NASA had turned down a request by Sen. William Proxmire (D-Wisc.) to postpone the Apollo-Soyuz Test Project (scheduled for launch on 15 July) until the U.S.S.R. brought back the two *Soyuz 18* cosmonauts who had been docked with the orbiting *Salyut 4* space station since 26 May 1975. Sen. Proxmire said that the Soviet Union would be unable to maintain ground control of two manned missions simultaneously. He had based his judgment on closed-door testimony given him as chairman of the Appropriations Subcommittee by Carl Duckett, deputy director for science and technology at the Central Intelligence Agency. The *Post* quoted Proxmire as saying that the Soviet command center would be strained to handle two manned flights, particularly if one were in trouble in space. Sen. Proxmire said that it would be a simple matter to deorbit the two *Soyuz 18* cosmonauts and bring them home.

In a telephone interview with the *Post*, ASTP Technical Director Glynn S. Lunney said that the Soviet Union had two mission-control centers, one managing the Salyut flight, the other managing ASTP. Lunney said NASA had calculated that the two Soviet spacecraft would cross over the same Soviet tracking stations only twice: One overlap might last as long as 30 sec and the other 90 sec. Soviet officials had assured NASA that ASTP communications would have priority during these two short passes. (O'Toole, *W Post*, 3 July 75, A33)

- NASA was inviting U.S. and foreign corporations, universities, and government organizations to propose experiments for the *Ats 6* Applications Technology Satellite (launched 30 May 1974) after it completed its year of service to India [see 2 July]. After the Indian experiment, *Ats 6* would be moved westward to the west coast of South America where it would operate within range of North and South America and adjacent ocean areas. NASA's announcement of experiment opportunities for the comsat's third year of operation invited proposals in the fields of societal, communications, or technological disciplines. (NASA Release 75–187)
- The Energy Information Center at the Univ. of N. Mex. had compiled for NASA a bibliography, *Wind Energy Utilization*. The 496-page volume included foreign and domestic literature on wind-power plants, wind-power generators, wind machines, and wind-energy storage facilities. The center at Albuquerque, established in 1974 as a cooperative effort between the university and NASA, had also published comprehensive bibliographies on hydrogen and solar thermal energy. (NASA Release 75–186)

*7 July:* NASA had begun a demonstration of hydrogen-injection technology, a technique that could help reduce fuel consumption in large aircraft engines by 20 to 25% and lower emissions of pollutants from general-aviation engines.

Working with the Federal Aviation Administration and industry, NASA was studying minor engine modifications that would reduce characteristic engine emissions. A separate NASA-industry program was investigating emission reduction and fuel efficiency by means of more extensive changes in the basic design of aircraft engines. In a third program, NASA was investigating both the noise and the emission characteristics of small general-aviation engines. (NASA Release 75–198)

*9 July:* Action by the House of Representatives to defer for 1 yr the funds budgeted for the Pioneer–Venus project [see 19 and 24 June] would "jeopardize the possible success of Pioneer–Venus," Sen. John V. Tunney (D–Calif.) said in a letter to Sen. William Proxmire (D–Wisc.), chairman of the Senate Appropriations Committee. The House recommended the delay to permit a budget decision in 1977 between the Large Space Telescope and the Pioneer–Venus mission. Sen. Tunney's letter said that the probe was timed for the 1978 launch window when earth and Venus would be in unusually favorable relative positions; the next favorable window would not appear for almost 2 yr, during 1980, and Venus would be considerably further away from the earth at that time than it would be during 1978. A delay in funding would postpone the program for at least 18 mo, would require design changes, and might cost $50 million more than original estimates.

Sen. Tunney said that the case for continued funding was a strong one. Most scientists felt that investigation of the Venusian atmosphere would provide valuable meteorological information that might substantially enhance weather-prediction capabilities.

Another reason for continuation of the program is that, in an era of high unemployment, few sectors had suffered more devastating setbacks than the aerospace industry. The Pioneer–Venus program had created more than 1500 jobs, which would be threatened by termination or delay of the program.

Curtailment of an ongoing scientific concept like Pioneer–Venus would severely reduce credibility of the government's long-term commitments, with added time, difficulty, and cost for future projects as private contractors would seek to protect themselves.

Sen. Tunney pointed out that the House's juxtaposition of funding for Pioneer–Venus with funding for the Large Space Telescope was inappropriate; the programs were not competing, but complementary. Pioneer–Venus would provide information on our closest neighbor, whereas LST would collect deep-space data of unprecedented scope and detail. "Cutting one program for the sake of the other would mean sacrificing a distinguishable body of scientific knowledge which would not be replaceable." (Text, letter, Tunney to Proxmire, 9 July 75; *CR*, 26 July 75, S13882–83)

- The rocket-powered M2–F3 lifting body—a wingless vehicle that derives aerodynamic lift from its body shape—arrived for display at the Smithsonian Institution's National Air and Space Museum in Washington, D.C. In a flight program beginning in 1963 and ending in 1972, the three versions of the M–2 vehicle had demonstrated that a manned reentry vehicle could reenter from space, maneuver through the atmosphere, and safely make a deadstick landing.

M2–F3 was the third version of Ames Research Center's M–2 lifting-body design. The first, M2–F1, was a lightweight plywood glider towed to altitude by a C-47 and released, whereupon the pilot guided the craft to a landing. A heavier version, M2–F2, was launched from underneath the wing of a B–52 aircraft. After nearly a year of testing the unpowered vehicle, an XRL–11 rocket engine was installed in preparation for rocket-powered flight; on the second glide flight after engine installation, the vehicle crashed, badly damaging the vehicle and injuring the pilot. The lifting body was rebuilt, incorporating modifications, and designated M2–F3. The M2–F3 made more than 37 glide and powered flights with a jet reaction-control system that successfully demonstrated the feasibility of using a single system from orbit to landing.

A lifting body design that had been considered for the reentry vehicle in the Mercury, Gemini, and Apollo programs was rejected in favor of a ballistic reentry design. The use by Space Shuttle of the maneuver-to-a-deadstick-landing concept was possible largely be-

cause of the confidence level built up by 9 yr of lifting-body flight tests. (NASA Release 75−199; *A&A 62* through *A&A 74*)
- Marshall Space Flight Center announced award of two contracts for 356 aluminum hand forgings for Space Shuttle solid-rocket boosters. A $465 935 contract went to Aluminum Co. of America and $62 900 went to Weber Metals & Supply Co. for forward-skirt thrust post fittings, inboard aft-skirt actuator support brackets, aft-skirt splice fittings, and aft-skirt holddown posts. (MSFC Release 75−151)

*11 July*: NASA announced the appointment of Donald P. Hearth as Director of Langley Research Center, effective in September. Hearth, Deputy Director of Goddard Space Flight Center since 1970, would replace Dr. Edgar M. Cortright, who is leaving NASA to pursue other interests.

Prior to being at GSFC, Hearth had served 3 yr as Director of Planetary Programs at NASA Hq. He came to NASA in 1962 from the Marquardt Corp. where he managed research on hypersonic propulsion and flight systems. Before that, he had been an aeronautical research scientist at Lewis Aeronautical Laboratory of NASA's predecessor, the National Advisory Committee for Aeronautics. (NASA anno 11 July 75)
- Didier Ratsiraka, President of the Supreme Council of Revolution of the Malagasy Republic, said in a radio broadcast to Malagasy citizens that his government would ask the U.S. to close its satellite tracking station there if the U.S. did not "raise concrete and satisfactory proposals before July 14." Ratsiraka said that "To avoid possible maneuvers of sabotage, the government has taken necessary measures to place the station under control of the armed forces from this day." People's Republic of China's Hsinhua news agency quoted Ratsiraka as saying that the U.S. was reluctant to fulfill its obligations under the agreement signed in October 1963 by the U.S. and Malagasy governments, being in "flagrant violation of the written engagements." (FBIS−PRC, 15 July 75, A3)
- NASA announced the beginning of a cooperative NASA-university-industry program to flight-test a low-cost flight-control system that would improve ride quality, reduce pilot workload, and improve flight safety. The control system—developed and ground-tested by the Univ. of Kansas—was being installed on a Beechcraft 99 commuter airliner for a flight-test program under the direction of NASA's Flight Research Center. Nearly a third of the aircraft's control surfaces were controlled by the new automated avionics system, with two-thirds remaining under direct control of the pilot. Any failures in the automated control system could be easily overridden by the pilot, eliminating the need for redundancy in the attitude-control system and permitting mechanization at a much lower cost.

With automated controls off, the separate surfaces responded to the pilot's commands and the control system was essentially that of the Beechcraft airplane. The flight-test program would include functional tests, envelope expansion, system optimization, and quantitative and qualitative evaluation by the three test pilots. (NASA Release 75−207; FRC Release 22−75)
- The Senate passed resolution, S. Con. Res. 47, setting aside the week of 16−22 July as "United States Space Observance Week." (*CR*, 11 July 75, S12438)

- Marshall Space Flight Center announced award of a $5 768 612 cost-plus-incentive-fee contract to Sperry Rand Corp. for design, development, test, and delivery of 37 multiplexer-demultiplexers (MDM) for the Space Shuttle rocket boosters. The MDM, an electronic device that permitted sending or receiving more than one message, signal, or unit of information on a single communication channel, was housed in the solid rocket boosters' integrated electronics assembly. It would process signals for such functions as ignition, thrust vector control, release of nose cap and frustum, jettison of solid rocket motor nozzles, detachment of parachutes, and turn-on of recovery aids. (MSFC Release 75-154)
- The midwest's Central Telephone and Utilities Corp. Cengas Div. surveyed heat loss from homes and commercial buildings in five Nebraska and South Dakota communities using an aerial technique developed with NASA support. In the survey—taken on cold clear nights during winter 1974-75—an aircraft equipped with a thermal-infrared scanner flew at 488 m, making successive runs over strips three city blocks wide. The scanner surveyed the roof-surface temperatures, which were recorded on magnetic tape. After the flight, ground-based equipment converted the data to strips of film for printing on photographic paper. The resulting imagery, called thermograms, were silhouettes of individual buildings in which warm roofs appeared in light tones and cool roofs in dark tones. Property owners were notified of the results. (NASA Release 75-201)

*12 July*: Kuwait's Communications Minister Sulayman Hammud al-Khalid signed a contract with Siemens AG, a West German company, to import and install Kuwait's second ground station for satellite communications. (AR-RA'Y AL-'AMM, FBIS—Arabian Peninsula, 18 July 75, C1)

*13 July*: Preparatory work was nearing completion on the Satellite Instructional Television Experiment (SITE), a joint India—NASA program to broadcast educational TV via the *Ats 6* Applications Technology Satellite launched 30 May 1974. Installation of 400 TV sets had been completed, and the buildings to house them had been wired for electricity. (Delhi Domestic Service, FBIS—India, 18 July 75, U2)

*14 July*: Soviet cosmonauts Aleksey A. Leonov and Valery N. Kubasov, relaxing on the eve of their launch for the Apollo-Soyuz Test Project, viewed the film *White Sun of the Desert*. The film, following a tradition set by the cosmonauts' predecessors in space, was a Russian action movie set during the civil war after the Bolshevik revolution in 1917. (Wren, *NYT*, 16 July 75, 1)

- Sen. William Proxmire (D—Wisc.) inserted into the *Congressional Record* a summary of the Central Intelligence Agency's report on the Apollo-Soyuz Test Project. The summary indicated that "although the Soviet have experienced many space problems the prospects for a successful mission are good." The report continued that the Soviet preparations were more extensive and thorough than for previous efforts, and that past failures had occurred at a phase during the mission which would not jeopardize U.S. astronauts should there be a recurrence. The report did point out that the Soviets had encountered severe problems in space. (*CR*, 14 July 75, S12511)
- NASA closed its tracking station on Madagascar after the agency was unable to reach agreement with the government of the Malagasy

Republic for continued operation of the station. Functions of the station would be taken over by the *Ats 6* Applications Technology Satellite launched 30 May 1974.

The Madagascar station had opened in 1964 under an agreement that called for station operation by NASA and Malagasy personnel. Data collected by the station were made available to the Malagasy Republic. The agreement had not called for an exchange of funds between the cooperating governments. The agreement expired in December 1973, and a new one could not be reached when Madagascar claimed rent for the station retroactive to 1964. (AP, *NYT*, 15 July 75; NASA Release 63—279; NASA Contracts Div., Phillips, interview, 2 Aug 77)

- The U.S.S.R. had turned down a U.S. invitation to participate in another joint manned space mission in 1976, Thomas O'Toole of the *Washington Post* reported a U.S. space official as saying. O'Toole reported that in 1974 NASA Deputy Administrator Dr. George M. Low had suggested to Soviet Academy of Sciences President Vladimir Kotelnikov that two Soviet cosmonauts and three American astronauts dock with a Soviet Salyut space station and work together in earth orbit for up to 2 wk. However, the Soviets had replied that they would undertake a repeat of the Apollo-Soyuz Test Project (scheduled for 15 July launch) but could not prepare to fly an Apollo-Salyut in any reasonable length of time. The *Post* quoted Dr. Low as saying, "Any second American—Russian space flight will have to wait until the 1980s."

  The *Post* also quoted Dr. Low as saying that NASA had suggested an Apollo-Salyut mission as early as 1971, but the Soviets had requested a change back to an Apollo-Soyuz mission during a presummit meeting in April 1972 just before President Nixon's Moscow visit. Dr. Low said that the Soviets had said that it was too expensive to modify Salyut to accept both a Soyuz and an Apollo spacecraft at the same time. (O'Toole, *W Post*, 14 July 75, A1)

- NASA announced award of a $150 000 10-mo study contract to The Boeing Co. to determine the basic design, costs, and environmental impact of potential power-generating satellite systems in space. Boeing's Aerospace Co. would investigate solar and nuclear families of space-based power generators, studying two different energy converters for each; define the satellites; study their cost; and determine potential environmental effects. One converter was to be a thermal engine converter like a steam turbine that transformed heat into a usable form of energy. The other, a thermionic converter, would produce electrically charged particles to power a passive electrical-generation system. Power generated by either method would be transformed into microwaves and beamed to earth for reconversion into electricity. (MSFC Release 75—156)

*15—24 July:* The joint U.S.-U.S.S.R. Apollo-Soyuz Test Project was successfully completed when NASA's *Apollo* spacecraft, carrying three American astronauts, and the Soviet *Soyuz 19*, carrying two cosmonauts, docked in orbit for 2 days and performed joint scientific investigations. The two crews also shared meals and exchanged mementoes of their flight.

*Soyuz Launch*: *Soyuz 19*, carrying cosmonauts Aleksey A. Leonov and Valery N. Kubasov, was launched into sunny skies from Baykonur

Cosmodrome at 5:20 pm local time (8:20 am EDT) 15 July. The spacecraft entered orbit with a 221.9-km apogee, 186.3-km perigee, 88.5-min period, and 51.8° inclination.

Foreign correspondents, barred from the launch site, watched the launch on color TV sets in a Moscow press center. The first Soviet launch to be televised live, it was transmitted to viewers throughout the Soviet Union, the U.S., and eastern and western Europe. President Ford watched from a U.S. State Dept. auditorium with Soviet Ambassador to the U.S. Anatoly P. Dobrynin and NASA Administrator James C. Fletcher, before Dr. Fletcher and Ambassador Dobrynin flew to Kennedy Space Center to watch the Apollo launch.

On the third orbit the *Soyuz 19* crew established contact with U.S. mission control in Houston, putting into operation the global Moscow–Houston–Soyuz-Apollo communications system. On the fifth orbit the cosmonauts made the first of two maneuvers to place *Soyuz 19* into a circular docking orbit. New orbital parameters were 231.7-km apogee and 192.4-km perigee. The spacecraft was spin-stabilized at 3° per sec with all systems operating normally.

*Apollo Launch*: At 3:50 pm EDT 15 July—7 hr 30 min after the *Soyuz* launch—a *Saturn IB* flawlessly lifted the Apollo spacecraft from Kennedy Space Center's launch complex 39, carrying Apollo commander Thomas P. Stafford, command-module pilot Vance D. Brand, and docking-module pilot Donald K. Slayton. The spacecraft entered orbit with a 173.3-km apogee, 154.7-km perigee, 87.6-min period, and 51.8° inclination. The spacecraft's launch-vehicle adapter was jettisoned at 9 hr 4 min ground elapsed time (9:04 GET, counted from the *Soyuz 19* launch) and the crew maneuvered the Apollo 180° to dock with the adapter and extract the docking module. These events were videotaped and transmitted to earth later via *Ats 6* (NASA's Applications Technology Satellite launched 30 May 1974). A maneuver 2 hr later at 7:35 pm circularized the orbit at 172 km. The *Saturn S–IVB* stage was deorbited into the Pacific Ocean 1 hr 30 min later.

A second *Soyuz 19* circularization burn of 18.5 sec at 8:43 am EDT 16 July placed that spacecraft in a circular orbit of 229 km, with all systems functioning normally.

*Rendezvous and Docking*: A series of Apollo maneuvers, with the final braking maneuver at 8:51 am EDT 17 July, put the Apollo spacecraft in a 229.4-km circular orbit matching the orbit of *Soyuz 19*. A few minutes later Brand reported, "We've got Soyuz in the sextant." Voice contact was made soon after. "Hello. Soyuz, Apollo," Stafford said in Russian. Kubasov replied in English, "Hello everybody. Hi to you, Tom and Deke. Hello there, Vance." All communications among the five crew members during the mission were made in the language of the listener, with the Americans speaking Russian to the Soviet crew and the Soviet crew speaking English to the Americans. Contact of the two spacecraft 51 hr 49 min into the mission (12:09 pm 17 July) was transmitted live on TV to the earth, and Stafford commented, "We have succeeded. Everything is excellent." "Soyuz and Apollo are shaking hands now," the cosmonauts answered. Hard docking was completed over the Atlantic Ocean at 12:12 pm—6 min earlier than the prelaunch flight plan—watched by millions of TV viewers worldwide. "Perfect. Beautiful. Well done, Tom. It was a good show. We're looking forward to shaking hands with you in board [sic]

Soyuz," Leonov said. Tass later reported that Kubasov told Moscow ground controllers that "we felt a slight jolt at the moment of docking" but that all went according to plan.

*Joint Activities*: At 3:17 pm hatch 3 opened; Apollo commander Stafford and Soyuz commander Leonov shook hands 2 min later. "Glad to see you," Stafford told Leonov in Russian. "Glad to see you. Very very happy to see you," Leonov responded in English. "This is Soyuz and the United States," Slayton told TV viewers around the world. Both Soviet Communist Party General Secretary Leonid I. Brezhnev and President Ford congratulated the crews and expressed their confidence in the success of the mission. Stafford then presented Leonov with "five flags for your government and the people of the Soviet Union" with the wish that "our joint work in space serves for the benefit of all countries and peoples on the earth." Leonov presented the U.S. crew with Soviet flags and plaques. The men signed international certificates and exchanged other commemorative items. After nearly 4 hr of joint activities, including a meal aboard the Soyuz, the Americans returned to the Apollo and the hatch was closed at 6:51 pm.

An integrity check of the hatches indicated an atmospheric leak on the Soviet side. Ground controllers later attributed the indication to temperature changes in the sealed docking module that were detected by the sensitive Soviet instrumentation. Future integrity checks of the hatches would be more rigorous, however.

Following a sleep period, the crews prepared for another day of joint activity. Kubasov described the mission to Soviet TV viewers while the rest of the crews performed experiments in their respective spacecraft. At 5:05 am 18 July Brand entered the Soviet spacecraft; Leonov joined Stafford and Slayton in Apollo, greeting them with "Howdy partner." Kubasov gave American TV viewers a tour of his Soyuz, and Stafford followed with a tour of the Apollo. Then both Kubasov and Brand videotaped scientific demonstrations for transmission to earth later. Kubasov and Brand ate lunch in the Soyuz while Leonov ate with Stafford and Slayton in Apollo.

During a third transfer, Stafford and Leonov went into the Soyuz and Kubasov and Brand joined Slayton in Apollo. Brand gave Soviet viewers a Russian-language tour of the eastern U.S. as seen from space. Further speeches and exchanges of commemorative items were made for both U.S. and Soviet viewers before the final handshakes at 4:49 pm EDT 18 July, when the crews returned to their respective spacecraft. The hatches were closed after Brand told Leonov and Kubasov, "We wish you the best of success. I'm sure that we've opened up a new era in history. Our next meeting will be on the ground." Total time for all transfers and joint activities was 19 hr 55 min. Stafford had spent 7 hr 10 min aboard Soyuz; Brand, 6 hr 30 min; Slayton, 1 hr 35 min. Leonov spent 5 hr 43 min in the Apollo, Kubasov 4 hr 57 min. During nearly 2 days of joint activities, the five men carried out five joint experiments.

*Undocking and Separation*: The Apollo and Soyuz spacecraft undocked at 95:42 GET (8:02 am EDT 19 July). While the spacecraft were in station-keeping mode, the crews photographed them and the docking apparatus, transmitting the pictures live on TV to earth. The Apollo spacecraft then served as an occulting disk, blocking the sun from the Soyuz and simulating a solar eclipse—the first man-made eclipse. Leonov and Kubasov photographed the solar corona as the Apollo backed away from the Soyuz and toward the sun. The two spacecraft then redocked at 8:34 am EDT with the Apollo maneuvering and the Soyuz docking system active while good quality TV was transmitted to earth. The second docking was not as smooth as the first because a slight misalignment of the two spacecraft caused both to pitch excessively at contact.

Final undocking—also with the Soyuz active—went smoothly and was completed at 11:26 am. As the spacecraft separated, the two crews performed the ultraviolet atmospheric absorption experiment, making unsuccessful data measurements at 150 m and then moving to a distance of 500 and 1000 m, where data were successfully collected. The Apollo maneuvered to within 50 m of Soyuz and took intensive still photography of the Soyuz. Separation maneuvers to put the two spacecraft on separate trajectories began at 2:42 pm with a reaction-control system burn. With the maneuvers completed, Leonov told the Apollo crew, "Thank you very much for your very big job. . . . It was a very good show." Brand answered, "Thank you, also. This was a very good job."

*Soyuz Orbit and Landing:* *Soyuz 19* remained in orbit nearly 30 hrs after the undocking. The cosmonauts conducted biological experiments with microorganisms and zone-forming fungi. At 2:39 am EDT 21 July the Soyuz crew closed hatch 5 between their orbital vehicle and descent module and began depressurizing the orbital module. Braking burns of the descent engines began at 6:06 am when the spacecraft was 772 km from the Apollo. The 194.9-sec burn slowed the spacecraft to 120 km per sec. After another burn to stabilize the spacecraft, the orbital and descent modules separated over Central Africa. While Soviet viewers watched the first landing of a Soviet spacecraft televised in real time, the main parachute deployed at 7 km and jettisoned before the soft-landing engines fired. *Soyuz 19* landed about 11 km from the target point northeast of Baykonur Cosmodrome at 6:51 am EDT 21 July, after a 142-hr 31-min mission. The rescue helicopter approached the capsule immediately and specialists opened hatch 5. Kubasov stepped out waving to rescue-team members, followed by Leonov, both cosmonauts in apparent good health and spirits. The cosmonauts returned to Baykonur for medical checks and debriefings.

*Apollo Postdocking Orbital Activities:* Apollo remained in orbit while its crew continued U.S. science experiments begun during predocking. Searching for extreme ultraviolet radiation, the ASTP crew marked the birth of a new branch of astronomy when they found, for the first time, extreme ultraviolet sources outside the solar system; some scientists had believed that such sources could never be found. One of the newly discovered sources turned out to be the hottest known white dwarf star. The Apollo detector also revealed the existence of the first pulsar discovered outside the Milky Way. About 200 000 light years from earth's galaxy, in the Small Magellanic Cloud, it was the most luminous pulsar known to astronomers, 10 times brighter than any discovered so far. After repairing some malfunctioning equipment, the astronauts also mapped x-ray sources throughout the Milky Way.

The crew completed nearly all the 110 earth-observation tasks assigned. Coordinated investigations had been made simultaneously by six groups of scientists on the ground, on ships at sea, and in aircraft. The astronauts looked at ocean currents, ocean pollution, desert geography, shoreline erosion, volcanoes, iceberg movements, and vegetation patterns.

On 23 July the command-module tunnel was vented and the crew put on spacesuits to jettison the docking module. The command and service module undocked from the DM at 3:45 pm EDT, and a 1-sec engine firing put the CSM into a higher orbit (232.2-km apogee, 219.0-km perigee) so that the DM could move ahead. A second maneuver put the CSM in a 223.2-km by 219.0-km orbit. Deorbit began at 4:38 p.m. The command module and service module separated, the drogue and main parachutes deployed normally, and the Apollo splashed down at 224:58 GET (5:18 p.m. EDT 24 July) in the Pacific Ocean 163°W and 22°N, 500 km west of Hawaii. This was the last ocean landing planned for U.S. manned space flights; future flights on the Space Shuttle would be wheeled touchdowns at land bases.

The CM landed in "stable 2" position (upside down) 7.4 km from the prime recovery ship, U.S.S. *New Orleans*. After swimmers from the rescue helicopter righted the spacecraft and attached a flotation collar, the Apollo was lifted by crane on to the deck of the recovery ship and Stafford, Brand, and Slayton stepped out to the cheers of the ship's crew. President Ford telephoned congratulations. During the welcome, the crew was evidently experiencing eye and lung discomfort; subsequent conversations and spacecraft data revealed that, during reentry, the earth landing system had failed to jettison the apex cover and drogues as scheduled and had had to be fired manually, without first disabling the reaction-control system thrusters. With the CM oscillating, the thrusters began firing rapidly to compensate, and combustion products — including a small amount of nitrogen tetroxide— entered through the cabin-pressure relief valves. As soon as the RCS system had been disabled, fresh air was once again drawn into the cabin. The crew members told flight officials that they had put on oxygen masks once the spacecraft had landed, and then activated the postlanding vent system.

Because of the crew's discomfort, further shipboard ceremonies had been canceled and the crew had been sent to sick bay and then to Tripler Hospital in Hawaii for observation until 8 Aug.

Primary ASTP mission objectives were to evaluate the docking and undocking of an Apollo spacecraft with a Soyuz, and determine the adequacy of the onboard orientation lights and docking target; evaluate the ability of astronauts and cosmonauts to make intervehicular crew transfers and the ability of spacecraft systems to support the transfers; evaluate the Apollo's capability of maintaining attitude-hold control of the docked vehicles and performing attitude maneuvers; measure quantitatively the effect of weightlessness on the crews' height and lower limb volume, according to length of exposure to zero-g; and obtain relay and direct synchronous-satellite navigation tracking data to determine their accuracy for application to Space Shuttle navigation-system design. The objectives were successfully completed, and the mission was adjudged successful on

15 Aug. (NASA MORs M−966−75−01, 7 July 75, 15 Aug 75; mission transcripts 15 to 24 July 75; change of shift press briefing transcripts 15 to 24 July 75; *W Post*, 14 to 25 July 75; *NYT*, 14 to 24 July 75; FBIS−Sov, 14 to 25 July 75; NASA Release 75−118; NASA *Apollo-Soyuz Test Project: Information for Press*, 1975; Froelich, *Apollo-Soyuz*, NASA EP−109; Ezell, *et al.*, *The Partnership: A History of the Apollo-Soyuz Test Project*; NASA SP−4209, 317−49)

*15 July:* President Gerald R. Ford sent a prelaunch message to both the U.S. and U.S.S.R. crews of the Apollo-Soyuz Test Project: "In a few short hours, you will be opening a new era in the exploration of space. Although others have gone before you, you will be blazing a new trail of international space cooperation." In the less than 2 decades since Yuri Gagarin and John Glenn orbited the earth, realizing the dreams of those early rocket pioneers who had believed firmly that man could fly in space, spaceflight had expanded considerably our knowledge of earth, moon, planets, and universe. "Your flight represents another stage in man's efforts to further his understanding of his environment. It has already demonstrated something else—that the United States and the Soviet Union can cooperate in such an important endeavor. Since the Apollo-Soyuz project was agreed to 3 years ago, crews, scientists, and specialists of both countries have worked diligently and productively, and in a spirit of cooperation, to bring us where we are today. I am heartened by the example of dedication and cooperation you have displayed. I am confident your efforts and example will lead to further cooperation between our two countries." (*PD*, 21 July 75, 750)

• Western Union Corp.'s Westar satellite system, which included *Westar 1* and *2* (launched 13 Apr. 74 and 10 Oct. 74), began commercial video service with the transmission of Apollo-Soyuz Test Project mission coverage from Houston to Chicago. TVN Independent Newservice, the first customer, used a transponder on *Westar 1* to transmit coverage for 75 clients in the U.S. and Canada. The two satellites, in synchronous orbit, were linked to earth stations near major cities. (*WU* Release, 15 July 75)

• The upper stage and instrument unit of the Saturn IB rocket that carried the U.S. astronauts into orbit for the Apollo-Soyuz Test Project reentered the earth's atmosphere at 9:45 pm EDT. The 14 100-kg stage and instrument unit, empty of 320 cu m of liquid hydrogen and liquid oxygen, broke into pieces as they reentered over the Pacific Ocean. Reentry was activated by ground commands; excess fuel was dumped through the engine, slowing the stage and ensuring that disintegration came in a remote ocean area. (MSFC Release 75−160)

• The *Washington Post* reported that, after launch of the Apollo-Soyuz Test Project, more than 1800 persons would lose their jobs. With the liftoff came the end of the Apollo and Saturn programs which had begun 25 May 1961 with President Kennedy's request for a national commitment to a lunar-landing program. Layoffs would cut the Kennedy Space Center workforce to less than 8500, down from a peak of 28 000. Of those who would lose their jobs, 1100 worked for NASA contractors— Rocketdyne Division of Rockwell, Boeing Co., Chrysler Corp., and McDonnell Douglas Corp.—who built and mated the stages of the Saturn launch vehicle to the Apollo spacecraft. Another 700 worked for contractors such as International Business Machines Inc., TRW Inc., Grumman Corp., Aerojet General Corp., and Rockwell International Corp.

Only 100 launch-operations contract employees would remain to help plan the launches of the Space Shuttle. The Shuttle launch force would begin to increase in size in 1977, but would never reach the size of the Saturn team because of increased automation. (O'Toole, *W Post*, 15 July 75, A9)

- NASA awarded a $5.5 million cost-reimbursable contract to the Univ. of Hawaii for construction of an infrared telescope facility on the summit of Mauna Kea, an extinct volcano on Hawaii Island. The contract covered the second part of a $6-million two-phase project covering design, construction, installation, and assembly of the facility.

  The 3-m infrared telescope, scheduled for 1977 completion, would be the world's largest. The Univ. of Hawaii would operate it in support of NASA's planetary exploration programs, particularly the 1977 Mariner mission to Jupiter and Saturn. The facility would provide a national capability of observing interstellar dust, exploding galaxies, and galactic nuclei in the middle and far infrared portions of the electromagnetic spectrum. (NASA Release 75-209)

*16 July:* At a Moscow exhibition of space art to commemorate the joint U.S.-U.S.S.R. Apollo-Soyuz Test Project, Soviet art glorified man's role in space, whereas U.S. art emphasized technology, the Baltimore *Sun* reported.

U.S. artists, commissioned by NASA and the National Gallery of Art to record the space program, showed basically factual treatments of blastoffs, launch towers, and mazes of steel gridwork. Artist Alden Wicks depicted Kennedy Space Center's Vehicle Assembly Building as "The New Olympus." Men played subordinate roles in U.S. art; they swarmed through control centers or, as in Jamie Wyeth's "Firing Room," sat by the dozens in front of individual TV monitors. Engineers ministered to launch vehicles like priests before an idol. Even in the pen-and-ink drawings of individual astronauts, faces were diminished by the detailed minutiae of spacesuit and life-support paraphernalia.

By contrast, man dominated the Soviet art. Artist Anatoly Yakushin depicted pioneer rocket theoretician Konstantin Tsiolkovsky against a geometric maze, but man controlled the idea. A painting by Boris Okorokov showed a young flight controller commanding abstract equipment, in contrast to Wyeth's passive TV-monitor watchers.

The irony of the two artistic visions was that they "reverse the technological visions of the nations' manned space programs," the *Sun* said. In reality, U.S. flights left much of the flying up to the astronauts, while the Soviets had kept their cosmonauts as passive as possible.

Also included in the exhibition were paintings by Apollo-Soyuz Test Project cosmonaut Aleksey A. Leonov, the only one of the artists who had seen outer space. He painted cosmic scenes in brilliant colors: His deep space was not a flat black, but a composite of black and indigo, streaked with magenta and stars with violet coronas. The Leonov earth as seen from space was deep marbled blue with atmospheric sheaths surrounding it in hard-edged layers, pale hues close to the planet, darkening in stages to the night of space. (B *Sun*, 16 July 75, A4)

- Department of Defense said the Navy had understated the long-term cost of building the new F-18 carrier aircraft by at least $1.6 billion, the *Washington Star* reported. The *Star* said the Navy had chosen the F-18 under pressure from DOD and Congress, after the Air Force selected the F—16 for its low-cost fighter instead of the fighter version of the F—18. At the time the Navy estimated the cost of each F—18 as $7.8 million, or $2.6 billion to develop and produce the first 128 F—18s; however, the *Star* reported, a DOD analysis found the 5-yr costs would be closer to $4.3 billion.

The future of the program was under discussion. One alternative would be to raise projected Navy budget to accommodate the increased costs of the F—18. Another would be for the Navy to drop F—18 and go back to industry to design and develop a lower cost fighter that would fit into the original budget and, in the interim, to let the Navy buy fighter-bombers from McDonnell Douglas to meet its fleet-modernization needs in the early 1980s. (Finney, *W Star*, 17 July 75, A16)

*17 July*: *Pioneer 10*, launched 2 March 1972 and traveling out of the solar system at 43 200 km per hr, was operating normally despite a report that the spacecraft was making unexplained orientation changes [see 11 June]. The apparent changes were caused when, during a routine change in spacecraft orientation on 6 May, the spacecraft's star sensor made an undetected shift in its lock from the star Betelgeuse to the star Sirius. This shift twisted the frame of reference for measuring direction, and for changing the direction of spin-axis pointing. As a result, onboard and ground computers had recorded a gradual northward shift of the spin axis relative to the earth's motion around the sun. Also, when the thrusters were fired to move the spin axis with the earth's east-west course, "southward" impulses were necessary.

No loss of data had occurred, and techniques were being designed to verify star reference during intervals when the star sensor was used. (ARC *Astrogram*, 17 July 75, 1)

- Two Ames Research Center research aircraft would be stationed in Hawaii for scientific research missions.

  A U—2 earth-resources survey aircraft would probe the stratosphere up to 20 km to measure the distribution and extent of ozone, nitric oxide, and aerosol particles in the area between the equator and 40° N latitude. Conducted by ARC to aid a national study, the sampling program was part of a semi-global study of the effect of ozone and nitric oxide on the world's climate over a long period of time. Similar missions had been flown over higher latitudes and polar regions. Another objective of the U—2 flights was to collect remotely sensed data on the Hawaiian Islands in support of several investigations, including one of the decline and die-back of Ohia and Manami trees in the forested areas of Hawaii.

  NASA's Kuiper Airborne Observatory, a C—141 jet aircraft equipped with a 91.5-cm infrared telescope, would make 16 flights to collect previously unobtainable data from the region of the galactic center, a highly complex intense source of infrared emission. (ARC *Astrogram*, 17 July 75, 1)

- The International Telecommunications Satellite Organization's board of governors had authorized issuance of a request for firm fixed-price proposals for design, development, manufacture, and test of seven Intelsat V satellites, with options for up to eight additional satellites.

  In 1979 Intelsat V satellites would replace the Intelsat IV—A series to be deployed in September 1975. Intelsat V satellites would have increased capacity by a fourfold use of the 6/4 ghz frequency band, use of both antenna-beam separation and dual-polarization concepts; introduction of the 14/11 ghz band for limited coverage in high-traffic regions; and a maximum effective bandwidth of 2280 mhz. A single satellite design would be used in all three ocean regions of the Intelsat global systems. Intelsat V satellites were expected to have a capacity of 12 000 two-way voice circuits plus one color TV channel. (INTELSAT Release 75-8)

- Lewis Research Center had awarded a contract to Rockwell International Corp. Rocketdyne Div. for a laser-heated rocket thrust chamber for spacecraft. The concept included transmitting laser energy from a ground laser-generating station to a spacecraft; propellant carried aboard the spacecraft would convert laser energy into propulsive thrust. The potential improvement in performance included a 400% increase in payload capability for a Space Tug.

  Rockwell would build an experimental thruster, to be powered by a 10-kw laser beam and tested at LeRC using an in-house carbon-dioxide laser system, and would provide a preliminary design for a 5000-kw thruster. (Rocketdyne Release RD-16)

- NASA announced appointment of Gerald D. Griffin as Deputy Associate Administrator for Operations in the Office of Manned Space Flight, effective 1 Aug. In his new position Griffin would plan for the most economical and flexible operation of the Space Shuttle and Spacelab. Griffin, who was Assistant Administrator for Legislative Affairs before this appointment, would be replaced by Dr. Joseph P. Allen, an astronaut assigned to the Astronaut Office at Johnson Space Center. (NASA anno 17 July 75)

*18 July:* The *Los Angeles Times* discussed the Apollo-Soyuz Test Project mission in an editorial: At the instant the Apollo and Soyuz docked, "the two ships ceased to be two and became one. And at that moment, the common fate of the five adventurers was indissolubly linked for the next 44 hours, symbolic of the shared destiny of the 4 billion passengers on the mother spaceship 140 miles below." That the passengers were from two different countries was of no significance. "They communicated in each other's language, and they saw the lovely blue earth across the vistas of space through the same eyes, with the same perception. It is a vision that is desperately needed by earthbound statesmen." (*LA Times*, 18 July 75)

*19 July:* Japan's National Space Development Agency and NASA signed a contract for launch of three Japanese geosynchronous satellites in 1977 and 1978. NASA would launch the satellites on a Delta launch vehicle on a cost-reimbursable basis.

The satellites—GMS, a meteorological satellite; CS, an experimental medium-capacity comsat; and BS, an experimental medium-scale broadcast satellite—would participate in the worldwide Global Atmospheric Research Project (GARP) sponsored by the United Nations World Meteorological Organization. (Kyodo, FBIS-Japan, 22 July 75, C3; NASA MOR M-492-101-77-01, 12 July 77)

*20 July:* The Apollo-Soyuz Test Project docking system functioned well even though "conditions were not too favorable in the second docking," spacecraft designer Vladimir Syromyatnikov said at a Moscow press conference. Although he said he was not yet familiar with all the details, Syromyatnikov said there had been a great deal of pitch between the two craft. "We know that the Soyuz system, after capture, is turned off, while the manual operation of the Apollo ship. . . should have been turned into a different mode." At point of capture Soyuz began to turn on its axis at approximately 3° per sec, and the shock absorbers on the Soyuz "were very strained in order to absorb this strain." All the pitch was "quelled" after about 40 sec, Syromyatnikov added.

When a reporter asked why this happened, flight engineer Sergei Tsibin said, "We know that at one moment on the Apollo there was a folding of the gyroscopic system. We also know that at the moment of locking and latching the Apollo and Soyuz were rolling at approximately 1 degree a second. This should not have taken place."

The *Los Angeles Times* reported on 21 July that the Russians were worried about a pressure drop in Soyuz and had voiced concern over possible damage to the Soyuz by the hard docking. However, the *Times* quoted NASA officials in Moscow as saying that the docking had been within prescribed limits though conceding that "it may have been a little harder" than the first one. (USSR press briefing transcript, 20 July 75; Toth, *LA Times*, 21 July 75)

- Pope Paul VI told a crowd of 5000 at the papal palace near Rome that he was enchanted by the unique encounter of the Apollo and Soyuz crews in space [see 15–24 July]. He said, "We too, our eyes dazzled by sky light or wide open toward the depths of nighttime space, will shout out, long live the heroic men of such a fantastic feat. Man has won and we cannot refrain from foretelling more wonderful advances in the dominance of nature beyond the heavenly sphere. Man will win." (ASTP air-to-ground transcripts, 20 July 75)
- The fierce rivalry between the U.S. and U.S.S.R. was not confined to ground, air, and sea, but extended to space, the People's Republic of China's *The People's Daily* said in an article entitled "Competition in Space and Hardship on Earth." Both superpowers had invested huge amounts of money, manpower, and materials to turn cosmic space into an arena for arms expansion and war preparations as well as espionage. U.S. and U.S.S.R. military satellites, which accounted for most satellite launchings, were used mostly to spy on each other. In recent years their rivalry in space had become more and more acute, disturbing tranquillity in space. (Peking NCNA, FBIS–PRC, 21 July 75, Al)

*21 July*: The 14-nation Arab League, joined by 6 other African and Mid-Eastern states in a 20-nation satellite organization, had finished initial technical and economic feasibility studies for an Arab satellite-communications network and were beginning to define specific technical requirements for the system, *Aviation Week and Space Technology* reported.

Program goals were to meet telecommunications requirements of Arabic-speaking states in North Africa and the Middle East for police work, weather broadcasting, civil aviation communications, and shipping, and to expand Arab technology into space. The Arab States Broadcasting Union estimated immediate need of 15 radio channels for direct broadcast of educational and cultural programs to remote areas, and 2 TV channels each for general and educational broadcasting. Estimated need of the Arab countries by 1976 would be as many as 1060 telephone, teletypewriter, and telegraph channels, and up to 6200 channels for this traffic by mid-1984.

*Av Wk* reported that, although the first satellites and initial ground stations would be bought from Western countries, the long-term goal was to shift operation and control to Arab nationals after 6 mo.

The system, which would provide communications and broadcast coverage for 95% of the Arab world, would include a main orbiting satellite, a standby orbiting satellite to replace the prime satellite in case of failure, and a reserve satellite kept on the ground ready for quick launch in case of satellite failure. (Ropelewski, *Av Wk*, 21 July 75, 56-59)

- A recommendation by Dr. James C. Fletcher, NASA Administrator, to transfer Skylab backup hardware to the Smithsonian Institution was agreed to by Rep. Olin E. Teague (D-Tex.), Chairman of the House Science and Astronautics Committee. In a letter to Dr. Fletcher, Teague said he understood the transfer would "in no way affect the status of the Saturn launch vehicles and Apollo spacecraft being retained for possible use." Teague said he shared a "belief that to display the Skylab hardware to the public will foster a better appreciation and increasing support for our national space program."

  Other equipment was being stored to permit restoration if additional space missions should become desirable [see 9 Apr.]. (Letter, Teague to Fletcher, 21 July 75)

- Cosmic rays—the most energetic particles in the universe—probably originated in supernovas within our own galaxy, Dr. Floyd W. Stecker, Goddard Space Flight Center scientist, wrote in an article in *Physical Review Letters*. Dr. Stecker had based his conclusions on data from NASA's *Explorer 48* (Small Astronomy Satellite-B, launched 16 Nov. 1972).

  Scientists had been speculating on the source of cosmic radiation since its discovery at the turn of the century, when it was thought to be identical with gamma rays. With the advent of radio astronomy and the rapid development of new experimental techniques and instrumentation, supernovas had been considered a possible source of cosmic rays; however, other scientists believed cosmic radiation had extragalactic origins, such as quasars. *Explorer 48*, measuring cosmic-ray intensity without interference from earth's atmosphere, had been able to obtain the gamma-ray information needed to complement ground-based radio-astronomy data.

  The SAS-B scientific team at GSFC had measured a large amount of gamma radiation coming from the Milky Way galaxy, the most intense coming from the center of the galaxy. As gamma rays resulted primarily from collisions of cosmic rays with interstellar gas, and the detailed *Explorer 48* data had identified the direction from which most of the gamma rays came, Dr. Stecker could pinpoint an area of intense activity midway between the galactic center and the earth. The active area was a region containing numerous interstellar gas clouds—but not enough to account for all the gamma-ray production—as well as a large number of supernova remnants. Dr. Stecker deduced that cosmic rays were to be found in the same regions as supernova remnants and in proportion to their number. (Stecker, *Physical Review Letters*, 21 July 75, 188-90; NASA Release 75-221)

*22 July*: The National Research Council's Space Applications Board, directed by the National Academy of Sciences and a NASA contract, released a report entitled "Practical Applications of Space Systems," with particular emphasis on approaches to socioeconomic benefits. The report stated that present institutional arrangements were "not adequate" to encourage future applications of space technology to promising areas such as natural resource exploration and management, telecommunications for education and health care, and long-range weather and climate forecasting. The report added: "There exists at present no institutional mechanism that permits the large body of potential users. . .to express their needs and to have a voice in matters leading to the definition of new systems."

Recommendations by the board included establishment of a national space applications council to direct policies affecting nonmilitary space applications; to set priorities for meeting user needs; to provide for exchanges between users and providers of space technology; and to encourage non-Federal investment in the application of space systems. The council could operate as an interagency group, with representation from state and local governments.

To meet the launch needs of future applications satellites, the board recommended that plans for the Space Shuttle provide an early opportunity to orbit payloads in either polar or geosynchronous orbits. The Shuttle-launched payloads would have applications in hazardous weather warnings and long-range weather and climate predictions; land-use planning; agriculture, forest, and range management; exploration for food, water, energy, and mineral resources; and environmental monitoring.

Other specific recommendations included lower cost, more accessible ground data-collection and readout stations; long-term observations of climatic factors—such as albedo, heat control of mixed layer in ocean, cloud distribution, and climatically significant changes in vegetation, land use, and snow and ice cover—to help provide a sound basis for long-range weather prediction; a resolution of 3 to 10 m for land-use applications; greater emphasis on Federal research and development programs for spaceborne sensors of water resources; continued development of earth physics techniques, including a system to measure relative displacement of tectonic plates within 3 cm; and vigorous continuation of the SEASAT-A and Nimbus-C programs.

The board had based its findings on the deliberations of 110 potential users of space-derived information and services who had attended a July 1974 study group in Colorado. The participants represented Federal agencies, state and local governments, industrial and business communities, and educational institutions. ("Practical Applications of Space Systems," NAS publ, July 75)

- Although more than 26 million Americans watched the Apollo-Soyuz Test Project handshake in space, TV coverage was "adequate without being particularly outstanding," TV critic Lee Winfrey said in the *Philadelphia Inquirer*. Two obstacles were insurmountable. NASA "is doggedly efficient, it got us to the moon first and it did the job again this time. But it is as colorless as a pane of glass and invariably capable of making the highest excitement look like the dullest routine." NASA had an unshakable air of the 1950s about it: all white, all male, crewcut, and straight-shouldered.

  The second obstacle was that the mission was "another goodie that Richard Nixon dreamed up for us. Which means it was all show and no substance." The U.S. paid $250 million for an aerial handshake.

"Nixon made us buddies with Brezhnev and so, through shadow shows like Apollo-Soyuz, we are supposed to remain." (Winfrey, *P Inq*, 22 July 75)

- Jackson M. Balch, director of NASA's National Space Technology Laboratories (NSTL)—formerly the Mississippi Test Facility—announced plans to retire. As head of the test facility for 10 yrs, Balch managed testing of Saturn V rockets for both the Apollo and Skylab programs. (NASA anno, 22 July 75)

*22–23 July:* Nearly 200 NASA managers, engineers, and scientists— headed by Dr. Edgar M. Cortright, Langley Research Center Director, and Viking Project Manager James S. Martin, Jr.—met at Kennedy Space Center for the Viking launch-readiness review. Those attending—including Dr. John E. Naugle, NASA Deputy Associate Administrator, and Dr. Noel Hinners, Associate Administrator for Space Science—heard status reports on the launch vehicle, spacecraft, and launch and tracking preparations. Everything was "in excellent shape" for the 11 Aug. launch of the nation's first mission to land unmanned spacecraft on Mars, said Martin. "We still have three weeks of work ahead, but time to do it in an orderly fashion, and no significant open items."

Viking–A would be mated with the Titan–Centaur launch vehicle on 28 July, followed by specialized tests, including an operational readiness test 6 Aug. during which the entire NASA and contractor Viking team would go through the countdown and several hours of simulated flight.

Problems discussed during the review included the x-band radio on the Viking–B Orbiter in which a defective coaxial cable had been replaced. A leak discovered in the Viking–A gas chromatograph mass spectrometer after sterilization was found to be so small— 0.000864 cubic centimeters per day—that it would not affect the mission. (NASA Releases 75–210, 75–216; KSC Release 156–75)

*24 July:* The Senate Committee on Appropriations reported out H.R. 8070, the Dept. of Housing and Urban Development–Independent Agencies Appropriations Bill for FY 1976.

### NASA FY 1976 Budget
### Comparison of Request, Authorization, and Senate Recommended Appropriations

(millions)

| Item | Budget Request | Authorization (diff. from req.) | Senate Appropriations Committee (diff. from req.) |
|---|---|---|---|
| Research & Development | 2678.4 | 2687.2 (+8.8) | 2685.4 (+7.0) |
| Construction of Facilities | 84.6 | 99.1 (+14.5) | 82.1 (−2.5) |
| Research & Program Management* | 776.0 | 776.0 (0) | 775.5 (−0.5) |
| Totals* | 3539.0 | 3562.3 (+23.3) | 3543.0 (+4.0) |

*Does not include a supplemental request for pay increases, introduced as H.R. 13172 and signed into public law P.L. 94–303 on 1 June 1976

The committee appropriation allowed for no new starts but did permit an increase of $400 million in funding for the Space Shuttle. The committee supported manned NASA activities such as an international cooperative space-docking mission built on the successes of Apollo and Skylab programs, and the development of the Space Shuttle as an economical, versatile transportation system to give a wide variety of users access to space. The committee also endorsed a space science flight program furthering knowledge of the earth, atmosphere, moon, sun, planets, interplanetary space, and stars; a research and development program to identify and demonstrate useful applications of space techniques in areas such as weather and climate, pollution monitoring, earth-resources survey, earth and ocean physics, communications, and space processing; U.S. leadership in aeronautics and space programs; worldwide tracking and data acquisition support for NASA's manned and unmanned programs; a program to assist in the development of national energy self-sufficiency; and a technology utilization program to speed dissemination of technological and engineering information gained during NASA programs to government, industry, and other users.

The committee restored $48.4 million for the two-spacecraft Pioneer–Venus mission scheduled for launch in 1978; the House had cut the funds requested by NASA in the original budget. The Senate agreed with NASA that the project had high scientific priority, and that delaying it to a less opportune launch window when Venus was further from the earth would require redesign of the spacecraft and would cost $50 million more than originally planned.

The committee also restored $1 million cut by the House from the $5 million requested for definition studies and advanced technological development of the Large Space Telescope. The study was necessary for NASA to define an optimum design at minimum cost.

The committee added $7 million to the $7 million NASA request for research and development to help understand and monitor physical and chemical processes in the upper atmosphere, emphasizing the need for studies of the depletion of stratospheric ozone.

Because some committee members had had reservations about the high cost and safety hazards of the joint U.S.–U.S.S.R. Apollo-Soyuz Test Project, the committee said it would like to be notified well in advance of any future joint space mission to permit an accurate assessment of potential benefits and costs.

The committee's recommendation of $82.1 million for construction of facilities, $25 million less than the request, would allow continuation of previous work including Space Shuttle facilities, facility rehabilitation and modification and minor construction, and facility planning and design. Not only did the committee, agreeing with the House action, cut the $2.5 million requested for an addition to the Lunar-Sample Curatorial Facility at Johnson Space Center, it also agreed with the House in denying funds to modify the 40- by 80-ft wind tunnel at Ames Research Center—not included in the original request but authorized by Congress—until the committee had had an opportunity to review the necessary funding in a formal budget request.

The recommendation of $775.5 million for research and program management, the same amount approved by the House, was $0.5

million less than NASA's request. The reduction reflected a 10% cut in General Services Administration rental charges.

The Senate committee recommended a total of $925 028 000 in funds for the transitional period 1 July–30 Sept. 1976.

### NASA Transitional Budget
### Comparison of Request, Authorization, and House-Recommended Appropriations

(millions)

| Item | Budget Request | Authorization (diff. from req.) | H.R. Appropriations Committee (diff. from req.) |
|---|---|---|---|
| Research & Development | 730.6 | 700.6 (−30.0) | 700.6 (−30.0) |
| Construction of Facilities | 14.5 | 10.8 (−3.8) | 10.8 (−3.8) |
| Research & Program Management* | 213.8 | 213.8 (0) | 213.7 (−0.1) |
| TOTALS* | 958.9 | 925.2 (−33.8) | 925.0 (−33.9) |

*Does not include a supplemental request for pay raises introduced as H.R. 13172 and signed into P.L. 94–303 on 1 June 76

Recommended funding was $33 872 000 below the total budget request and $122 111 less than the authorization. (Sen Com Rpt 94–326; Budget Chron Hist FY 1976, NASA Off of Budget Operations, 16 June 75)

- "We have just witnessed another flawless Apollo splashdown and the successful completion of the world's first international [manned] spaceflight," Dr. James C. Fletcher, NASA Administrator, said at an ASTP postrecovery press briefing. "At this time it is difficult, if not impossible, to assess the full significance of the ASTP mission, or even its ultimate influence on international affairs." However, "to a world beset by tension and suspicion it must, indeed, be a heartening sign to see Russians and Americans cheering each other. . . . By going into space together, we have shown a sometimes skeptical world that perhaps there is a real chance of world unity." (ASTP Transcript PC–55)

- The Apollo-Soyuz Test Project reflected the "progress made in the relations between our two nations and . . . the successes of the policy of peaceful coexistence, and promotes a further improvement of the international situation, and the strengthening of mutually rewarding contacts," Acting President of the Soviet Academy of Sciences Vladimir Kotelnikov said at a press briefing.

  General Vladimir Shatalov, head of cosmonaut training, told the newsmen, "While placing great emphasis on the importance of this joint flight, the Soviet Union also attaches great importance to its national space program, which at present centers on the establishment of long term manned orbital stations of the Salyut type." He said that the success of ASTP along with that of *Salyut 4* "testifies to the maturity of the Soviet space program, to the fact that it is able to man and launch several spacecraft at the same time and control their flight."

ASTP cosmonaut Aleksey A. Leonov, who had been promoted from Soviet Air Force colonel to major general, said he had thought language would be the biggest problem during the joint session but "we understood each other very well." He added, "we didn't have problems during our joint operation."

When asked about future joint projects, Willis Shapley, NASA Associate Deputy Administrator, said, "For the future I see accomplishments in two directions. . . . On one hand we must jointly plan and execute further joint technical projects. At the same time we must further develop the areas of political understanding and cooperation between our two countries. . . . Preliminary discussions are already underway between specialists. . . from both sides." (Soyuz crew press conf. transcript, Apollo news center, Houston, 24 July 75)

- Walter Cronkite, U.S. broadcaster who had covered the space program from its beginnings to date, reminisced about the space program during an interview with Kennedy Space Center's *Spaceport News*. Cronkite recalled that news coverage in the early 1950s was done from the beach with very little official information. "All we had were tips from engineers and others we met in bars." By Alan B. Shepard's 5 May 1961 launch, things had improved only slightly; coverage took place from the back of a station wagon with a microphone, monitor, and very little incoming information. Later launches had seen drastic changes. "We've got so much information today that it's the traffic cops' job to take care of the flow, whereas before it was more imagination and fill-in-the-blanks."

  Asked if he would be interested in riding aboard the Space Shuttle, Cronkite said, "I think newsmen will go on the Shuttle. . . . Just as soon as we prove out the equipment. . . . I would guess that somehow or other we're going to see that a pool man gets aboard, and I hope it's me."

  Regarding the apparent apathy of the American public toward the space program, Cronkite stated that it would be difficult in this age to sustain the "high intensity of feeling about the space program, the race to the moon, [and] landing on the moon. . . . We don't have an attention span in the modern world, with so much going on, that permits us to remain at that high level of excitement." (*Spaceport News*, 24 July 75, 6)

- A terrestrial solar-energy recording system now installed at Marshall Space Flight Center would provide accurate data on sunfall for use in developing solar-energy measuring systems, MSFC announced. Designed and built by International Business Machines Corp. as part of the center's earth-based solar-power activities, the monitor would use a sun tracker and stationary sensors to measure and record solar energy. The 203-cm by 178-cm instrument included a pyrheliometer to measure energy coming directly from the solar disc and 2 pyranometers to measure direct and scattered energy from any direction in its hemispheric field of view. The system could also evaluate the ability of developmental devices to convert or reflect solar energy. Collected data were recorded on tape for computer analysis. (MSFC Release 75−168)

- NASA announced completion at Lewis Research Center of one of the largest indoor echo-free chambers in the U.S. The 1416-cu m chamber—called the Engine Fan and Jet Noise Facility—would aid the center's research into jet-aircraft noise by permitting engineers to test noise characteristics of quiet fans for advanced aircraft engines and to evaluate new ways of reducing the rumble of jet nozzles. All surfaces of the 16-m wide, 17-m long, 5-m high facility had been treated with anechoic (echo-free) 76-cm fiberglass wedges that absorb sound. Acoustic tests had shown that the facility could absorb essentially all the sound in the region of interest for aircraft-engine fan models up to 51 cm in diameter, or jet nozzles up to 10 cm in diameter. Engineers could study noise coming from either the front or the rear of the fan. The facility's control room was linked directly to LeRC's central computer, permitting instant analysis of much of the data. (NASA Release 75—212)

*25 July:* At an Apollo-Soyuz Test Project crew-status briefing, Richard S. Johnston, Director of Life Sciences, told reporters that Houston control could not make voice contact with the crew during Apollo reentry [see 15—24 July]. The first medical bulletin on the astronauts' condition, received at Houston several hours later, had informed Houston officials of the crew's eye and lung irritation. After splashdown, astronauts Thomas P. Stafford and Donald K. Slayton had noticed that fellow crewman Vance D. Brand had passed out. He had revived after about 1 min when Stafford adjusted his oxygen mask and ensured a proper flow of oxygen.

Johnston also said that preliminary diagnostic procedures had indicated that the crew did not have "any apparent medical problem at that time" but that x-rays had shown increased hylar markings originating from the center of the chest. The doctors elected to move the crew to sick bay and keep them under observation. The astronauts had been given cortisone to decrease the inflammation of their lung tissue and relieve the symptoms of cough and substernal discomfort.

During the briefing, word came that the astronauts had arrived at Tripler Hospital in Hawaii where they would enter the intensive care unit. "They walked in smiling and they looked happy," said JSC Public Affairs Officer Robert V. Gordon. (ASTP Transcript PC—56)

- The House of Representatives' $48-million cut in NASA's Pioneer—Venus program [see 19 June] had taken the agency by surprise, a *Science* editorial reported. "Not only had the program been scrutinized and endorsed by House and Senate authorization committees, but construction of the spacecraft, for which planning began 4 years ago, is now well advanced." The potential loss was even more bitter because it was the "child of the scientific community."

Although the House action was intended not to eliminate Pioneer—Venus altogether but rather to delay it, to consider the relative priority between the project and the Large Space Telescope, "they are not in any way comparable in their scientific objectives." The House Appropriations Committee's action seemed to be based on "a misunderstanding of the physical constraints of launching a spacecraft to Venus." Once the 1975 launch window had passed, the next opportunity would be in 1980, when the relative position of the two planets would require spacecraft approach at higher velocities that would call for a larger retro-propulsion motor and more fuel. These requirements would reduce the scientific payload, necessitate redesign, and waste much of the $40 million already spent.

The subcommittee's decision seemed to have been made abruptly "without any real attempt at investigating the consequences." The subcommittee staff had never asked NASA whether the mission could be delayed.

Even more remarkable, said *Science*, was the motivation given for the action in the committee report: "Some astronomers have been critical of NASA's Space Science Program because they contend that a disproportionate level of NASA dollars have been used on planetary astronomy missions, while little or no funds have been allocated to deep space astronomy which is the principal mission of the Large Space Telescope." The public bickering over funds among scientists tended to give "the budget cutters . . . a field day."

To many space scientists, the deferral raised questions about the credibility of the long-range planning process for large science projects. Scientists argued that "cutting the mission after it has already been approved by Congress and after money and considerable talent have been committed to its development may jeopardize efforts to design efficient spacecraft and plan a logical program of planetary exploration." (Hammond, *Science*, 25 July 75, 270−71)

- NASA announced the development of an improved inorganic paint, a potassium silicate zinc-rich coating with a silicon additive that could provide long-term protection from salt spray, fog, heat, and rapid temperature changes. Produced at Goddard Space Flight Center as an anticorrosion coating for use in the space program, the paint had shown no deterioration after 5300 hr of continuous exposure to a 3% brine spray. The paint had also been applied to a 2-m panel mounted under the Golden Gate Bridge in San Francisco for further testing. (NASA Release 75-213; *Newsweek* (int'l ed), 18 Aug 75)

- Comsat General Corp. of the Communications Satellite Corp. announced its intent to pay MCI Communications Corp. and Lockheed Aircraft Corp. $1.5 million each for their common stock in the CML Satellite Corp. Restructured in 1972 by Comsat General Corp. and with thirds owned by Comsat, Lockheed, and MCI, CML was to have entered the domestic-communications-satellite business but had stopped short because of the financial difficulties of MCI and Lockheed.

  On 3 July 1974, Comsat and International Business Machines Corp. had announced plans to buy out the shares in CML owned by Lockheed and MCI, subject to approval by the Federal Communications Commission. Restrictions imposed in a 23 January 1975 ruling by the FCC—concerned that the combination of the two companies would create a monopoly—required that no one partner own less than 10% or more than 49% of the stock. Under this "balanced CML" ruling, Comsat and IBM were seeking a third partner; to that end, Comsat was acquiring all of the stock so that it could be reapportioned in the proposed new structure. (Comsat General Release 75−45)

*26 July:* By a vote of 73 to 7, the Senate passed H.R. 8070, the Dept. of Housing and Urban Development−Independent Agencies' Appropriations Bill for FY 1976, as reported out by the Appropriations Committee on 24 July. NASA was provided a total of $3.543 billion for FY 1976, and $925 million for the transition period from 1 July to 30 Sept. 1976. (*CR*, 26 July 75, S13863−98)

- The lung condition of the Apollo-Soyuz Test Project astronauts, who had inhaled nitrogen tetroxide fumes during Apollo reentry [see 15− 24 July], had improved, NASA Flight Surgeon Dr. Arnauld Nicogossian told reporters at a crew-status briefing. The astronauts had been transferred from the intensive-care unit of Tripler Hospital to other hospital quarters. Astronaut Donald K. Slayton's x-rays showed a 70 to 80% improvement in the condition of his lungs; astronauts Vance D. Brand and Thomas P. Stafford had improved, but not as much. (ASTP Transcript PC−58)

- The Senate passed a resolution, S. Res. 222, congratulating NASA and the Soviet Academy of Sciences on the outstanding success of the joint Apollo-Soyuz Test Project [see 15−24 July]. (CR, 26 July 75, S13914)
- People's Republic of China launched its third satellite into an earth orbit with a 455-km apogee, 182-km perigee, 90.9-min period, and 69.0° inclination. PRC's Hsinhua news agency reported 27 July that the satellite had been launched "under the guidance of Chairman Mao's [Tse-tung] proletarian revolutionary line . . . promoting production and other work and preparedness against war." Hsinhua also reported that the instruments on board the satellite were operating normally.

The satellite was PRC's first in 4 yr. Baltimore *Sun* correspondent Edward K. Wu attributed the apparent slowdown in the PRC space program to an estimated 25% cutback in military spending. PRC's first two satellites had been successfully launched 24 April 1970 and 2 March 1971. (Wkly *SSR*, 25−30 July 1975; FBIS−PRC, 28 July 75, E1; Wu, B *Sun*, 28 July 75, A4; *W Post*, 28 July 75, A14; *A&A 70*; *A&A 71*)

*27 July:* The Apollo-Soyuz Test Project crew was making "excellent progress" recovering from lung ailments that developed when the crew inhaled nitrogen tetroxide during Apollo reentry [see 15−24 July], Johnson Space Center Director of Life Sciences Richard S. Johnston told a crew-status press briefing at JSC. Johnston told reporters that, if the crew continued to progress, they would be released from Tripler Hospital on 29 July. However, their activities would be restricted for an additional 10 days to protect them from undue exposure to large crowds, to ensure continued improvement, and to avoid respiratory complications. (ASTP Transcript PC-59)

*28 July:* Soviet engineers had been working on a "spaceplane" to carry cosmonauts to and from orbit as part of a long-term project to establish a major space platform in the 1980s, the *Christian Science Monitor* reported. Western observers had expected the reusable spacecraft to be launched by an expendable booster, but *CSM* reported that the rumored aim was to fly it from the back of a ramjet-powered boost vehicle that would return to base. The project was thought to be linked with development of a large multiman modular space station that the Soviets expected to assemble in orbit before 1980. *CSM* reported that U.S. Air Force reconnaissance satellites had noted preparations at the Baykonur Cosmodrome to retest the huge SLX−14 booster, scheduled to launch the station. The SLX−14 had been delayed for 5 yr by a series of mishaps: Two launch complexes had been built for the SLX−14, one of which burned during a 1969 fueling test. Intelligence reports indicated it had been rebuilt.

In an August 1975 issue of *Sputnik* magazine, reprinted from the Soviet *Nauka I Zhin*, U.S.S.R. Chief of Cosmonaut Training Vladimir A. Shatalov confirmed Soviet interest in a reusable space vehicle.

Although "the future of cosmonautics is largely connected with long-term orbital stations," he said, the fact that carrier rockets were extremely expensive and could be used only once was "becoming a serious brake on the development of cosmonautics and relevant research." Neither the airplane nor the spaceship had proven capable of flying into orbit and then reentering. But there is every reason to believe that in the not-so-distant future aviation and cosmonautics would "draw appreciably closer together." Shatalov said Soviet and U.S. scientists had already made important high-altitude and high-speed experiments. Shatalov visualized "a new-type spaceship" consisting of a smaller one launched from a larger one after reaching required speed. The carrier craft would return to a landing field, and the smaller craft would orbit, accomplish its mission, and then land. "Creation of an apparatus with greater orbital maneuverability would ... make it possible ... to deliver satellites on orbits in the spacecraft's freight compartment, serve and repair them in space and bring back to earth samples of explorations and observations and even the satellites themselves if they go out of order." Although "creation of this new type of space apparatus" posed a host of technical problems, "modern science and technology can solve them" and "the time was not far off when such a plane would make its maiden flight." (Gatland, *CSM*, 28 July 75, 11; *Sputnik*, "Plane for Outer Space," Aug 75, 68–71)

*29 July:* Dr. James C. Fletcher, NASA Administrator, and Donald P. Hearth, Deputy Director of Goddard Space Flight Center, gave in their testimony before the Subcommittee on Space Science and Applications of the House Committee on Science and Technology a preview of the new NASA report "Outlook for Space." The report had been submitted to NASA by a 21-member study group established by Dr. Fletcher in June 1974 to develop an unconstrained listing of desirable and practical civilian space activities; to group these around goals, objectives, and themes; to define R&D tasks required by potential commercial and operational uses of space; to identify social and economic challenges that could benefit from space-reported data; and to relate space goals and objectives to national ones.

All space objectives defined by the study group were evaluated in terms of their contribution to national interests, which included expansion of human knowledge, physical benefits to the U.S. and humanity, contribution to the vitality of the nation, exploration of the unknown, national prestige and self-esteem, and international cooperation.

The report suggested categorizing space missions as either earth-oriented or extraterrestrial. Earth-oriented programs for the rest of the century should help increase food production by improved weather and water-availability forecasting, improved crop forecasting, improved detection and monitoring of diseases and infestations in vegetation. The consequent economic benefits and stabilization of the commodity market could gain hundreds of millions of dollars per year.

Other earth-oriented space programs should emphasize improved forecasting of seasonal and climatic trends, to aid in the managing of food and energy resources and associated transportation requirements. Benefits from improved climate forecasting could reach hundreds of millions of dollars annually.

The space program should also address itself to the challenge of the unknown by adding to man's knowledge of cosmic and human evolution and destiny. Spacecraft could take instruments above the "dark and dirty glass" of earth's opaque atmosphere, and should continue to carry instruments to other planets, setting up laboratories there as well as bringing samples back to earth for analysis. Extraterrestrial missions should include textural, chemical, and isotopic planetary studies and comparisons of atmospheric trends for all planets and their satellites.

Plans should also continue for human activities in space, using both the space shuttle and free-flying laboratories to repair and maintain orbiting spacecraft systems and perform research in space. Although the study panel had found it difficult to assert that manned orbiting or planetary bases should be undertaken in the next 25 yrs, it found that man was an integral and necessary part of the more creative space-exploration programs. If the U.S. was to avail itself of the potential benefits of space, the next logical step for NASA's manned program could be development of a permanent space facility in which crews could work for extended periods. (H.R. transcript)

- Communications Satellite Corp. had received initial orders to lease terminals on tankers and other commercial vessels for communications via the Marisat maritime-satellite system. Exxon International Corp. had ordered five shipboard terminals to provide communications with Exxon tankers at sea. The U.S. Maritime Administration had ordered terminals for six U.S.-flag ships as part of a cooperative cost-sharing program between USMA and the vessel owners.

The Marisat system, scheduled for commercial service later this year, would provide high-quality telex, data, and telephone communications 24 hrs a day to ships and offshore facilities. (ComSatCorp Release 75–48)

*30 July:* Rep. Olin E. Teague (D–Tex.) and Rep. Charles A. Mosher (R–Ohio) introduced H.R. 9058, a revised bill to establish national science policy and organization, incorporating the most promising features of H.R. 4461 [see 6 March] and the Administration bill H.R. 7830 [see 9 and 11 June].

The proposed bill included a statement of national policy similar to that in H.R. 4461, but with language recommended during an earlier 3 wk of hearings. The proposal to establish an Office of Science and Technology Policy was based on the Administration bill with three major changes: First, H.R. 9058 would give the President the power to appoint up to four assistant directors, giving him the potential of a three- to five-man council as he chose; this would give later Presidents more flexibility of organization. Second, the President-appointed Director and assistants would be subject to confirmation by the Senate. Third, H.R. 9058 described the duties and functions of the new office more precisely than had the Administration bill.

H.R. 9058 would establish, within the Executive Office of the President, a Federal Science and Technology Survey Committee to survey the Federal science effort, including missions, goals, personnel, funding, organization, facilities, and general activities. The 5 to 12 committee members would report findings and recommendations to the director of the new advisory office, with particular attention to organizational reform, science information systems, technology innovation and transfer, Federal-state and Federal-industry liaison and cooperation, and improved science budgeting process. Upon completion of the report, the committee would cease to exist. (CR, 30 July 75, E4265)

- NASA announced selection of Martin Marietta Aerospace for a $22.8 million cost-plus-award-fee contract for checkout, control, and monitoring of subsystem hardware for the Space Shuttle launch processing system (LPS). Martin would design, fabricate, test, and install hardware for the subsystem, composed of hardware assemblies communicating directly with computers and with flight and ground equipment of the Space Shuttle. The LPS was designed to use automated and modular techniques to meet Space Shuttle launch, maintenance, and refurbishment requirements. (NASA Release 75-219)
- At the Air Force Flight Dynamics Laboratory's 50 mw open-jet wind tunnel, engineers had been able to more precisely measure the material and shape endurance of model nosetips for spacecraft and missiles. Using a newly developed flared nozzle with a smoothly decreasing pressure field, and a new technique for reentry simulation, engineers could gradually bring the sample nosetip up to the nozzle mouth, simulating the transition from moderate to maximum temperature and pressure. Previous techniques could show only variations between distinctly different materials; use of the new nozzle could detect subtle differences between very similar materials. (AFSC Release OIP 172.75)

*During July:* The world's news media commented on the joint U.S.-U.S.S.R. Apollo-Soyuz Test Project [see 15-24 July].

A 15 July article in West Germany's *Frankfurter Allgemeine* said that "technically the Apollo-Soyuz enterprise will show nothing much." The reason for the mission had been political since it was agreed upon in 1972. It was "fully in line with the detente and rapprochement policy of the two superpowers." The astronautic field was particularly useful for propaganda because it could mark the "beginning of an intellectual and scientific exchange of opinion of advantage to both sides." However, "if one looks...at rapprochement in other fields, such as SALT..., one must remain skeptical regarding 'atmospheric' improvements to be achieved outside the atmosphere of the world."

Milika Sundic, a Yugoslavian commentator, said concerning ASTP, "Nobody doubts its success, or...the good intentions and will of the two superpowers." The technical aspect of the mission was least important, and the political side "constitutes its essence." The joint undertakings both on earth and in space reflected the superpowers' belief in a "full equilibrium of forces rather than trust in one another." This lack of trust was the consequence not only of different systems but different interests which went beyond national borders. The fact that the U.S. and U.S.S.R. were leaders of two military alliances could not lead to greater trust between them. But, "without blocs and without ambitions going beyond national borders, the situation would be...more favorable both for the superpowers and for the rest of the world." One hoped that nobody intended to divide up space along bloc lines and that nobody would misuse it for solving this world's problems. "Third World countries may not be present at today's rendezvous in space, but they have a claim on the part played by science in the Soviet-American undertaking." Regarding their rights on the earth, "the Third World has no intention of renouncing them, and even less of abandoning the fate of peace to the big powers, for peace is indivisible both on earth and in space."

Hungary's 15 July issue of *Nepszabadsag* said the joint space flight was a scientific undertaking inseparable from "normalization" between the two states and reflecting the advance of peaceful coexistence. The joint cosmic attempt called attention to the vast possibilities inherent in dynamic scientific cooperation between the two countries in other, no less important, areas.

Bulgaria's 15 July *Rabotnichesko* said in an editorial that ASTP not only would mark "a new stage in the development of space research and will open up new prospects for cooperation between the U.S.S.R. and U.S.A. but will also be a telling example to be followed by all other states and peoples on the earth. . . ."

A 26 July commentary by a Peking correspondent in the People's Republic of China said that the U.S.-U.S.S.R. handshake in space—despite the show and all the money spent to advertise it—could not "cover up their fierce struggle on earth." In fact, the two superpowers "regard each other as the enemy in space. They have been contending for military superiority in space ever since the first satellite was launched." The handshake was but a contention with "each trying to cheat and outwit the other."

Finland's *Karjalaimen* called the joint mission "historic" from the point of view of both space technology and world politics. The superpowers had found it necessary to begin together to regulate the world's conflicts to prevent a world conflagration or wars that could endanger the interests of the superpowers. ASTP was one of the public symbols of this policy. (NCNA, FBIS-PRC, 28 July 75, A1; FBIS-Bulgaria, 16 July 75, C1; FBIS-Hungary, 16 July 75, F1; FBIS-Yugoslavia, 16 July 75, I1; FBIS-Finland, 16 July 75, P1)

- Preparations continued for August launch of the two Viking spacecraft. Following the mating of Viking Lander-Capsule 2 (VLC-2)—with Orbiter 2 (VO-2)—to become the Viking-A spacecraft—cabling connections between the two were made and precountdown test was run on 3 July. Propellants were loaded on 9-10 July and final encapsulation inside the shroud was completed on 11 July. Viking-A was mated to the Titan IIID launch vehicle on 28 July.

  The Lander (VLC-1) and Orbiter (VO-1) of Viking-B were mated on 21 July and the spacecraft was encapsulated on 24 July, to be kept in a planned holding mode until it was ready for mating with its launch vehicle. (Viking status bulletins 8 & 9, 8 July 75, 29 July 75)

- The Air Force had announced operational concepts for Space Shuttle launches at Vandenberg Air Force Base. Construction, planned to begin in mid-1978, would cost $50 million less than original estimates through maximum use of existing facilities such as Space Launch Complex No. 6—constructed in the 1960s for the Air Force's manned orbiting laboratory cancelled in June 1969—and existing roadways and railroad tracks. An "integrate on the pad" concept and modification of surplus Saturn IB rocket transporters to move the Orbiter and external tanks to the launch pad had eliminated the need for a new highway and railroad system. A study coordinated with the Navy showed that the solid-rocket booster that dropped off the main tank into the ocean could be successfully recovered with a minimum of new facilities by using an existing Navy dock and large stationary crane.

Operations facility requirements would be programmed in two phases: Phase one, construction of a complete ground-support system, would allow the DOD to operate Space Shuttle missions from Vandenberg by December 1982. Phase two called for completion of a second launch pad in late 1986. (AFSC *Newsreview*, July 75, 16)

- The Space and Missile Systems Organization had announced award of a $2 059 000 contract to Raytheon Co. Equipment Div. for a self-repairing onboard satellite computer system. The 11-kg computer would carry two or more spares for each of its basic subunits, and a configuration-control unit (CCU) to monitor the computer's operating subunits and diagnose problems; if a subunit failed, the CCU would diagnose the problem and switch to a spare. The computer would also perform satellite housekeeping including power control, antenna positioning, temperature control, attitude control, and navigation, jobs normally performed by a computer on the ground. (AFSC *Newsreview*, July 75)

- Arnold Engineering Development Center was studying the feasibility of the secondary burning—called external burning—of fuel-rich exhaust to increase the efficiency of solid rocket motors. The concept called for the diversion of the exhaust of burning propellant through ports that ring the rear end of the motor casing, and its injection at right angles into the airstream. Miniature nozzles in each port would accelerate the exhaust gas to sonic speed. Turbulent mixing of the fuel-rich sonic gas and the supersonic airflow would cause a secondary combustion, increasing pressure at the base of the motor and overall thrust of the propulsion system. (AFSC *Newsreview*, July 75, 12)

- Potential propellants for the Space Shuttle had been tested at an underground facility at Arnold Engineering Development Center, the Air Force Systems Command reported. Testing was part of a selection process to find a compatible propellant for the auxiliary motors that would separate the Orbiter from the two solid-rocket boosters.

  Three candidate propellants were tested at a simulated altitude of 40 000 m. Reusable ballistic test-evaluation system motors, each loaded with 45 kg of propellant, were fired seven times, burning for 2 sec and generating 53 400 newtons of thrust. Exhaust plumes were directed at 15- by 15-cm tiles of thermal protective material at various distances and angles; after the tests, the tiles were checked for erosion of the surfaces and edges. Pressure and temperature were measured in the rocket plumes, and surface temperatures and heating rates were measured on 21 selected material samples. All data and samples were sent to Marshall Space Flight Center for analysis. (AFSC *Newsreview*, July 75, 3)

- NASA awarded a 1-yr contract to Boeing Co. to study a proposed Large Lift Vehicle (LLV) made up of Space Shuttle engines, fuel tanks, and avionics. Launched without the Space Shuttle Orbiter, the LLV could carry tons of freight into space, or could rocket to the moon with a second stage attached. It would be less expensive than the *Saturn V* because the avionics package could be retrieved in orbit and the twin boosters were recoverable. In a 1974 interview with the *Huntsville Times*, John H. Disher, NASA Director for Advanced Programs, had said that no definite mission for the LLV was planned, but that NASA was looking to the future. (Casebolt, *Huntsville Times*, 12 Aug 74; NASA Hq Adv Progs Off, Fero, interview, 10 Aug 75)

## August 1975

*1 August:* The Indian Space Research Organization began transmitting instructional programs to inexpensive ground receivers in more than 2400 isolated villages throughout India, using NASA's *Ats 6* Applications Technology Satellite launched 30 May 1974. As part of the 1-yr satellite instructional television experiment (SITE), the transmissions would stress improved agricultural techniques, family planning and hygiene, and school courses pertinent to Indian villagers' needs. After the experiment, *Ats 6* would be repositioned over the western hemisphere.

During its first year in operation, *Ats 6* had transmitted medical and educational programs to remote communities in Alaska, the Rocky Mountains, and Appalachia. During the 15–26 July Apollo-Soyuz Test Project mission, the satellite had helped relay communications from the Apollo spacecraft to ground stations, increasing coverage from the usual 17% to 55% for each orbit. (NASA Release 75–221; NASA MOR M–966–75–01, 7 July 75; Borders, *NYT*, 3 Aug 75, 10)

- Sen. William Proxmire (D–Wisc.), a critic of the joint U.S.–U.S.S.R. Apollo-Soyuz Test Project, reported that the Soviet Union had bugged the conversations of U.S. astronauts and technicians at the Moscow space center before the July space flight. Using information received from two sources and confirmed through official channels, Sen. Proxmire said that the bugs were discovered while the astronauts and technicians were watching a televised Soviet hockey game. Wanting a closer view of the game, U.S. astronaut Robert F. Overmyer moved his chair nearer the TV screen, causing a wire leading from the chair into the floor to snap. The *Washington Post* quoted a spokesman for NASA as confirming the incident, but saying there was no evidence the wire "had anything to do with a listening device." (AP, *W Post*, 2 Aug 75; *Av Wk*, 11 Aug 75, 17)

- Kennedy Space Center announced award of a $5 137 000 contract to Mayfair Construction Co. to modify the Vehicle Assembly Building (VAB) for Space Shuttle operations. The contract provided for modifications to High Bay 3's extensible work platforms, installation of new checkout cells in High Bay 4, and modifications to the north door of the VAB transfer aisle and related support facilities. (KSC Release 160–75)

*2 August:* *Landsat 2*, launched 22 Jan. 1975, developed a noise problem with one of its two image tape recorders, *Aviation Week & Space Technology* reported. The satellite's mission was not in danger because the backup recorder was functioning properly. (*Av Wk*, 11 Aug 75, 13)

*3 August:* Space expenditures resulted in tangible economic benefits, according to a report, "The Economic Impact of NASA R&D Spending," being prepared for NASA by Chase Econometric Associates, Inc., the *Philadelphia Inquirer* reported. Using methods developed for regular national economic forecasts, Chase predicted that, if NASA's research and development budget were increased by $1 billion for the

1975–84 period, the U.S. gross national product (GNP) would swell by $23 billion or 2% over the normal rate of growth. Labor productivity in the nonfarm areas of the economy would rise more than 2% over the normal growth rate, and more than one million jobs would be created, reducing the unemployment rate by nearly 0.4% by 1984.

According to Chase, the key to NASA's domestic economic importance was in the agency's widespread technological advances that benefited a wide range of industries. Although the technological benefits were readily visible, NASA's influence on the nation's economy was more subtle, taking about 5 yr to work its way through the system. (Text, NASA Final Rpt, *CR* 144351, April 76; Holland, *P Inq*, 3 Aug 75)

- Appointment of Dr. S. Ichtiaque Rasool, special assistant to the NASA Deputy Associate Administrator, to be the Deputy Associate Administrator for Space Science became effective. In that position Dr. Rasool would be primary adviser to Dr. Noel W. Hinners, Associate Administrator for Space Science. He would also serve as chairman of the Space Science Steering Committee, responsible for drawing skills and resources of the nation's scientific community into NASA programs. Dr. Rasool had joined NASA in January 1965 as senior research scientist at the Goddard Institute for Space Studies; in 1971 he was named Deputy Director for Planetary Programs, serving until 1974 when he became special assistant to the Deputy Associate Administrator. (NASA anno, 22 Aug 75)

*3 August – 6 November:* Ariel 5, a U.S.–U.K. cooperative satellite launched 15 Oct. 1974 to study galactic and extragalactic x-rays, detected weak cosmic x-ray emissions from the constellation Orion on 3 Aug. The emissions, subsequently confirmed by NASA's *Explorer 53* (SAS 3 Small Astronomy Satellite launched 7 May 1975), steadily increased in intensity until they were five times greater than any observed to date. *New York Times* reporter Walter Sullivan quoted Dr. Terry Matilsky, Massachusetts Institute of Technology scientist, as saying that when the emissions were first observed "we couldn't believe it."

Sas 3—operated by Goddard Space Flight Center—was able to pinpoint the location of the emissions to within 1 or 2 arc-min. With this clue scientists hoped to explain the phenomenon using ground-based optical and radio telescopes. Observatories around the world had been alerted and were trying to pinpoint the source.

The *Philadelphia Inquirer* reported MIT's Dr. Saul A. Rappaport as speculating that the emissions probably had been caused by masses of material falling from a large star into an extremely dense neutron star or black hole. The falling matter would heat up enough to emit bursts of x-rays and visible light.

Dr. Noel Hinners, NASA Associate Administrator for Space Science, reported 6 Nov. to the House Committee on Science and Technology's Subcommittee on Science and Applications that the *Explorer 53* observations had identified the optical counterpart of the x-ray emissions as a faint, otherwise normal-looking star that had brightened 100 times over previous observations. Dr. Hinners repeated Dr. Rappaport's theory and told the subcommittee that, because a similar brightening of this star had been observed in 1917, it was being classified as a recurrent nova. (1977 NASA authorization hearing transcripts, vol 1, part 1; Sullivan, *NYT*, 17 Aug 75, 21; AP, *P Inq*, 31 Aug 75)

4 *August*: NASA announced completion of a program conducted by Jet Propulsion Laboratory to measure the constituent gases of the stratosphere to help determine the effects on the earth's ozone layer of gases released from aerosol spray cans. Using a Fourier interferometer onboard a U−2 aircraft, JPL scientists had measured the distribution of hydrogen chloride molecules at altitudes between 10 and 21 km during 6 separate flights in May. Hydrogen chloride—produced by the breakdown of aerosol gas molecules in the upper atmosphere and released naturally, in small amounts, from the ocean surface and volcanic eruptions—was one of several trace gases thought to play a major role in controlling the equilibrium of the protective ozone layer. JPL had found traces of the chloride beginning at 15 km, and reaching a maximum of almost 1 part per billion at 20 km. Although the results did not represent a direct effect of aerosol gas, or Freons, on stratospheric ozone, JPL scientists recommended continuing measurements to detect any future buildup.

The ozone measurements, part of NASA's Stratospheric Research Program, had begun when a JPL-designed instrument was flown on the Anglo−French Concorde supersonic transport in 1973. Data collected on these flights included the first detection of nitric oxide, along with new information on the geographic and vertical distribution of water, carbon monoxide, carbon dioxide, and methane, and furnished a base for the 1975 study. (NASA Release 75−223; Miles, *LA Times*, 21 Aug 75)

- The Vertical Motion Simulator (VMS), an aeronautical facility designed to reproduce up-and-down and sideways motions of aircraft during takeoff and landing, was being built at Ames Research Center, NASA announced. Costing about $3.5 million, the VMS would move as much as 18 m in height and 12 m sideways, accurately simulating flare and touchdown. Practice in the simulator would enable pilots to cope better with the complexities of flying sophisticated aircraft such as short takeoff and landing (STOL) and vertical takeoff and landing (VTOL) types. (NASA Release 75−224)
- Michael Collins, Director of the Smithsonian Institution's National Air and Space Museum, and Santiago Astrain, Secretary General of the International Telecommunications Satellite Organization, signed an agreement to transfer three early Intelsat communications satellites to the Museum as part of its 1976 inaugural display. INTELSAT was providing the museum with a backup model of the INTELSAT's *Early Bird 1*, launched 6 Apr. 1965 as the first commercial comsat. A backup model of the second-generation Intelsat II and an engineering model of third-generation Intelsat III satellites would also be sent for display. Fourth-generation satellites would not be represented because they were too large for display.

  The new Air and Space Museum, located in Washington, D.C., was scheduled to open 4 July 1976. (INTELSAT Release 75−9)
- Westar communications satellites would soon be used for communications to and from offshore oil drilling platforms and exploratory vessels in waters adjacent to the continental U.S., Western Union Telegraph Co. announced. The offshore facilities previously had depended for communication on microwave relay networks and cabling, methods that had become increasingly expensive and impractical as the facilities were moved farther and farther off shore. (Westar news release, 4 Aug 75)

- "Technology will be available for manned flights well before the year 2000, allowing even better research of the planet's [Mars'] composition," Dr. George Sands, associate Viking Project scientist at Langley Research Center, stated in an interview with the Newport News *Times Herald*. "The Viking missions [scheduled for August launch] are just the prelude of things to come." Dr. Sands continued, "The results of [the] probe to Mars this month will be very significant for the future of planetary exploration and additional trips to Mars." (Biggins, *Times Herald*, 4 Aug 75)
- Marshall Space Flight Center was seeking industry proposals for procurement of parallel definition studies of the atmospheric, magnetospheric, and plasmas in space (AMPS) payload, a reusable research facility to be integrated with Spacelab. The proposed 12-mo studies were to define an overall AMPS program with special emphasis on ground- and flight-support systems and subsystems, systems engineering and integration, and ground/orbital operations with the Space Shuttle and Spacelab.

    AMPS, a single laboratory system flown with the Space Shuttle, would perform experiments and observations in atmospheric science, magnetospheric physics, and plasma physics. (MSFC Release 75−172)

*4−6 August*: The American Institute of Aeronautics and Astronautics, at its 1975 aircraft systems and technology meeting in Los Angeles, Calif., presented the following aerospace awards.

| Award | Recipient (affiliation) | Purpose |
| --- | --- | --- |
| Aircraft Design | Walter E. Fellers (Northrop Corp.) | For 30 yr of outstanding achievement in design and development of fighter aircraft to meet rapidly changing requirements |
| Octave Chanute | Alan L. Bean<br>Jack R. Lousma<br>Dr. Owen K. Garriott<br>(NASA Skylab 3 astronauts) | For outstanding contributions to manned space flight and scientific research. Using the most sophisticated instruments ever flown in space, crew performed scores of scientific experiments and observations vastly expanding man's knowledge and enhancing quality of life on earth |
| 38th Annual Wright Brothers Lecture | Henri Ziegler (Airbus, France) | Topic: International Cooperation in Aerospace Projects: cooperation between European industries and between Europe and United States |

(AIAA Release, 22 July 75)

*5 August:* The Saturn IB, used to launch three U.S. Apollo-Soyuz Test Project astronauts into space [see 15−24 July], had experienced no unscheduled holds whatever during countdown, making it the most perfect launch of the Saturn series, Ellery B. May, manager of the Saturn Program Office at the Marshall Space Flight Center, said. Saturn 210, manufactured in 1967, was 10th of the IB series and had the oldest engine used. All Saturn launches had been successful and had met their objectives; although some had anomalies, such as leaks or faulty wiring, no major configuration changes were ever necessary.

The ASTP mission marked the end of the Saturn series, but its impact on technology would continue. "Space Shuttle main engines (SSME) are an outgrowth of technology from Saturn engines," May explained. The major difference between the two was that the Shuttle must have much higher pressure systems than ever before required and must be reusable. (MSFC Release 75-174)

- The wingless unpowered X-24B lifting body had made its first landing on a concrete runway at Edwards, Calif., to demonstrate maneuver and safe landing of an unpowered reentry vehicle on a conventional runway. After launch from a B-52 aircraft flying at 14 000 m, test pilot John N. Manke, chief Flight Research Center pilot for the X-24B project, had ignited a small rocket engine, propelling the X-24B to a speed of 1381 km per hr and altitude of 18 300 m. Manke then shut off the engine and glided to a perfect 300-kph landing on the 4600-m runway.

    The 11.3-m-long X-24B was a part of a joint NASA-Air Force program to study transonic flight characteristics and landing ability of a vehicle designed for hypersonic speeds. (FRC Release 24-75; *LA Times*, 6 Aug 75)

- Sen. John Tunney (D-Calif.) defended the nation's space program during a tour of Space Shuttle facilities at Rockwell International Corp. in Downey, Calif. According to Sen. Tunney, the most important justification for the space program was to provide new technology in weather mapping and control, to help increase the amount of land available for food production, and to locate additional stocks of natural resources in this country. Sen. Tunney stated, "I quite frankly think without a space program...mankind is going to have a rendezvous with destiny which would be catastrophic...." But with this kind of program, a world food supply and adequate resources can "keep this country going indefinitely." (*Pasadena Star News*, 6 Aug 75)

- An Ames Research Center wind tunnel used primarily for testing the Space Shuttle had been shut down after a steel flange failed, setting off a high-powered explosion of compressed gas. The blast scattered hundreds of hot aluminum oxide pebbles over a wide area, causing several fires but no serious injuries. Space Shuttle testing had been postponed until the facility, which generated pressures up to 126 $kg/cm^2$ (1800 psi) in testing models at 14 times the speed of sound, could be declared operational again. An investigation board was formed to determine the cause of the accident and recommend actions to prevent recurrence. (ARC *Astrogram*, 14 Aug 75, 1; ARC Experimental Fluid Dynamics Br, interview, 27 June 77; UPI, *W Post*, 7 Aug 75, A17)

- Two Boeing Co. scientists had proposed construction of a 64-million-kg solar power satellite, with 57 sq km of mirrors, to collect and concentrate solar energy, together with thermal engines to convert solar power into electricity, the *Christian Science Monitor* reported. A microwave transmission system would convert the electricity to a form suitable for transmission to earth. *CSM* quoted Boeing as saying that within 2 decades "Powersat" could provide up to 10 000 mw of useful power, twice the hydroelectric capacity of Grand Coulee Dam. (AP, *CSM*, 6 Aug 75, 7)

- Kennedy Space Center announced award of an $18 749 million contract to Blount Brothers Construction Co. for modification of Launch Complex 39 Pad A, to accommodate all early Space Shuttle missions. The complex had been the site of all but one of the historic Saturn V launches; Pad B, from which the Skylab Orbital Workshop was launched, would be modified later. The contract included conversion of the mobile launcher for Shuttle operations. (KSC Release 164-75)

*7 August:* Marshall Space Flight Center announced selection of United Technologies Corp. for negotiation of a fixed-price contract of $1.775 million for solid-propellant booster separation motors (BSM) for use on the first 6 development flights in the Space Shuttle program, beginning in 1979. The BSMs would separate the two reusable solid rocket boosters (SRB)—each SRB requiring eight BSMs, four forward and four aft—approximately 110 sec after launch. (MSFC Release 75–175)

- European Space Agency announced a postponement in the launch of COS–B, scheduled for a 6 Aug. 6:56 pm PDT launch from Western Test Range, because of hydraulic-valve failure in a telemetry aircraft needed to monitor third-stage ignition of the launch vehicle. The launch had been rescheduled to 8 Aug. at 6:48 pm PDT. COS–B would be ESA's first satellite launch since the new agency's establishment on 30 May. (ESA Release, 7 Aug 75)
- Scientists were calling for names for the thousands of mountains, craters, and chasms being found on planets faster than they could be named, the *Christian Science Monitor* reported. The two Viking spacecraft scheduled to reach Mars in 1976 and the two Soviet Venera spacecraft scheduled to approach Venus in October would no doubt uncover a host of new unnamed features. Names for known features on Venus, Mars, and the moon would be officially chosen at the August 1976 meeting of the International Astronomical Union (IAU), world clearinghouse for solar-system nomenclature. (Jones, *CSM*, 7 Aug 75, 2)

*8 August:* NASA launched *Cos-B* scientific satellite for the European Space Agency at 3:48 pm PDT from Western Test Range on a three-stage thrust-augmented Thor-Delta booster. Launch had been delayed for 2 days because a telemetry aircraft was grounded with hydraulic-valve failure. The satellite entered the planned highly elliptical polar orbit with a 99 886-km apogee, 344-km perigee, 36.7-hr period, and 90.13° inclination. NASA's objective for the launch was to place *Cos-B* into a highly elliptical near-polar orbit with sufficient accuracy to allow the satellite to conduct its planned scientific experiments.

Purpose of the *Cos-B* mission was to study extraterrestrial gamma radiation in the energy range 25 mev to 1 gev in order to establish the intensity of the average gamma-ray flux; examine large-scale anisotropy of the radiation over angular regions corresponding to galactic features; study quasars, x-ray, and radio sources of small angular size in supernova remnants; measure energy spectra of various galactic radiations; and study time variations of radiation from pulsars, x-ray, and radio sources.

On 11 August, ground controllers sent low-voltage commands from ESA's space operations center (ESOC) in West Germany to check out spacecraft and experiment performance. After a second switch-on at high voltage, a series of inflight tests further validated all onboard systems and experiment performance. On 15 Aug. a complete review of spacecraft operational status showed all air and ground systems were operating normally; by 19 Aug., the first data had been returned to earth.

*Cos-B*, first satellite launched for the newly organized ESA, was the eighth developed by ESA's predecessor, the European Space Research Organization [see 30–31 May]. Under the direction of the European

Space Research and Technology Center (ESTEC) and prime contractor Messerschmitt-Boelkow-Blohm GmbH, *Cos-B* had been designed and built by organizations in 7 of the 10 ESA member countries. The single experiment—a 115-kg spark chamber—had been assembled from five primary and two subsidiary experiment units supplied by six institutes in four European countries. After separating from the launch vehicle, *Cos-B* transmitted data in real time to ground stations of the European Space Tracking Network with central control at the European Space Research Operations Center in West Germany.

For NASA, Goddard Space Flight Center had the responsibility for the Thor-Delta launch vehicle under the direction of the Office of Space Science. GSFC also had responsibility for limited tracking support. (NASA MOR S−492−301−75−01, 22 July 75, 10 Sept. 75; NASA Release 75−214; Releases 7−19 Aug 75)

- *Science* magazine reported that U.S. scientists had reprogrammed NASA's *Landsat 1* earth resources satellite (formerly called *Erts 1*, launched 23 July 1972) to photograph large areas of the Soviet Union, although not specifically requested to do so by Soviet officials. The magazine reported that the intent of the U.S. scientists was to aid Soviet scientists, after discussions at a series of 1972 joint working group meetings—convened for geological studies of comparable land sites in each country—indicated that the Soviets had nothing available to them equivalent to this kind of satellite imagery. Because the Soviet government was sensitive about U.S. satellites flying over the U.S.S.R., the Soviet scientists who were to benefit had never been directly informed about the reprogramming decision. Resulting images had never been directly handed over to the Soviet scientists lest doing so should violate the 1972 joint space agreement signed by both countries. *Science* quoted Goddard Space Flight Center's Dr. William Nordberg, scientist responsible for the reprogramming, as saying, "They knew where they could buy them." *Science* also reported that the Soviet scientists had ordered the Landsat images from the U.S. Geological Survey's Sioux Falls distribution center. Dr. Nordberg denied that any of the reprogrammed imagery had been done for "economic espionage." (*Science*, 8 Aug 75, 441−2)

- Photographs from *Landsat 1* (launched 23 July 1972 as *Erts 1*, Earth Resources Technology Satellite) indicated that a continental collision had been in progress between India and Eurasia, pushing China eastward at the rate of 2.54 centimeters a year, Massachusetts Institute of Technology scientists reported in *Science*. This eastward movement could account for China's unusual pattern of earthquakes—widespread instead of along narrow fault lines, the MIT geologists suggested. Despite the current theory that the Indian subcontinent was sliding under Eurasia, the MIT scientists had concluded that rigid Indian plates of rock were causing China to be pushed into the Pacific. (Molnar *et al.*, *Science*, 8 Aug 75, 419−425; *NYT*, 10 Aug 75, 49)

*10 August:* During a panel discussion at a 4-day meeting of the Astronomical Association of Northern California, scientists rejected the idea that inhabitants of other planets would send manned spacecraft or unidentified flying objects to establish contact with earth. Such a venture would require enormous expenditures of energy and funds. The *Washington Star* quoted Ames Research Center scientist John Billingham as saying, "The most logical way to establish contact is radio communication." Billingham was devoting 2 yr to the subject of intelligent extraterrestrial life, and how to reach it if it existed. He said that several persons around the globe had been constantly trying to make radio contact with outer space by aiming a radio telescope at a suitable star, hoping that an inhabited planet might be orbiting. (UPI, *W Star*, 11 Aug 75, A16)

- Contrary to the impression of peaceful cooperation created by the U.S.-U.S.S.R. Apollo-Soyuz Test Project docking mission [see 15-26 July], both countries had increasingly emphasized the military potential of manned space flight, the *Philadelphia Inquirer* reported the Center for Defense Information as saying. The Center, a private organization critical of defense spending, charged that the 1967 United Nations treaty which reserved outer space for peaceful uses had not prevented either country from exploiting space for military purposes. Although the Pentagon had listed its space spending at $2.25 billion, it had acknowledged the existence of another $667 million in Air Force "miscellaneous" requests for secret space programs. The Center said that, although the Soviet Union's space program was 15 yrs behind the U.S., frequent launches indicated that they were attempting to catch up. The Center concluded that, contrary to the spirit of the outer space treaty, space technology and systems were being developed that had the potential for new confrontations between the superpowers. (UPI, *P Inq*, 10 Aug 75)
- Launch of the Viking—A spacecraft to Mars was postponed because of a faulty thrust-vector control valve found in the solid booster stage of the Titan-Centaur launch vehicle during countdown tests. The launch team decided to remove and replace the valve, rescheduling the launch for 14 Aug. (Viking Status Bulletin No 10, 20 Aug 75; O'Toole, *W Post*, 12 Aug 75; McElheny, *NYT*, 12 Aug 75)
- Dr. Harrison H. Schmitt, Assistant Administrator for Energy Programs since May 1974, would resign effective 30 August. Previously, Dr. Schmitt had been chief of the Astronaut Office, Science and Applications Directorate, at Johnson Space Center. Selected as a scientist-astronaut by NASA in 1965, he had been the lunar module pilot on *Apollo 17* (7-19 Dec. 1972) and the only civilian scientist to walk on the moon. Dr. Schmitt planned to return to his home state of N. Mex. to pursue geological consulting and personal activities. (NASA anno, 11 Aug 75)
- *Westar I*, first U.S. domestic commercial communications satellite (launched by NASA for Western Union Telegraph Co. 13 April 1974), had transmitted its first live sports event—the Milwaukee Brewers vs. Texas Rangers baseball game—within the 48 contiguous states, WU announced. WU Vice President Dow C. Pruitt noted that the low-cost transmission reflected "the power of technology to change basic living patterns." (*Westar News*, 11 Aug 75)

*12 August:* In the general purpose laboratory (GPL) at Marshall Space Flight Center, investigators from MSFC and Johnson Space Center were conducting experiments in high-energy cosmic-ray astronomy, superfluid helium, optical x-ray astronomy, and life sciences. The objective of the experiments, designated concept verification test (CVT) 5, was to verify operational and integration concepts for their applicability to Spacelab, the reusable manned space laboratory being built by the European Space Agency as a payload for the Space Shuttle.

The purpose of the high-energy astronomy experiment was to demonstrate the use of minicomputers for experiment operation, control, and data management, and the real-time interaction of payload specialists with experiment operations. The superfluid helium experiment examined characteristics of helium droplets at temperatures near 2.2 K (−512°F), which would help in developing cryogenic techniques for use in Spacelab. The optical astronomy experiment, which demonstrated remote experiment operation by obtaining data from a Ferson 40.6-cm telescope 9.2 kilometers away, gave practice in investigator-experiment interaction and helped in developing computer control techniques, display requirements for telescope operation and data management, and astronomy training requirements for crew members. The four life-sciences experiments included studies of sporophore formation, the vestibular function, a clinical diagnostic system, and muscle-like contractile protein. Investigators also participated in an experiment to demonstrate the use of a computer in real-time mission planning, scheduling, and applications.

By involving investigators in the CVT simulations, NASA would be able to plan the most cost-effective use of Spacelab. (MSFC Release 75−178)

- The development of space law, which had begun about the time of the launch of U.S.S.R.'s *Sputnik 1* in 1957, was a unique process. "The problems are both literally and symbolically out of this world," NASA General Counsel Neil S. Hosenball said in an interview with the Orlando *Sentinel Star*. Does any state have the right to claim the natural resources of celestial bodies? Were earth's criminal and civil laws applicable in space? What was the difference between establishing a planetary base and planting a nation's flag? Which court would rule on extraterrestrial squabbles? These were some of the questions lawyers and politicians would have to consider in the space age. Few precedents existed, and no judge had yet heard a case in a legal dispute originating outside the earth's atmosphere. In fact, a world court, recognized by all major powers, did not exist.

  In 1962, a United Nations ad hoc committee on peaceful uses of outer space had developed principles later incorporated into the 1967 Outer Space Treaty. These included a ban against military activity in outer space and against any nation's claiming a section of space or a planet as its own. Other treaties dealt with specific points such as the rescue of astronauts and the registration of spacecraft and space stations. Hosenball stressed that most charters and treaties were drawn up ahead of time in anticipation of problems likely to occur as space exploration expanded. The *Sentinel* quoted G. P. Zhukov, chief legal counsel of the U.S.S.R. Academy of Sciences, as saying that détente between the U.S. and U.S.S.R. had been assisted by the success of "legal regulation in space. . . . Man's foot had not yet stepped on the surface of the moon and we had already formulated what man could and could not do on the moon. The law was ahead of science." (Orlando *Sentinel Star*, 12 Aug 75)

- NASA commemorated the 15th anniversary of the 12 Aug. 1960 launch of *Echo 1*, world's first communications satellite. Visible to millions around the world, the 30-m gas-inflated mylar (0.0013-cm thick aluminum-coated plastic) balloon served as a passive radio reflector in space, bouncing signals from one point on earth to another. Its first transmission sent the recorded voice of President Eisenhower from

Goldstone, Calif., to Holmdel, N.J.; during its first year in orbit, it served in approximately 150 experiments, including voice, teletype and facsimile communications. *Echo 1*'s large size, small mass, and extreme sensitivity to solar radiation pressures and aerodynamic drag helped scientists confirm orbital behavior theory.

Despite its original 1-yr life expectancy, *Echo 1* remained in orbit for nearly 8 yr until, leaking gas through its skin bombarded by meteorites, it reentered the atmosphere and burned up 23 May 1968. (NASA Releases 61–177, 75–217; *A&A 68*, 23 May)

*13 August:* Dr. Wernher von Braun—German-born rocketry pioneer and former Marshall Space Flight Center Director and NASA Deputy Associate Administrator—had undergone successful surgery at Johns Hopkins University Hospital in Baltimore for removal of a malignant tumor, the *Washington Post* reported. (AP, *W Post*, 13 Aug 75)

- The launch of Viking A had been postponed for the second time [see 11 Aug.] when Jet Propulsion Laboratory technicians discovered that the batteries had dropped from their normal charge of 37 volts to 9 volts. The batteries had been drained because a motorized rotary switch that should have remained off until 7 min before launch had accidentally turned on. The dangerously low voltage decided officials to remove the entire spacecraft from the launch vehicle and return it to the spacecraft assembly and encapsulation facility for troubleshooting. At the same time, the Viking-B spacecraft, in a planned holding mode before mating with its launch vehicle, would be moved to the launch complex to replace its ailing sistercraft and become the Viking-A mission. The launch had been rescheduled for 20 Aug., with the launch window opening at 5:22 pm. (Viking Status Bulletin No 10, 20 Aug 75; NASA Audio News, 14 Aug 75; O'Toole, *W Post*, 14 Aug 75, A23; McElheny, *NYT*, 14 Aug 75)

*14 August:* A team of four physicists—Dr. Paul B. Price and Dr. Edward K. Shirk of the Univ. of Calif., and Dr. Weyman A. Osborne and Dr. Lawrence S. Pinsky of the Univ. of Houston—reported finding evidence of a particle with only one magnetic pole. The discovery of the monopole, if confirmed, could have a major impact on physics, with practical applications such as new medical therapies, new sources of energy, extremely small and efficient motors and generators, and new particle accelerators of much higher energy than any yet built.

The monopole, representing the basic indivisible unit of magnetism, would exist in two forms of opposite polarity, the north pole and the south pole. This would be analogous to the basic electric charge, the electron, which existed in the positive and negative charges, the proton and electron. Existence of the monopole had been predicted in 1931 by Paul Dirac, Fla. State Univ. professor.

The four physicists detected tracks of the particle that had apparently penetrated a sandwich of more than 33 layers of plastic and emulsion photographic film suspended from a balloon floating at 39 600 m. (Sullivan, *NYT*, 15 Aug 75, 1; AP, *B Sun*, 15 Aug 75, A3; Joffee, *W Post*, 15 Aug 75, B1)

- U.S. intelligence reports indicated that the Soviet Union had fired 22 missiles and orbited 9 satellites in July, making it one of the most active months in Soviet space history, the Baltimore *Sun* reported. In an impressive display on 2 July, the Soviets fired 11 ballistic missiles from four different sites within an hour, launched eight SSN-6 submarine-based rockets at 8-sec intervals, and launched an SS-7, SS-8, and SS-11 all within 2 min. Other tests included submarine missile launchings by northern and far eastern fleets, medium- and intermediate-range missile firings from Kapustin Yar, and a test flight of the new SS-17 missile which could carry four warheads. The month's only failure was the malfunction of the second stage of a SS-19 missile, a vehicle believed to be capable of delivering six nuclear warheads to separate targets over ranges of 9260 km.

The nine satellite launches included six Cosmos satellites, a Molniya communications satellite, a Meteor meteorological satellite, and the manned *Soyuz 19*, the Soviet spacecraft participating in the joint Apollo-Soyuz Test Project. (Corddry, B *Sun*, 14 Aug 75, A2)

- The first 10 automatic landings of a powered-lift short-takeoff-and-landing (STOL) aircraft equipped with a STOLAND automatic flight-control system had been completed by Ames Research Center personnel at the Naval Auxiliary Landing Field in Calif. The STOLAND system, a versatile onboard digital computer, provided the pilot with navigation, guidance, and control information. Automatic controls, including control of the aircraft's lift with automatic flaps and augmenter nozzles, permitted steeper climbouts, approaches, and landings, offering potential relief for airport traffic congestion. Half the aircraft's lift was provided by engine thrust, permitting flight at much lower speeds than conventional aircraft.

   The STOLAND system was developed by Sperry Flight Systems for use in the joint NASA-Dept. of Transportation operating experiments program and the augmentor-wing handling qualities and flight-control programs. (NASA Release 75-226)

*15 August:* Flight Research Center pilot Einar Enevoldson, in a ground-based instrumented cockpit, guided the remotely piloted research vehicle (RPRV) to its first unpowered landing, at FRC. Following air-launch from a B-52 aircraft flying at 14.6 km, the 7.3-m three-eighths-scale vehicle performed maneuvers to measure control effectiveness during spinning flight. The spin was continued until 4.6-km altitude when Enevoldson deployed the landing gear and performed a series of stability and control maneuvers. He initiated final approach at 1524-m altitude. The RPRV landed on the dry lakebed using small steel skids individually mounted on the ends of three conventional automobile shock absorbers installed inside the aircraft's fuselage.

The RPRV had been developed by FRC to provide a more economical and far less hazardous means of flight-testing new vehicles. Earlier versions of the RPRV had been recovered in midair using a parachute and helicopter. (FRC Release 26-75; FRC Proj Off, Rezek, interview, 12 Sept 75)

- Kennedy Space Center deactivated its hyperbaric decompression chamber after 6 yr of operation. The chamber had never been used for its intended purpose—to save the lives of astronauts in the event of a malfunction during atmospheric chamber tests—but had treated 12 local civilian divers for the bends, a painful crippling illness caused by formation of nitrogen bubbles in the bloodstream when a diver moves too quickly from high-pressure water depths to the surface.

The $100 000 chamber intended for the manned space program had been opened to public use by the center director. With the end of the Apollo program and the next manned flight 4 yr in the future, the decision was made to deactivate the chamber. The Orlando *Sentinel Star* quoted Russell P. Lloyd, KSC engineer in the Support Operations Div., as saying the possibility existed that the chamber would be reactivated for the Space Shuttle program, but for the time being it would remain closed. (Lloyd, interview, 19 Sept 77; Upchurch, Orlando *Sentinel Star*, 17 Aug 75)

- NASA and the U.S.S.R. Academy of Sciences had agreed on a joint tour of the U.S. and the Soviet Union by the 5 crew members of the Apollo-Soyuz Test Project mission [see 15–26 July]. U.S. astronauts Thomas P. Stafford, Vance D. Brand, and Donald K. Slayton would join Soviet cosmonauts Aleksey A. Leonov and Valery N. Kubasov on 21 Sept. in Moscow for a 2-wk tour of the U.S.S.R. Plans for the U.S. tour were not disclosed. (NASA Release 75–230)

*18–20 August:* A workshop on "Operational Applications of Satellite Snowcover Observations" was held at Lake Tahoe, Nev., to discuss recent progress in extracting meaningful snowcover information from satellites. Albert Rango, Goddard Space Flight Center scientist, reported that, with the cooperation of nine Federal and state agencies, NASA had developed an applications systems verification test (ASVT)—using data from satellites and instrumented aircraft, conventional ground information, and prediction models—to quantitatively determine the usefulness of remote-sensing technology in an operational applications system.

In another paper Rango reported that low-resolution meteorological satellite data and high-resolution earth-resources satellite data had been used to map a snow-covered area over Wyoming mountains. Predictions based on satellite data of the April through June 1972 streamflow were within 3% of the actual total. Also, composite results from 2 yr of data over the mountains indicated that Landsat snowcover observations could be useful in predicting runoff and seasonal streamflow.

James C. Barnes of Environmental Research and Technology, Inc., reported that the earth resources experiment package (EREP) flown aboard the 1973–74 Skylab Orbital Workshop missions had provided the first opportunity to examine reflectance characteristics of snowcover in several spectral bands from the visible to the near infrared. Results indicated that near-infrared methods could distinguish between snow and water droplet clouds and could detect areas of melting snow. (NASA *Hq WB*, 28 Feb 76, 2; NASA SP–391, text)

*18–21 August:* Evidence of a 10th planet existing in earth's solar system 16 million yrs ago was presented at the 146th annual meeting of the American Astronomical Society in San Diego by Dr. Thomas C. Van Flandern, U.S. Naval Observatory astronomer. Studying the orbits of 60 long-period comets, including 1974's Kohoutek, Dr. Van Flandern had concluded they had shared a common origin 16 million yr ago.

The evidence was consistent with calculations made in 1971 by Canadian astronomer Dr. Michael W. Ovenden. Dr. Ovenden's studies of irregularities in the orbits of planets, especially Uranus and Neptune, suggested that a major disruptive event had occurred in the solar system approximately 16 million yrs ago. He had arrived at this time by age-dating carbon-bearing meteorites that had fallen to earth, and theorized that the event could be explained by the explosion of a huge planet 90 times the size of earth.

During the meeting Dr. Stuart Bowyer, Univ. of Calif. astronomer, surprised the 740 astronomers attending the meeting by reporting that a Univ. of Calif. telescope flown aboard the Apollo spacecraft during the joint U.S.-U.S.S.R. Apollo-Soyuz Test Project [see 15–26 July] had picked up the extreme ultraviolet (EUV) radiation of a star. EUV, a small segment of the spectrum between normal ultraviolet and x-rays, had been considered by astronomers as practically invisible because of probable absorption by interstellar dust and gas. Dr. Bowyer said that the EUV had come from a white dwarf star 300 light yrs away in the constellation Coma Berenices. (NASA Gen Mgmt Review Rpt, 18 Aug 75, 9; Alexander, *LA Times*, 26 Aug 75)

*19 August*: The Soviet Union might attempt to land an automatic microbiological laboratory on Mars before the U.S. Viking spacecraft, whose twice-delayed launch is now scheduled for 20 Aug., could arrive, the *Christian Science Monitor* reported Finnish astronomer Heikka Oja as telling the British Interplanetary Society. By launching a probe between 17 Sept. and 3 Oct. and using higher speeds and shorter trajectories, the Russians could reach Mars in May 1976, 2 mo before the scheduled Viking arrival. Between 1971 and 1973 the Soviets had made 4 attempts to land probes on Mars; only *Mars 3* (launched 28 May 1971) succeeded in softlanding a capsule (on 2 Dec. 1971). Landing in a dust storm, it ceased transmitting after 20 sec. Oja said the Soviets hoped to repeat this experiment, this time in a region near the south polar cap of Mars where the thin, mostly carbon dioxide atmosphere might contain moisture. (Gatland, *CSM*, 19 Aug 75)

- Marshall Space Flight Center announced the award of a $1.9-million firm-fixed-price contract to the Martin Marietta Corp. for fabrication, acceptance testing, and delivery of 322 pyrotechnic-initiator controllers (PIC) for the Space Shuttle solid rocket boosters. (MSFC Release 75–182)

*20 August*: *Viking 1* was successfully launched at 5:22 pm EDT from Eastern Test Range's Launch Pad 41 on a Titan III-Centaur launch vehicle. The launch, first scheduled for 11 Aug., had been postponed until 14 Aug. to replace a faulty thrust-vector control valve on the launch vehicle [see 11 Aug.]. A second delay had occurred when the spacecraft's internal batteries had discharged to 9 volts, well below safe limits [see 13 Aug.]. The spacecraft, Viking A, had been removed from the launch vehicle and replaced by Viking B. The range safety officer had issued a waiver of standard procedures and permitted removal and installation of the spacecraft without detanking and depressurizing the launch vehicle, thus allowing the launch to be rescheduled for 20 Aug.

*Viking 1*, consisting of a 2360-kg orbiter attached to an 1180-kg lander, was placed in a trajectory to Mars within the designed 3-sigma limits. Nearly 35 min after trajectory insertion, the Centaur stage separated from the spacecraft. The solar panels were deployed and oriented toward the sun, and the biocap—a container used to hermetically seal the lander while in earth's atmosphere—was jettisoned. On

27 Aug. at 2:30 pm EDT a trajectory-correction maneuver targeted the spacecraft to its Mars orbital insertion point. By 28 Aug. all systems aboard the spacecraft were operating normally and in a cruise mode. Travel time to Mars was estimated at 10 mo, with a possible Mars landing on 4 July 1976, the height of the U.S. Bicentennial. A second spacecraft, Viking B, had been scheduled for a 1 Sept. launch to arrive at Mars in 12 mos.

The Viking missions supported NASA's planetary program goals of exploring the solar system and gaining knowledge of its origin and evolution, and of the origin and evolution of life and of the dynamic processes that shaped man's terrestrial environment. Specific purpose of the two Viking missions was to increase knowledge of Mars by making observations from Mars orbit and by direct measurements in the atmosphere and on the surface, with particular emphasis on biological, chemical, and environmental data on past or present existence of life on the planet.

*Viking 1* was targeted to land in the Martian equatorial region, in a very low area resembling the fluvial plain of a dry river bed—an ideal site to search for organic material. Alternate sites for both Viking landers had been selected and there would be some flexibility until insertion into Mars orbit, after which the available band of landing sites would be narrowed by orbital geometry. Once in orbit, the orbiter would survey prospective landing sites; when the decision on where to land had been made, retrorockets would separate the lander from the orbiter and the lander would descend to the surface, sequentially braked by its aeroshell's drag, by a parachute, and finally by its retrorockets. After landing, the science instrumentation would be activated and exploration would begin; data would be transmitted back to earth by the lander radio or through a radio-relay link with the orbiter.

The orbiter's main body, an octagonal ring 46 cm high with alternating 140- and 51-cm sides, consisted of 16 modular compartments. The entire structure was 10 m across the tips of the extended solar panels and 3.3 m high from the lander attachment points to the launch vehicle. Besides propulsion, navigation and control, communications, power, computer, and data-storage subsystems, the orbiter carried its science instrumentation mounted on a scan platform that could move with 2 degrees of freedom. The three science instruments were twin high-resolution, slow-scan TV cameras; an infrared atmospheric water detector; and a high-resolution infrared thermal mapper for detecting surface-temperature variations.

The lander was a six-sided aluminum and titanium box 46.2 cm deep and enclosed top and bottom by cover plates; it measured 3 m wide and 2 m high from the footpads to the top of the S-band antenna. The six sides measured 109 cm and 56 cm alternately, with the three landing legs attached to the three narrower sides. The three retrorockets were mounted at 120° intervals on the lander; generating up to 2838 newtons (638 lbs) thrust, they would be ignited at an altitude of 1220 m to slow the lander's descent from 222 kph to 9 kph at landing. Four small engines provided attitude control.

In addition to subsystems for thermal control, power, and communications, the lander carried six instruments: twin scanning TV cameras, a sophisticated biology laboratory, a gas chromatograph mass spectrometer, an x-ray fluorescence spectrometer, a meteorology instrument, and a seismometer. Used with numerous temperature, pressure, and magnetic sensors, the instruments would perform entry, landed, and radio experiments.

NASA's investigation of Mars had begun in 1964 with the launch of *Mariner 4*. That spacecraft's flyby on 15 July 1965 had revealed the planet to be moonlike with a dry, barren, cratered surface showing little evidence of wind or water erosion and no magnetic field. The better instrumented *Mariners 6* and *7* that flew by the planet during the summer of 1969 reported a chaotic view of jumbled ridges and valleys unlike anything found on the earth or moon. The spacecraft also showed wide featureless expanses where craters had been somehow eroded, suggesting that Mars had been geologically active. *Mariner 6* and *7* photographs showed a thin layer of snow—probably carbon dioxide—and instruments measured a thin atmosphere of mostly carbon dioxide.

High-quality photographs from *Mariner 9*, placed in Martian orbit on 13 Nov. 1971, provided evidence of fluid erosion, glacial action, and volcanoes. The Viking missions had continued the Mariner explorations of Mars.

Viking was managed by Langley Research Center under overall direction of NASA's Office of Space Science; LaRC also had management responsibility for the lander system designed and built by Martin Marietta Corp. Jet Propulsion Laboratory designed and built the orbiter system. The Titan-Centaur launch vehicle was the responsibility of Lewis Research Center; LaRC managed launch and flight operations executed by Martin Marietta, Kennedy Space Center, and JPL. Goddard Space Flight Center was responsible for tracking and data systems, and JPL managed the mission control and computing-center system. (NASA MORs S−815−75−01/02, 1 Aug 75, 28 Aug 75, 16 Sept 75; NASA Release 75−183; LaRC Launch and Mission Operations Status Bulletins Nos 9−11; *Viking: Mission to Mars* (NASA SP−334); *W Post*, 18−22 Aug 75; *NYT*, 18−22 Aug 75)

* Doctors had rechecked x-rays of Apollo-Soyuz Test Project astronaut Donald K. Slayton's lungs taken before the 15 July launch and found that they also had shown a shadow revealed by postflight x-rays, Dr. Arnauld E. Nicogossian, NASA's ASTP physician, said at a Johnson Space Center press briefing. Dr. Nicogossian said that the spot had not been noticed before because the x-rays were difficult to read and the spot had been covered by normal vasculature and bony structure. Only a 6 Aug. tomogram [see 25 Aug.] had verified that the shadow was a lesion. Dr. Nicogossian said the nitrogen tetroxide inhaled by the ASTP astronauts during reentry [see 15−26 July] had not been shown to have caused or increased the size of the lesion.

Dr. Nicogossian added that, if cancer were diagnosed during the surgery, the usual medical procedure was to remove a part of the lobe, but he did not think that would prevent Slayton from flying in the future.

Slayton said that he felt "pretty damn lucky" that the lesion had been discovered when it was. "This thing could have been discovered before the flight. . . . And I could have easily gotten jerked off the flight, so that would have been bad. Secondly, if we hadn't had the gas in the cockpit on reentry, . . . [the doctors] could very easily not have picked this thing up until my next annual physical which is 6 or 8 months off." Then it would have been a "lot tougher" to fix.

Slayton, one of the seven original U.S. astronauts, had been grounded in 1962 because of an erratic heart rate, first detected in 1959, but had been returned to flight status March 1972 in time for assignment to ASTP. He told the briefing that he hoped to be out of the hospital within 7 days and back to work in 2 wk, and on flight status again within a month. (Transcript, ASTP PC−63, 20 Aug 75)

- NASA announced that the U.S. tour by the five Apollo-Soyuz Test Project crewmembers had been postponed to 13 Oct. because of astronaut Donald K. Slayton's scheduled lung surgery. The crew's tour of the U.S.S.R., scheduled to begin 22 Sept., would continue on schedule [see 15 Aug.]. (NASA Release 75−236)
- Lt. Col. Michael Love successfully completed the 28th flight of the X−24B lifting body, making its second and final runway landing [see 5 Aug.] at Flight Research Center. The purpose of the flight was to land on a concrete runway, survey body pressure, study a lefthand fin tuft, check out the thermal protection system, and perform stability and control maneuvers with the rudder bias at 5° toe-out. Launched from a B−52 aircraft at 13 700 m, Col. Love ignited the vehicle's rocket engine and the X−24B reached a speed of mach 1.53 and an altitude of 21 900 m before engine shutdown and unpowered glide and landing.

  Flights of the 11-m delta-shaped vehicle were part of a joint NASA-Air Force program to study transonic flight characteristics of an aerodynamic shape that could be the forerunner of future hypersonic cruise vehicles. (X−24B flt rpt, NASA Prog Off; FRC Release 25−75)
- The Large Area Crop Inventory Experiment (LACIE), a low-budget cooperative NASA−Department of Agriculture−National Oceanic and Atmospheric Administration program designed to assess U.S. wheat-crop yields and to forecast production, could develop into a worldwide crop program, J. F. ter Horst reported in the *Los Angeles Times*. If elevated in status and budget priority by President Ford, LACIE could accurately measure a worldwide grain feast or famine each year, identifying regions of big harvests and potential shortages, vital in allocating world food supplies to meet the needs of growing world population. Ter Horst quoted an advisor of President Ford as saying that, if the LACIE system were in worldwide operation, "we would know what the Russian grain crop was likely to be, what they would be buying, and just how much of our crop we could afford to sell them."

  Ter Horst added that intelligence sources indicated the Russians had already measured U.S. crops by satellite. "If true, that could explain the confidence of Russian grain-purchasing moves this year. It also indicates an accelerated U.S. political need to play catch-up in this phase of the space race." (ter Horst, *LA Times*, 20 Aug 75)
- A *Today* editorial had commented on the planned Viking flight to Mars: "Success of the Viking flight could mean additional Congressional support for more sophisticated flights in the future. . . . It will be almost another year before we will have the answers to the Viking experiments. . . . At that time, we will be celebrating the 200th birthday of our country—and what better way to celebrate the birth of one nation than with the discovery of another?" (*Today*, 20 Aug 75)

*21 August*: The Titan-Centaur launch vehicle for the Viking—B had been moved to the pad at Launch Complex 41 just 24 hr after the successful launch of *Viking 1*. Technicians would mate the spacecraft to the launch vehicle on 27 Aug., preparing for the scheduled launch on 1 Sept. (KSC Release 174—75)

- The Space Shuttle was, in many ways, the "ultimate recycling program," Robert Anderson, president and chief executive officer of Rockwell International Corp. said on Transportation Day at the Canadian National Exhibition in Toronto. Anderson said that the Shuttle program, with its international scope and emphasis on reusability, was the first major step toward economical and effective use of space.

  More advanced Shuttle-launched satellites would benefit the world by aiding crop control; locating new sources of minerals and fossil fuels; monitoring weather, pollution, and localized disasters such as oil slicks and forest fires; mapping oceans and urban areas; and improving communications. "For the job of transportation is not just to move people and goods from point to point—it is to also move history forward by enabling the pioneers to explore the new frontiers and by enabling society to capitalize on their discoveries." (Text; Rockwell Intl Release R—42)

- NASA announced appointment of Herbert J. Rowe, chairman of the board of PEMCOR, Inc., as Associate Administrator for External Affairs. Rowe would coordinate activities of the Office of Industry Affairs and Technology Utilization and the Office of Public Affairs, the offices that oversee NASA interactions with external organizations and individuals and disseminate information about NASA programs. (NASA anno, 26 Aug 75)

- NASA announced retirement from military service of NASA *Skylab 4* (16 Nov. 1973—8 Feb. 1974) astronauts Gerald P. Carr and William R. Pogue effective September 1. Carr, a Marine Corps colonel, would remain with NASA as a civilian astronaut. Pogue, an Air Force colonel and an astronaut since April 1966, would leave NASA to become a vice president of High Flight, an interdenominational evangelistic foundation founded by former astronaut James R. Irwin. As crew members on *Skylab 4* Pogue and Carr, along with Dr. Edward G. Gibson, share the world record for individual time in space of 2017 hr 15 min 32 sec. Pogue's departure would reduce the number of NASA astronauts to 31. (NASA Release 75—233)

- The Knights of Columbus, a U.S. Catholic fraternal organization, had agreed to pay for worldwide satellite coverage of three live 90-min Vatican events, the *Washington Post* reported. Four Intelsat satellites would telecast the midnight mass at Christmas, the Stations of the Cross from the Roman Colosseum on Good Friday, and the Pope's Easter Sunday message. Each broadcast would cost an estimated $25 000. (*W Post*, 21 Aug 75)

*22 August:* Defense, space, and energy accounted for most of a record increase in Federal research and development funding in FY 1976, the National Science Foundation's report *An Analysis of Federal R & D Funding by Function, Fiscal Years 1969—76* stated. The largest dollar increase, $1860 million, was for national defense; the second and third largest increases, $343 million each, were for space and energy.

Relative increases in FY 1976 were largest for education, 102%; energy, 37%; national defense, 20%; and space, 13%.

Total Federal obligation for R&D in the 1976 budget of $21.7 billion was a record high. The $2.7-billion increase over FY 1975 was also a record increase for any one year, enough to indicate a real rise in R&D activity, allowing for a reasonable inflation factor. The upward change contrasted with an average annual decline of 2.7% in constant dollars over the period 1969—75. (NSF *Highlights*, 25 Aug 75, 1)

- NASA announced selection of McDonnell Douglas Astronautics Co. for negotiation of a $14.8-million fixed-price incentive contract to develop, build, and deliver Space Shuttle solid rocket booster (SRB) structures, and to design and build the tooling necessary to produce them. The SRB structures would support the Space Shuttle on the launch pad, transfer thrust loading to the Orbiter and external tank, and provide structural support for the SRB recovery system, electrical components, and thrust-vector control system. (NASA Release 75—238)

- A group of 28 scientists, engineers, sociologists, and economists concluded a 10-wk (16 June—22 Aug.) NASA—Stanford University study at Ames Research Center by recommending that the U.S. adopt a space colonization program using available technology. The "city in space" envisioned by the study group could be a 1.5-km-wide wheel-shaped habitat for 10 000 persons positioned on the moon's orbit at a point 385 000 km from both earth and moon. Costing an estimated $100 billion, the 454-million-kg wheel, or torus, would rotate around its hub at 1 rpm to simulate earth's gravity. The rim would house inhabitants as well as shops, schools, light industry, and closed-loop agriculture; heavy industry could be located outside to take advantage of weightlessness and high vacuum in space.

  A major commercial activity of the first colony would be to construct solar-power satellites. Placed in geosynchronous orbit above the earth, the satellites would collect and convert sunlight into energy and beam it to earth as low-density microwaves.

  The space colony would have several advantages that might make it self-supporting: weightlessness for manufacturing and transportation, massive use of lunar minerals, and continuous natural sunlight for increased agricultural productivity. The study group had considered social, cultural, safety, and ecological difficulties of a space colony, but had found "no unsurmountable problems that would prevent humans from living in space." (NASA Releases 75—229, 75—249; ARC Release 75—41; Dunstan, *W Post*, 23 Aug 75)

- The press commented on the mission of the Viking spacecraft to Mars [see 20 Aug.]. The *Washington Post* said that the essence of the space program had been "to provide mankind with new knowledge, not in hopes that this knowledge will be useful immediately here on Earth but in hopes that it will expand our understanding of the universe in which we live and . . . enable us to reach better solutions to our philosophical and political, as well as practical, problems." The Viking missions had opened a new era in which the search for knowledge was to be done largely by machines; the rewards promised to be rich, beyond measure in strictly monetary terms.

  The *Christian Science Monitor* agreed, commenting that Viking represented "an opportunity for mankind's self-awareness to take a greater stride away from earth-centered thinking than was afforded by Neil Armstrong's historic step on to the moon." Discovering organic life on Mars would strengthen the conviction that life existed abundantly throughout the universe and that we were not alone. Some had questioned whether the U.S. should spend a billion dollars on such a program when resources were hard pressed on earth, but a new discovery, that life was not a meaningless chance but part of a grand design, could give new inspiration to humanity's efforts to deal with earthly problems.

Here was a challenge for the architects of détente, the *Monitor* asserted. The U.S. and Russia had been needlessly duplicating scientific efforts, their cooperation amounting to little more than "arm's-length information exchange." The two countries had much to share. "By abandoning costly competition for a truly joint program, the two countries could gain from each other's capabilities, minimize the cost to each of them, and pursue this cosmic outreach on behalf of all mankind." (*W Post*, 22 Aug 75, A24; *CSM*, 22 Aug 75, 28)

*23 August:* An editorial in the *Washington Star* commented on the Viking mission to Mars: "Perhaps the most miraculous thing about the Viking spacecraft . . . is that the American people, pinched as they are by hard times, would more or less cheerfully dispatch a billion dollars on a one-way . . . journey into the trackless void." Still, enough might have been learned in the making of this remarkable vehicle to "constitute fair recompense for all the money it represents," and whatever it discovered at journey's end might be worth twice the cost. (*W Star*, 23 Aug 75)

*24 August:* A planet's size and bulk might be more important than its location in the solar system in determining whether it can support life, Dr. Robert Jastrow, Director of NASA's Goddard Institute for Space Studies, said in a *New York Times* article. Conventional scientific thought had always been that the chance of life's evolving on a planet was narrowly restricted by the planet's distance from its parent star. However, planetology—the study of comparative geology of the planets, made possible by space probes such as Pioneer, Mariner, and Viking—provided evidence for the newer theory.

The most important single factor in the geology of earth-like planets was the amount of internal heat left over from their birth. Planets as large as earth conserved heat, losing it very slowly through volcanic action over long periods of time. Molten volcanic material carried gases, including water vapor, that had been trapped below the planet's surface; the vapors condensed into oceans, other gases formed an atmosphere, and life began. Mariner's photos of Mars had confirmed the existence of volcanoes; but, because of the planet's relatively small size, these exhausted their internal heat supply and died out rapidly about 100 million yr ago. During its active period, Mars must have had in its atmosphere large volumes of volcanic gases, including water vapor; Mariner photos had shown what were apparently dry river beds. When the volcanoes became extinct, the water and atmosphere leaked away into space and the small planet was left dry and nearly airless. Any life forms found on Mars by Viking or future space probes would probably be fossils.

The study of lunar material added another example to the comparative geology of earth-like planets: The moon rocks brought back by

Apollo astronauts had revealed volcanic activity on the moon; because the moon was smaller than Mars, this activity had stopped even earlier and the gases and moisture escaped even more quickly, leaving another dry, airless, and lifeless body.

Scientists concluded that Mars was cold and lifeless not because it was farther from the sun than earth but because it was smaller and geologically inert. If the earth were moved out to the orbit of Mars, the average temperature of the earth would drop somewhat but its insulating atmosphere, continually replenished by volcanic gases for some billions of years longer, would maintain a livable temperature over large areas of the surface. (Jastrow, NYT, 24 Aug 75, E7)

- Soviet ground controllers altered the trajectory of *Luna 22*, the unmanned lunar probe launched 29 May 1974, bringing it closer to the moon's surface. Resulting parameters were 1286-km apogee, 100-km perigee, 3-hr period, and 21° inclination. Tass reported 3 Sept. that all onboard systems were operating well and that the spacecraft was continuing to transmit to earth information about the moon's surface. (Tass, FBIS—Sov, 3 Sept 75, U1)

*25 August:* Apollo-Soyuz Test Project astronaut Donald K. Slayton entered Texas Medical Center to undergo exploratory surgery for a small lesion on his left lung. NASA physicians had discovered the lesion in x-rays taken during the astronaut's postflight recovery from gas inhalation [see 25 July]. Slayton's x-rays had shown a complete clearing of lung infiltrates but revealed a 4-mm discrete shadow. Specialized x-rays, called tomograms, taken 6 Aug. had confirmed that the shadow was indeed a lesion and not a part of the normal lung structure. NASA physicians, conferring with chest specialists at the Texas Medical Center, recommended surgery; all the doctors agreed that the lesion had not resulted from the gas inhalation. (JSC Release 75—69)

- The People's Republic of China had confirmed that its three satellites—*China 1*, launched 24 April 1970; *China 2*, launched 3 March 1971; and *China 3*, launched 26 July 1975—had earth-observation capability, *Defense Space Business Daily* reported. Although the PRC's *People's Daily* had published cloud photos taken by one of the satellites, and the implication was that the satellites were forerunners of meteorological satellites, their performance suggested the development of higher resolution observation-reconnaissance satellites. All three satellites had been placed into orbits with perigees at reconnaissance altitudes sufficient to cover all targets of interest to the PRC. (*SBD*, 25 Aug 75, 290)

- NASA's use of the firefly's two light-producing chemicals, luciferin and luciferase, to test for the chemical presence of adenosine triphosphate (ATP), an energy-storage compound present in every living cell, had increased interest in luminescence as a tool for medical research, the *New York Times* reported. Research done in 1948 by Johns Hopkins Univ. scientists had demonstrated that ATP was the third essential ingredient—along with luciferin and luciferase—necessary for the firefly's glow, and that the amount of light generated was directly proportionate to the level of ATP present. The Hopkins scientists had also developed a simple test for life itself, since luciferin and luciferase added to any living matter could reproduce the firefly's glimmer. NASA applied this information, with more sophisticated light-detection instruments, to develop a test for life on Mars for use aboard the Viking spacecraft (launched 20 Aug.)

NASA scientists working with the firefly chemicals at Goddard Space Flight Center had developed many applications, including a method of detecting bacteria in water; this permitted speedy diagnosis of urinary infections, as well as testing the effect of various antibiotics on particular infections.

Univ. of Calif. scientists continued the research, adapting the techniques for related work in measuring creatine phosphokinase (CPK) in the bloodstream. CPK, present in all human blood, was produced in abnormally high quantities during muscle-cell degeneration that accompanied cardiac arrest or muscular dystrophy. CPK could be treated to produce ATP and, with the application of firefly chemicals, to produce a glow proportionate to the amount of CPK in the bloodstream. Researchers had used the measurements as a quick blood test to tell whether a patient had suffered a heart attack. (NYT, 25 Aug 75)

*26 August—3 September:* NASA launched *Symphonie 2*, France and West Germany's second experimental communications satellite at 9:42 pm EDT 26 Aug. from Eastern Test Range. A three-stage thrust-augmented Thor-Delta launch vehicle boosted the spacecraft into a synchronous transfer orbit with a 37 974-km apogee, 413-km perigee, 678.3-min period, and 13.2° inclination. On 29 Aug. at 11:48 am EDT, ground controllers activated the onboard liquid-fueled apogee motor to circularize the orbit at geosynchronous altitude. By 31 Aug. *Symphonie 2* was in an orbit with a 35 870-km apogee, 35 364-km perigee, 23-hr 47-min period, and 0.0° inclination. NASA's primary objective for the mission—to launch the *Symphonie 2* into a synchronous transfer orbit with sufficient accuracy to enable the satellite to accomplish its operational mission—had been successfully completed, and the mission was adjudged successful on 3 Sept.

When finally positioned at 11.5° west longitude over the equator, the 402-kg *Symphonie 2* would provide 1200 telephone, 8 voice, and 2 TV channels for experimental communications between Europe and the African and South American continents.

Second of two experimental comsats developed by the French-West German Consortium Industriel France-Allemand pour le Satellite Symphonie (CIFAS), under the direction of Germany's Gesellschaft für Weltraumforschung GFW and France's Centre National d'Etudes Spatiales (CNES), *Symphonie 2* would expand the French-West German experimental satellite TV and telegraph communications program begun with the launch of *Symphonie 1* on 18 Dec. 1974.

In October 1973 NASA had agreed to provide the launch vehicle and services on a cost-reimbursable basis, and in June 1974 had signed a launch services contract with the consortium. Estimated cost of hardware and services was $12 million. Goddard Space Flight Center, under the direction of the Office of Space Science, was responsible for the launch vehicle and for limited tracking during initial activities. When the satellite separated from the vehicle third stage, the Symphonie Project Operations Group in France and West Germany would assume operational responsibility. (NASA MORs S−492−204−75−02, 25 July, 3 Sept; GSFC *SSR*, 31 Aug 75; GSFC *Wkly SSR*, 21−27 Aug 75; NASA Release 75−234)

- A prototype of the YC—15 advanced medium short-takeoff-and-landing (AMST) cargo transport successfully completed its first flight. Airborne for 2 hr 26 min, the YC—15 was flown from Long Beach municipal airport to Edwards Air Force Base, Calif., achieving a peak altitude of 5500 m and a top speed of 485 kph. During the flight the pilots evaluated the aircraft's flight controls, handling and slow-speed-flight qualities, and speed advance to the assigned maximum.

    Built by McDonnell Douglas Corp. for the Air Force, the YC—15 was the first large transport aircraft to include both the NASA-developed supercritical wing for improved flight performance and reduced fuel consumption, and externally blown flaps for powered lift. This aircraft and a second nearing completion would be evaluated, during a 12-mo flight-test program, against established Air Force performance goals and against two YC—14 prototype AMSTs built by Boeing Co. (Fink, *Av Wk*, 11 Aug 75, 18—20; *NYT*, 27 Aug 75, 58; AFSC *Newsreview*, 16 Oct 75, 16)

*27 August*: Astronomers had picked up microwave signals that dated back 10 billion yrs to the creation of the universe, Sir Bernard Lovell, director of the Jodrell Bank radio telescope station, told the 137th annual meeting of the British Association for the Advancement of Science. First picked up accidentally by equipment testing space communications, and later monitored by a sounding rocket, the signals apparently had originated in the cataclysmic explosion—the "big bang"—caused when a single primeval fireball exploded to form the universe. The observed radiation was "a relic of the high-temperature phase of the universe," Lovell stated, "perhaps within a second or so of the beginning of the explosion."

Lovell urged scientists to reexamine their responsibility to society during their quest for knowledge of the universe and life on other planets; he questioned whether man could "survive for long the consequences of the probing of scientists," saying that the search might produce answers too overwhelming for the mind of man to comprehend. He warned that extensive military involvement might lead to great human disaster, recalling how man's quest for knowledge had led to the development of nuclear weapons. (AP, *W Star*, 28 Aug 75)

- NASA announced that astronaut Thomas P. Stafford would leave the NASA astronaut corps effective 1 Nov. to become commander of the Air Force Flight Test Center at Edwards Air Force Base, Calif. Serving as a NASA astronaut since September 1962, Stafford made the first rendezvous in space when he flew the *Gemini 6* mission (launched 15 Dec. 1965) to meet the already orbiting *Gemini 7* crew. He also commanded *Gemini 9* (3—6 June 1966), which rendezvoused with the previously launched augmented target docking adapter. Stafford was commander of *Apollo 10* (18—26 May 1969), first lunar-orbital mission to use the complete Apollo spacecraft; during the mission he and crew member Eugene A. Cernan flew the lunar module to within 15 km of the surface while John W. Young orbited the moon in the Apollo spacecraft. In July Stafford headed the three-man Apollo crew for the joint U.S.-U.S.S.R. Apollo-Soyuz Test Project mission (15—24 July 1975); he had accumulated a total of 507 hr 43 min in space and flown 6 space missions. (NASA Release 75—241)

*27 August—7 September*: Field operations of an ocean bathymetry expedition sponsored jointly by NASA and the Cousteau Society had been successfully completed in the Central Bahamas. Objective of the expedition was to evaluate the usefulness of Landsat satellite sensors for measuring water depth in shallow seas and for improving the accuracy of mapping ocean-bottom features. Thirteen satellites—including NASA's *Landsat 1* and *2*, *Sms 1*, and *Ats 3*; National Oceanic and Atmospheric Administration's *Noaa 3* and *4* and *Essa 8*; and six satellites of the Navy's Transit Navigation System, plus two research vessels—Cousteau Society's *Calypso* and the Johns Hopkins University's *Beayondan*—recorded bathymetric data at selected sites. Scuba divers measured ocean floor reflectivity and water transparency with sophisticated underwater instruments.

Participating in the expedition were Cousteau Society head Jacques Cousteau and his son Philippe; NASA Project Manager Dr. Enrico P. Mercanti and science monitor Ross McCluney, both of Goddard Space Flight Center; and Dr. Fabian Polcyn of the Environmental Research Institute of Michigan. NASA Director of User Affairs Russell L. Schweikart coordinated the project. Also participating was President Gerald R. Ford's son Jack, who accompanied the expedition for the first phase, assisting with several underwater and onboard experiments. (NASA Releases 75–240, 75–257)

*28 August*: The Air Force announced successful test flight of an F–111D equipped with a turbofan jet engine having all-composite third-stage fan blades 40% lighter than conventional titanium blades. This was the first military operational evaluation of a rotating structural engine component made of composite materials. (AFSC Release OIP 214.75)

- A red-tide detection program had been approved by Goddard Space Flight Center and the Florida Department of Natural Resources (FDNR). Caused by an ocean-borne phytoplankton, *Gymnodinium breve*, "red tide" had left thousands of dead fish rotting on beaches in coastal estuaries. Ocean-color scanners—mounted on *Landsat 1* and *2* (launched 23 July 1972 and 22 Jan. 1975) and on NASA's U-2 research aircraft—could detect subtle color variations in coastal waters, to warn of changes in concentration and species of marine phytoplankton populations and indicate possible red-tide invasions. Research vessels would verify the findings by on-site sampling. Program managers hoped the data would lead to the development of an early warning system that would allow crews to combat effects of the red tide. (NASA Release 75–242)

*29 August*: Successful lung surgery performed 26 Aug. on Apollo-Soyuz Test Project Astronaut Donald K. Slayton revealed no evidence of malignancy, the Johnson Space Center *Roundup* reported. During the 2.5-hr surgery, doctors at the Texas Medical Center removed a triangular wedge that included the 4-mm nodule detected during postflight x-rays [see 20 Aug.] plus a small amount of normal surrounding tissue. Adjacent lymph nodes were biopsied and found normal; doctors found no signs of any other lesions. Slayton would remain in the hospital for 7 to 10 more days. (JSC *Roundup*, 29 Aug. 75, 1)

- A backup Skylab Orbital Workshop and an airlock module and multiple docking adapter would be shipped by barge to Washington, D.C., for display at the Smithsonian Institution's National Air and Space Museum. Stripped of all equipment that could be used for future programs, the workshop would be cut into three sections for transport;

two doors would be added and the floor strengthened to accommodate the flow of tourists. The museum was scheduled to open 4 July 1976. (MSFC Release 75—189)

- Rockwell International Corp., prime contractor for the Space Shuttle, announced selection of Consolidated Controls for a $1.5-million contract to provide high- and low-pressure helium valves for the Shuttle Orbiter reaction-control system (RCS). The valves would control helium output in the RCS propellant tanks, serving as a manifold shutoff valve in the vernier engine and as a low-pressure helium shutoff valve. (Rockwell Release SP—29)

*31 August:* Communications Satellite Corp., U.S. representative in the International Telecommunications Satellite Consortium, transmitted its twelfth annual report to the President, covering the activities of ComSat from 1 July 1974—30 June 1975.

Two more Intelsat IV satellites had been orbited, *Intelsat IV F—8* 21 Nov. 1974 over the Pacific Ocean and *Intelsat IV F—1* 22 May 1975 over the Indian Ocean. A third satellite, Intelsat IV F—6, failed to achieve orbit when the launch vehicle malfunctioned.

By 30 June 1975 a network of 112 antennas—27 more than last year—at 88 earth stations was providing 379 ComSat pathways, with 107 countries leasing satellite services, an increase of 7 over the previous year.

Comsat General Corp., a subsidiary, had entered into a joint venture with the European Space Agency and the Government of Canada on 3 Dec. 1974 to provide satellite capacity for an Aerosat (aeronautical satellite) communications test and evaluation program. Comsat General and International Business Machines Corp. had notified the Federal Communications Commission that they would seek to enter the domestic satellite business with one or more partners. ComSat's earth stations in Connecticut and California neared readiness for Marisat (maritime satellite) communications services in the Atlantic and Pacific Ocean areas.

ComSat reported a net income of $44 918 000, or $4.49 per share, up from $36 299 000 or $3.63 per share in 1973. (Text)

*During August:* "The importance of space to our defensive military effort can only increase in the future," Gen. Samuel C. Phillips, retiring commander of the Air Force Systems Command and NASA's Apollo Program Director from 1964—69, said in an interview. Gen. Phillips speculated that aircraft, including the B—1 bomber under development by the U.S., would be an important part of the military force "for as many decades ahead as anyone cares to project." However, he saw energy scarcities increasing the need for training simulators and alternative aircraft fuels. (AFSC *Newsreview*, Aug 75, 1)

## September 1975

*1 September:* NASA's Small Astronomy Satellite (*Sas 2*, launched as *Explorer 48* 16 Nov. 1972) had observed a pulsar in the Vela constellation that generated two bursts of gamma rays for each burst of radio waves, the *Astrophysical Journal* reported. No other pulsar had been found to exhibit such properties. The new observation indicated that pulsars—small, dense, rapidly spinning stars thought to be remains of a huge stellar explosion or supernova—were much more complex than had been believed.

Detection of gamma rays from Vela meant that pulsars were a likely source of cosmic rays. Many astrophysicists had believed that only very young pulsars could produce cosmic energies, but the *Sas 2* observations cast doubt on this conclusion: Scientists believed that the Vela pulsar was more than 10 000 yrs old. (D.J. Thompson *et al.*, *Astrophysical Journal*, 1 Sept 75; GSFC Release G–75–20)

- Appointment of Lt. Gen. William J. Evans as commander of the Air Force Systems Command became effective. Gen. Evans, who had been deputy chief of staff for research and development at Air Force Hq, was replacing retiring Gen. Samuel C. Phillips. AFSC had been responsible for developing Air Force aerospace technology. (AFSC *Newsreview*, Aug. 75)

*2 September:* Marshall Space Flight Center announced the award of a $4 409 000 cost-plus-incentive-fee contract to Bendix Corp. for design, development, test and evaluation, and fabrication of integrated electronic assemblies (IEA) for Space Shuttle solid rocket boosters on the first six Shuttle developmental flights. Each booster would require two IEAs: The forward IEA would initiate release of the nose cap and frustum, jettison the solid-rocket-motor nozzle, detach the parachute, and turn on recovery aids. The aft IEA would interface with the Orbiter, the forward IEA, and other avionic systems. Delivery of the IEAs would begin in 1976 and continue through 1 Apr. 1979. (MSFC Release 75–190)

*3 September:* Jet Propulsion Laboratory scientists had released "Automobile Power Systems Evaluation Study," urging adoption of a billion-dollar program to mass-produce, by 1985, a fuel-saving pollution-free replacement for the standard internal combustion engine, Richard Witkin reported in the *New York Times*. Funded by a $500 000 grant from the Ford Motor Co. with the understanding that the study would be independent and unbiased, JPL's report recommended accelerated, parallel development of the Brayton gas turbine and Stirling external combustion engine until a clear choice could be made between them. The study found that either engine not only was virtually pollution-free and cut fuel consumption by 30 to 45% but also could be mass-produced at a cost differential small enough to be easily recovered through fuel savings by the first owner.

The JPL engineers defined the fuel economy, emissions, and costs of vehicles powered by nine other engines as well as by improved versions of the conventional internal combustion engine. Six sizes of cars were analyzed, covering the range of consumer preferences. Electric and hybrid power systems were also evaluated.

Aside from development of the Brayton and Stirling engines, the report recommended other programs to reduce fuel consumption 15 to 35% by improved vehicle packaging, weight reductions, and transmission improvements. The report also recommended that the U.S. work in the interim to improve the fuel economy and emissions level of conventional car engines. (JPL LabOratory, 1975/5, 21; Witkin, *NYT*, 4 Sept 75, 18)

- Marshall Space Flight Center was requesting proposals—due 22 Sept.—from the aerospace industry for a study leading to design and development of an Atmospheric Cloud Physics Laboratory (ACPL) as a payload aboard Spacelab, MSFC announced. ACPL experiments would give scientists a knowledge of weather processes important for long-range weather prediction and weather modification. After evaluating the proposals, MSFC would award two parallel 12-mo firm-fixed price contracts. (MSFC Release 75−192)

- Dale D. Myers, president of Rockwell International Corp.'s North American Aircraft Operations, announced appointment of Bastian Hello as president of Rockwell's B−1 Bomber Div. Hello, who had been vice president of the division since 1973, was succeeding Robert E. Greer, who was retiring. Before joining the B−1 Div., Hello was vice president and general manager of Rockwell's Space Div.'s Apollo launch operations organization at Kennedy Space Center. (RI Release LA−2)

*4 September:* Former astronaut Alfred M. Worden, Chief of Ames Research Center's Application Div., announced that he would leave government service, effective 5 Sept. Worden was also retiring from the Air Force after 20 yrs. During the *Apollo 15* lunar landing mission (26 July−7 Aug. 1971), Worden was command module pilot and, with fellow crewmen David R. Scott and James B. Irwin, spent 12 days 7 hr 12 min in space.

Worden transferred to ARC in September 1972 as a research scientist in the Airborne Science Office. He was also Chief of the Systems Studies Div. before his appointment to the Application Div.

In his announcement Worden said he would spend one year lecturing on "Spaceship Earth" and writing a book commissioned by the High Flight Foundation, an interdenominational evangelistic organization based in Colo. (NASA Release 75−248)

*5 September:* A Viking Data Review Management Center had been established at Kennedy Space Center, the *Spaceport News* reported. The center would promptly disseminate data from the two Viking spacecraft to all concerned government agencies and contractors, and would aid in solving launch-related problems as quickly as possible.

During the *Viking 1* launch on 20 Aug., data and review teams of NASA and contractor personnel at KSC were responsible for rapidly analyzing incoming data for top management. Any anomaly that could affect the launch of Viking B had to be detected and resolved as quickly as possible. Center personnel had tracked some 20 problems from discovery to resolution. (KSC *Spaceport News*, 5 Sept 75, 2)

- Jet Propulsion Laboratory had proposed some economically sound and technically feasible alternatives to the automobile industry's inefficient and polluting internal combustion engine [see 3 Sept.], a *Los Angeles Times* editorial said. Development of a new economical and pollution-free engine on a mass-production basis would take time and money but "the auto industry can well afford the $150 million a year that would be needed." The JPL report voiced skepticism that industry could do the job alone. "Additional funding from the federal government is required, as well as a commitment that introduction of an alternative engine by 1985 is a national goal. JPL makes a persuasive case that all this is feasible and in the national interest. Given the potential in fuel savings and reduced air pollution, the money...would be a solid investment." (*LA Times*, 5 Sept 75)
- A group of 35 representatives from central Fla. planning agencies attended a 1-day workshop at Kennedy Space Center to assess the possibility of using satellite data for land-use planning. Edward J. Hecker of the KSC Office of Earth Resources told the group that the 4-cent-per-sq-mi cost of obtaining satellite data was one-third the cost of aircraft-obtained data. Data collected by "on foot" surveys cost $20 per sq mi, and the cost of using an automobile for this purpose ran about $2 per sq mi.

   Dr. Garland L. Thomas, coinvestigator for the earth-resources mission of *Landsat 2* (launched 22 Jan.) and employee of the Brevard County (Fla.) Planning and Zoning Dept., told the group that, using satellite data and photography, planners could determine which areas were developable and which were not, produce comprehensive land-use planning maps, monitor water quality, and trace the process of eutrophication in central Fla.'s many lakes. (KSC Release 181–75)

*8, 9, and 17 September:* Data collected by sounding rockets, satellites, instrumented balloons, and high-flying aircraft had supported the hypothesis that fluorocarbons such as those found in aerosol spray cans might be breaking down the earth's ozone shield, scientists told hearings of the Senate Committee on Aeronautical and Space Sciences' Subcommittee on the Upper Atmosphere. Ozone in the stratosphere screened the earth from harmful ultraviolet radiation that might increase incidence of cancer and other skin diseases and prove lethal to many life forms as well as damage eyes, crops, and livestock. An increase in atmospheric fluorocarbons could also have a greenhouse effect, permitting sunlight to pass freely but inhibiting the escape of heat in the form of infrared radiation. This would warm the lower atmosphere and the earth, altering aquatic and terrestrial ecosystems and environmental chemicals, and creating climate changes. An extreme greenhouse effect had been blamed for Venus' oven-hot temperatures.

Dr. Warren R. Muir, cochairman of the Interagency Task Force on Inadvertent Modification of the Stratosphere, told the subcommittee, "Based on our preliminary assessment there seems to be a legitimate cause for concern." If additional government research—with NASA coordinating—had proved more conclusively the danger to the environment, the government should take action for "prompt enactment of toxic substances legislation."

Dr. Wilson K. Talley of the Environmental Protection Agency agreed that fluorocarbons might pose a problem to the stratosphere, and that the total effect of releasing fluorocarbons and other halocarbon compounds into the atmosphere might not become apparent for 10 yrs. But, with evidence of public danger not strong enough to stand up to a court challenge by manufacturers, government regulation of the offending substances would be unenforceable. (Transcript; Sullivan, *NYT*, 10 Sept 75, 1)

*9 September: Viking 2* was successfully launched at 2:39 pm EDT from the Eastern Test Range on a Titan-Centaur launch vehicle. The spacecraft entered on a trajectory toward Mars accurate to within the 3-sigma tolerance. If flown without midcourse correction, *Viking 2* would reach Martian orbit 279 259 km from the target.

The launch had been rescheduled from 1 Sept., after the launch crew during a 28 Aug. precountdown test discovered a degradation in the detection threshold of the orbiter's S-band radio receiver. On 1 Sept., project officials decided to remove the spacecraft from the launch vehicle for additional evaluation. The launch was rescheduled for 9 Sept. after evaluation concluded that the anomaly had been caused by a faulty connector or conductor between the diplexer module and the high-gain antenna. Since the faulty part could not be isolated, the antenna and cabling were replaced, the orbiter and lander retested, encapsulated, and mated to the launch vehicle. After launch, all onboard systems and experiments were checked out. A trajectory correction maneuver on 19 Aug. targeted *Viking 2* for a rendezvous with Mars on 7 Aug. 1976.

*Viking 2* was the second of two identical planetary probes [see 20 Aug.], each consisting of an orbiter and a lander, launched toward Mars in 1975 for rendezvous and orbital and surface exploration in 1976. (NASA MOR S−815−75−01−02, 16 Sept 75; Viking Science Activities rpt no. 89; Mission Operations Status Bulletins 10−14)

- Japan's N booster launched, from Tanegashima launch site, its first payload—*Kiku*, an 83-kg spacecraft designed to measure launch vibrations and spacecraft temperatures in space. *Kiku* entered orbit with a 1104-km apogee, 977-km perigee, 106-min period, and 47° inclination. The launch vehicle consisted of a Rocketdyne MB−3 first-stage engine, modified from the U.S. Thor program, with three Thiokol Corp. Castor 2 solid-rocket strap-ons. The second stage, designated LE−3, had been designed by Japan's National Space Development Agency with Rocketdyne assistance. The third stage was a Thiokol TE−364−4 solid-rocket motor. (GSFC *Wkly SSR*, 4−10 Sept 75; *Av Wk*, 22 Sept 75, 50)

- A star tracker, STELLAR (Star Tracker for Economical Long Life Attitude Reference), was being developed by Jet Propulsion Laboratory for a variety of future manned and unmanned space missions. STELLAR used a solid-state silicon detector, called a charge-coupled device (CCD), for image sensing and a microcomputer to process and format the data. Previous star trackers had used high-voltage vacuum-tube image dissectors for image sensing; the tubes were difficult to produce, had a limited lifetime, developed errors, and could provide attitude-reference data from only a single star in any given observation frame. STELLAR, which substituted a single low-voltage 6-mm-sq CCD chip for the complex and expensive tube, offered the advantages of solid-state reliability and of simple programming to adapt to varying mission requirements. In addition to substantial savings in cost of flight hardware, the CCD imaging devices were expected to revolutionize TV camera design, permitting miniature low-cost home monitors within the next several years.

STELLAR, which had been developed under the management of NASA's Office of Aeronautics and Space Technology, would provide trackers for planetary Mariner spacecraft after the Mariner–Jupiter–Saturn mission in 1977 and for such Space Shuttle payloads as the proposed Shuttle infrared telescope facility. For the telescope facility, STELLAR would track as many as 10 stars simultaneously, providing position coordinates and star magnitudes for each. (NASA Release 75–251)

- The aft fuselage for Space Shuttle Orbiter 101 arrived at Rockwell International Corp.'s Palmdale, Calif., assembly facility after a 160-km truck journey from Rockwell's Downey plant. The aft fuselage joined the midfuselage, vertical tail, and wing panels delivered earlier by Rockwell subcontractors. The forward fuselage was due in Palmdale in October, with rollout of the first orbiter scheduled for the third quarter of 1976. Approach and landing tests would begin during the second quarter of 1977. (JSC *Roundup*, 12 Sept 75, 4)
- NASA had awarded a $3.358-million contract to RCA Corp.'s Astro-Electronics Div. to build a return-beam vidicon (RBV) two-camera TV system for the Landsat-C spacecraft. With twice the resolution offered by the cameras on *Landsat 1* and *2*—which viewed three identical ground scenes 160 km sq through separate spectral filters—the RBV camera system would produce side-by-side panchromatic pictures, each covering 80 km sq. (NASA Contract NAS 5–22–350; Raizen, GSFC Landsat Procurement Off, interview, 15 Sept 77; *SBD*, 21 Aug 75)
- NASA announced selection of the Boeing Co. and Martin Marietta Corp. for negotiations leading to a contract for delivery of a base module for the first of NASA's Applications Explorer Missions (AEM–A), also known as the Heat Capacity Mapping Mission (HCMM). The base module would be a platform for the heat-capacity-mapping radiometer and all support instrumentation, including receivers, transmitters, power system, and attitude-control system. The contract would call for delivery of the base module within 20 mo of the award.

    Planned for launch in 1978, AEM–A would gather thermal-inertia data from the earth's surface, data which should distinguish rock types, locate mineral resources, detect soil moisture, and read temperatures of vegetation cover. (NASA Release 75–253)

*10 September:* Dr. George M. Low, NASA Deputy Administrator, told the Senate Committee on Aeronautical and Space Sciences that the combined efforts of government and industry could reduce fuel consumption of civil air transport by 40 to 50% without degrading the environment or reducing aircraft safety. The conclusion was based on a study by NASA, Federal Aviation Administration, DOD, and industry at the request of the committee. Need for a vigorous fuel-conservation program had been underscored by estimates from the Federal Energy Administration and Air Transport Association of America that U.S. outlays for imported oil had increased from $3 billion in 1970 to $24 billion in 1974, and could reach $32 billion by 1977. Jet fuel costs had risen from 12 cents a gallon (4.4 cents per liter) in 1973 to 29 cents (10.5 cents) in 1975.

NASA studies had shown that, by the year 2005, the U.S. could save one million barrels of refined aircraft fuel per day by use of new technology in civil air transport. NASA's proposed fuel-efficiency plan would include six major programs of research on engine components, engine efficiency, aircraft design, turboprop performance, laminar flow control and drag reduction, and increased use of composite materials for aircraft parts. (Transcript, NASA Release 75−252)

- Engineers at Marshall Space Flight Center were completing tests seeking to refine the means of towing to shore for refurbishment and reuse recovered solid rocket boosters from Space Shuttle launches. These tests explored the question of whether the parachute recovery lines and riser lines on the three recovery parachutes to be attached to each booster would, when under tension during descent, stretch enough to develop sufficient latent energy to deploy the 52-mm-dia by 15-m long tow rope from a tray located in the booster's forward skirt. The tests employed a 10-m Saturn S−II-stage adapter ring, to which the parachute lines were attached. A crane hoisted the adapter ring until the stress on the rings was 12 000 kg, about one-half the stress expected on the parachutes lowering the booster. An explosive bolt was detonated, abruptly freeing the parachute lines; they contracted and the lines and fitting to which they were attached were yanked upward more than 60 m, quite enough energy to deploy the tow line.

  The recovery system parachutes would be fitted with flotation devices so they could be recovered. On the first six Shuttle developmental flights, the set of parachutes would be cut free from one booster in each set of two and both would be tested to determine which configuration made it easier to retrieve and tow the booster. (MSFC *Marshall Star*, 10 Sept 75, 2)

- *Apollo 17* Astronaut Harrison H. Schmitt, the only scientist astronaut to walk on the moon, had announced his candidacy for U.S. Senator from N. Mex., the *Washington Post* reported. A Republican, Dr. Schmitt would oppose Democratic Senator Joseph M. Montoya, who was expected to seek reelection. (*W Post*, 10 Sept 75)

- The appearance of a supernova on 5 July 1054 had been recorded in primitive art by ancient Indians on the walls of a N. Mex. cave, the *New York Times* reported. Goddard Space Flight Center astronomer Dr. John C. Brandt, leader of the team of astronomers who had discovered the cave art, said that the date of the pictograph—a crescent moon with a nearby object variously described as a circle, cross, or asterisk—coincided with records of the Chinese Sung dynasty. The Chinese had reported that, on the morning of 5 July 1054, a minor star in the constellation Taurus appeared to explode into extraordinary brilliance. Ancient Chinese astronomers had described the event— which modern astronomers identified as the supernova that had created the Crab Nebula—as visible in daylight for 23 days before fading and at night for another 633 days. Although the modern astronomers could not prove that the cave art depicted the Crab Nebula supernova, "What we have is circumstantial evidence that, to us, looks pretty good," Dr. Brandt said. No other supernova of known date had been visible from North America near a crescent moon.

  The supernova pictographs were not the only evidence of astronomical observations by American Indians; Navajo gourd rattles and sand paintings often depicted the Pleiades and Orion constellations, and

Hopi designs included representations of the Milky Way, constellations, sun, moon, and stars. Heaps of stones strategically placed centuries ago in Wyo. hills could have been used by Indians to sight stars and to predict and confirm the summer solstice. (Rensberger, *NYT*, 10 Sept 75, 39)

*11 September:* NASA launched a two-stage solid-propellant Nike-Hawk sounding rocket from Wallops Island, carrying an 84-kg payload to a 164-km altitude. The flight, to demonstrate successful mating of a surplus Hawk engine with the Nike M−88 first stage, also tested a newly developed recovery system for both the single-stage Hawk and the Nike-Hawk vehicle. The system, designed for both land or air recovery, could return up to 136 kg of payload by parachute. The sounding rocket carried a Univ. of Pittsburgh high-resolution photoelectron spectrometer to measure daytime photoelectron energy distribution with a very high-energy resolution. The payload impacted 178 km downrange and was retrieved by the research vessel *Annandale*. (WSC Release 75−10)

- NASA was considering resuming exploration of the moon in 1980, using a low-cost unmanned satellite to map the entire lunar surface from orbit, Goddard Space Flight Center scientist Marius B. Weinreb told United Press International in a telephone interview. The proposed satellite, together with a small communications relay satellite, would gather information needed by scientists to organize the vast amount of data collected during the Apollo lunar landing missions.

  One mission of the satellite would be to provide a detailed chart of the lunar gravitational field, including field variations, to tell scientists more about the composition of the moon and perhaps about its origin and evolution. Previous lunar orbiters had merely mapped and photographed the lunar surface. Although scientists had learned much from Apollo data about the gravitational field of the moon's near side, they knew little about the far side. The proposed satellite—equipped with instruments including x-ray and gamma-ray sensors—and its accompanying high-orbiting comsat, which would remain in range of both the satellite and the earth at all times, would send back to earth data on the entire lunar surface.

  Weinreb told UPI that NASA administrators had considered including the proposed project in the FY 1977 budget. Budget restrictions might prevent NASA from initiating new projects. (UPI, *NYT*, 11 Sept 75, 11; interview, Weinreb, 13 June 78)

*12 September:* Former astronaut William R. Pogue (USAF, Ret.) received the 1974 Thomas D. White Space Trophy during ceremonies at the National Geographic Society Headquarters in Washington, D.C. Pogue received the trophy—given annually to the military or civilian member of the Air Force who made an outstanding contribution to U.S. progress in aerospace—for his achievements as pilot of *Skylab 4*, third and final mission aboard the U.S.'s first manned orbiting workshop. The mission (16 Nov. 1973 to 8 Feb. 1974), longest manned flight in history, had established the foundation for future long-term missions in space. Pogue had retired from both the Air Force and the astronaut corps on 1 Sept. (DOD Release 456−75)

- Flight Research Center's F−111 Integrated Propulsion Control System (IPCS) aircraft had made its first flight with the new system operating, FRC's *X-Press* reported. The IPCS used digital electronics instead of conventional hydromechanical controls for actuating the aircraft's supersonic jet engine and engine inlet. The new system was expected to allow aircraft to fly at full high-speed performance limits without adverse interaction among the engine, the inlet, and the aerodynamic flow over the fuselage. During the 1-hr flight, the modified inlet and engine were operated entirely under computer control, with routine test transfers between digital and hydromechanical control at all flight conditions. (FRC *X-Press*, 12 Sept 75, 2)

- NASA had selected Brooks Air Force Base in San Antonio, Tex., as the site for a remote storage facility for 10 to 20% of the 382 kg of lunar material brought back from the moon by the Apollo crews, the Johnson Space Center *Roundup* reported. All lunar material had previously been stored in JSC's curatorial facility; the additional site would safeguard against any catastrophe that might affect the material at the main site.

  Samples at Brooks, a cross section of all lunar materials collected during the Apollo missions, would be held in "dead storage" to keep them in pristine condition for analysis by future generations. (JSC *Roundup*, 12 Sept 75)

- Charles M. Duke, Jr., an astronaut since 1966 and lunar mobile pilot on the *Apollo 16* lunar landing mission (16–27 April 1972), announced he would leave NASA, effective 1 Jan. 1976. He would also resign from the Air Force after nearly 19 yr of service. During *Apollo 16*, Duke logged 11 days 1 hr 43 min in space, including a 71-hr 14-min stay on the lunar surface. He and John W. Young made 3 surface explorations of the Descartes region for a total of 20 hr 15 min of lunar excursion.

  Since April 1973 Duke had served as technical assistant to the manager for Space Shuttle systems integration. (JSC Release 75–74)

*15 September:* In a letter to Rep. Don Fuqua, chairman of the House of Representatives' Committee on Science and Technology Subcommittee on Space Science and Applications, NASA Administrator Dr. James C. Fletcher summarized options for missions that could use the remaining Apollo-Saturn flight hardware. Options included an earth-orbital mission as an Apollo-Soyuz Test Project backup; a Skylab revisit; an Apollo geosynchronous mission; a lunar orbital mission; an earth-orbital mission to visit a Soviet space station; and a manned-satellite inspector. The hardware could also be used to launch a U.S. unmanned space station carrying a large U.S.S.R. payload, a large orbiting reflector, a space solar-power demonstration, a planetary probe, or multiple payloads. Another possibility might be to carry an Apollo spacecraft into space aboard the Space Shuttle to provide a crew-rescue capability or to serve as a habitable module, to demonstrate a Shuttle-serviced free-flying module, or to inspect and service orbiting satellites.

Dr. Fletcher's letter added that he did not foresee that funds would be available to support additional Apollo-Saturn missions in the period preceding the Space Shuttle operation. The subcommittee at a 25 March meeting had asked NASA to summarize the options when the agency had sought permission to dispose of surplus Apollo-Saturn equipment [see 9 April]. (Text, letter from Fletcher to Fuqua and attachment "Conceptual Flight Possibilities for Saturn-Apollo Hardware," 15 Sept 75)

- Norman Pozinsky's appointment as Deputy Associate Administrator for Tracking and Data Acquisition became effective. Pozinsky was replacing H. R. Brockett, who retired. Pozinsky previously had been Director of the Network Development and Engineering Div. in the Office of Tracking and Data Acquisition. (NASA anno 17 Sept 75)

*16 September:* Dr. James C. Fletcher, NASA Administrator, presented 23 awards to Marshall Space Flight Center employees and contractor personnel for contributions to the Apollo-Soyuz Test Project mission (15–26 July). Ellery B. May, former manager of the Saturn Program Office at MSFC, received NASA's highest award, the Distinguished Service Medal. Exceptional Service Medals were awarded to 14 MSFC employees; Public Service Awards went to six contractor employees active in the ASTP mission. The Group Achievement Award went to the ASTP Experiments Team and the Stress Corrosion Review Team.

ASTP astronauts Thomas P. Stafford, Vance D. Brand, and Donald K. Slayton, as well as NASA Deputy Administrator Dr. George M. Low, Associate Administrator for the Office of Space Flight John F. Yardley, and ASTP Program Director Chester M. Lee, attended the ceremony honoring MSFC workers who had provided the Saturn launch vehicle as well as 10 of the 27 onboard experiments for the mission. (MSFC Release 75–196; *Marshall Star*, 17 Sept 75, 1–2)

- NASA announced appointment of Gonzalo Fernandez as Assistant Associate Administrator for Center Operations, replacing Raymond A. Kline, appointed Assistant Administrator of Institutional Management. Fernandez, who came to NASA from the Air Force, would be responsible for the Offices of Administrative Data Processing Management, Information Systems, NASA Aircraft, Safety and Environmental Health, and Operating Systems. (NASA anno, 16 Sept 75)

*17 September:* One of the most highly detailed and instrumented models of the Space Shuttle ever constructed had been wind-tunnel-tested at the Arnold Engineering Development Center, the Air Force Systems Command announced. Located throughout the 92-cm model were 835 temperature sensors to measure heat levels of major components at jettison of the two expended solid-propellant boosters.

Two other Shuttle tests also had been completed for NASA at AEDC: A smaller model and computer-controlled systems were used to study aerodynamic forces during the separation of the expended solid boosters. During the test the Orbiter and external fuel-tank models were attached to the tunnel supports and a solid booster model was attached to the computer-controlled support. Measurements were recorded with a single booster model at various points below, beside, and behind the Orbiter-tank combination and with the booster and Orbiter-tank at various pitch and yaw angles.

In the third test, heating data were obtained on an Orbiter model positioned at a high angle of attack at mach 8. Special paints that changed from a solid to a liquid at specified temperatures were used to record heating histories on the surface of the model, and oil-coating techniques were used to examine airflow. (AFSC Release OIP 226.75)

*18 September:* President Gerald R. Ford presented the 1974 National Medal of Science, the nation's highest science award, to 13 distinguished American scientists. Among those honored during the White House ceremony was Dr. Rudolf Kompfner of Bell Telephone Laboratories Research Communications Div., who received the award for his invention of the traveling-wave tube and for major contributions to communications satellites and to optical communications.

Dr. Linus C. Pauling received the award for his contributions in the fields of structural chemistry, molecular biology, immunology, and genetic diseases. Dr. Pauling, two-time winner of the Nobel Prize, had been long bypassed for the national award, despite repeated recommendations of a scientific advisory committee, because of his outspoken criticism of much of American foreign policy. Victor Cohn of the *Washington Post* reported that the award signaled "an end to a White House war on unfriendly scientists conducted during the Johnson and Nixon administrations."

President Ford told the 200 persons attending the ceremonies that, looking back over 200 yr of the nation's history, we owed a great debt to science and to all the men and women who have "carried on the scientific enterprise of this country." The whole spirit of science—one that urged us to innovate, to explore the unexplored, and to answer the unanswered—was the true spirit of America. Although scientific priorities had changed over the years, the "Nation's commitment to that most basic of all inquiries—basic research—has not diminished. . . . Our Nation's future and that of the world depends on the creativity and the genius of men and women such as these we honor today." (*PD*, 22 Sept 75, 1027–28; Cohn, *W Post*, 18 Sept 75, A1; *NYT*, 19 Sept 75, 23)

- The National Aeronautic Association announced the selection of Clarence L. Johnson to receive its Wright Brothers Memorial Trophy for 1975. Johnson, retired from Lockheed Aircraft Co. after 42 yrs with the company, would receive the award for his "vital and enduring contributions over a period of 40 years to the design and development of military and commercial aircraft." As head of Lockheed's Advanced Development Projects, he had played leading roles in the development and design of more than 40 of the world's finest aircraft including the World War II fighter bomber, the P−38; the Constellation and Super Constellation; the first U.S. operational jet, the F−80; the high-altitude U−2 reconnaissance aircraft; and one of the first business jets, the Jet Star.

  The trophy had been awarded annually for significant public service of enduring value to aviation in the U.S. (NAA News, 18 Sept 75)

- The American Institute of Aeronautics and Astronautics announced that James B. Lazar, NASA's Chief of Electric Propulsion and Flight Experiments Branch in the Office of Aeronautics and Space Technology, and Roderick W. Spence of the Los Alamos Scientific Laboratory would receive the James H. Wyld Propulsion Award. The award was given annually in recognition of "outstanding leadership of this nation's advanced propulsion programs in the separate fields of nuclear and electric rocket propulsion."

  Lazar had been responsible for NASA's electric propulsion program, providing NASA with the option for a highly reliable, efficient, and flight-proven space propulsion system.

  Dr. Spence had directed design and development of the nuclear rocket reactors at the Los Alamos laboratory, providing the foundation of technology for the nation's nuclear rocket program. (AIAA Release 18 Sept 75)

- A giant washing machine capable of doing 318 kg of laundry would help to develop economical and efficient methods of cleaning and refurbishing the Space Shuttle solid-rocket booster parachutes, and to identify the problems encountered during the process, Kennedy Space Center's *Spaceport News* reported. The washer—2 m wide by 2 m high and 10 m long, and holding 23 668 liters of water—would wash a 23-m 318-kg 80-gore parachute. Once washed, the parachute, 136 kg heavier, would be taken by monorail to a drying area. A team of five technicians was required to complete the 4-hr process. (KSC *Spaceport News*, 18 Sept 75, 6)

*19 September:* *Salyut 4*, launched into earth orbit on 26 Dec. 1974, was continuing its flight, Tass reported. To date, the space station had completed 4200 revolutions of the earth and, in automatic mode, was collecting data and making astrophysical observations with the onboard instrumentation. All systems were functioning normally.

Since its launch, *Salyut 4* had been inhabited by two two-man crews. The first crew, launched 11 Jan. 1975 aboard *Soyuz 17*, lived and worked aboard the station for nearly 30 days. A second crew, launched 24 May 1975 aboard *Soyuz 18*, visited the station for a Soviet manned space flight record of 63 days. A third mission, launched 5 April, was aborted when the booster third stage malfunctioned. (Tass, FBIS—Sov, 23 Sept 75, U1; *A&A 74*)

- The American Institute of Aeronautics and Astronautics presented its Haley Astronautics Award to *Skylab 4* astronauts Gerald P. Carr, Edward G. Gibson, and William R. Pogue at the AIAA-American Geophysical Union Conference on the Exploration of the Outer Planets, in St. Louis, Missouri. The award was presented "for demonstrated outstanding courage and skill during the record-breaking 84-day Skylab mission." During the mission, which had begun 16 Nov. 1973, the crew had successfully completed 56 experiments, 26 science demonstrations, 15 detailed subsystem objectives, and 13 student investigations. They also collected extensive earth-resources data and logged 338 hrs of Apollo Telescope Mount data, making detailed observations of solar processes. (MSFC *Marshall Star*, 3 Sept 75, 4)

*20 September—5 October:* Apollo-Soyuz Test Project astronauts Thomas P. Stafford, Vance D. Brand, and Donald K. Slayton and their families visited the U.S.S.R., joining Soviet ASTP cosmonauts Aleksey A. Leonov and Valery N. Kubasov and their families for a goodwill tour of the Soviet Union. The two crews had not met since they said goodbye in space during their historic mission [see 15—24 July]. After greetings at the Moscow airport with wide grins, bear hugs, and back slaps, Stafford told the airport crowd in Russian, "We are very pleased to be back in the Soviet Union again. Now we can say our joint work has been fulfilled successfully. I am confident that our flight will serve progress and peace on earth." During the welcoming ceremonies Boris Petrov, U.S.S.R. Chairman of the Intercosmos Council, told the Americans, "The Soviet people were glad to see your success and are eager to welcome you on earth and in the Soviet Union."

The astronauts, cosmonauts, and their families visited Star City on 21 Sept. to thank the Soviet space community for its work on the ASTP mission. "For 2 years Star City was a second home to us," Stafford told the crowd of Soviet space workers. "Thank you for all your work." The astronauts' families laid flowers at the memorial to Yuri Gagarin, first man to fly in space. Stafford, Brand, and Slayton presented Star City with a plaque bearing the Soviet and American flags and exchanged other gifts with the Soviets.

On 22 September Leonid I. Brezhnev, General Secretary of the Communist Party Central Committee, met the ASTP astronauts and cosmonauts at the Great Kremlin Palace, greeting each as "hero" and congratulating them on their historic mission, saying it would "pave the way for generations to come to go on with this work." During the 35-min reception he told the crews of his hope for a treaty to limit nuclear arms. "Both sides—the United States and the Soviet side—should make every possible effort for good, neighborly relations, really good, in all aspects, from every point of view," he said.

On 23 Sept. the two crews and their families began a six-city tour of the Soviet Union that took them to Leningrad, Kiev, Volgograd, Novosibirsk, Sochi, and Tbilisi. Everywhere the crews were swamped by enthusiastic spontaneous crowds of up to 3000 persons waiting on corners and at airports for hours to cheer the spacemen. "The hospitality is fantastic," Stafford remarked at Volgograd. The *New York Times* reported that each appearance of the spacemen was greeted with applause and cheering and they responded with waves and autographs.

The astronauts ended their Russian tour on 4 Oct. with a farewell press conference back in Moscow. Stafford, Brand, and Slayton praised the "enthusiastic people with warm hearts" they had met during their tour. During the press conference—held exactly 18 yr after the U.S.S.R. had launched the first manmade satellite, *Sputnik 1*, on 4 Oct. 1957—Stafford encouraged continued cooperation between the U.S. and U.S.S.R. in such areas as lunar and planetary exploration, space medicine, space meteorology, and earth environment.

The three U.S. astronauts and their families flew to London for an overnight stay before flying back to Houston on 5 Oct. (B *Sun*, 21 Sept–6 Oct 75; *NYT*, 21 Sept–6 Oct 75)

*22 September:* NASA and the Air Force Systems Command's Space and Missile Systems Organization had jointly contracted for procurement of 12 advanced meteorological satellites, saving more than $37 million, AFSC announced. Both agencies had developed satellites to observe and report worldwide weather conditions and were negotiating procurement of improved, longer lived meterological spacecraft. The contracts for long-leadtime parts for the new satellites—nine for NASA and three for the Air Force—were a follow-on to an earlier Air Force procurement program for a meterological satellite identified as Block 5D. Although some of the systems differed, similar technical requirements permitted the basic spacecraft to serve both agencies. Use of an already developed basic design would save NASA about $33 million, and the joint production contract would save both agencies nearly $2.4 million in material and assembly costs. (AFSC Release OIP 186.75)

- The appointment of Lt. Gen. Duward L. Crow (USAF, Ret.) as Associate Deputy Administrator became effective. Gen. Crow succeeded Willis H. Shapley, who retired. Before joining NASA in September 1974 as Assistant Administrator for Dept. of Defense and Interagency Affairs, Gen. Crow had been Assistant Vice Chief of Staff of the Air Force since October 1973. He had previously been Comptroller of the Air Force.

Replacing Gen. Crow as Assistant Administrator for DOD and Interagency Affairs was Lt. Gen. William W. Snavely (USAF Ret.) who had been Air Force Deputy Chief of Staff for Systems and Logistics since January 1973. (NASA Release 75−260)

- A NASA-developed low-cost highly efficient solar thermal-energy collector and a ball bearing design with a life expectancy 20 times greater than that of present ball bearings had been selected to receive *Industrial Research* magazine's award for two of the 100 most significant new products developed during the year, NASA announced.

  Developed at Lewis Research Center in cooperation with Honeywell, Inc., the solar collector absorbed 20% more solar-radiated energy and produced a greater flow of high-temperature fluids. The flat-plate collector combined an efficient flow scheme, insulation, dual antireflective glass covers, and black chrome absorption coating to provide heating and cooling, with low cost and low pollution, for both commercial buildings and homes.

  LeRC developed the bearing with General Electric Co. and Industrial Tectonics, Inc., using computer analysis for maximum performance and reliability. Advanced techniques for producing and melting steel provided a cleaner, longer lived material for the bearing, and advanced fabrication and machining methods contributed greater accuracies and surface finishes to the bearing rings. The bearing was lubricated through holes in the bearing's split inner ring rather than by a conventional lubricant jet pointed at the bearing rings. This advanced bearing design was expected to replace mainshaft-thrust ball bearings presently used in large turbine engines, pumping systems, electric utilities, and petroleum pipeline operations. (NASA Release 75−258)

*23 September:* A Senate and House of Representatives conference committee reported out H.R. 8070, appropriating FY 1976 funds for the Department of Housing and Urban Development and sundry independent executive agencies, after resolving the disagreements of the two congressional bodies. Total NASA FY 1976 appropriation recommended by the committee was $3 551 822 000, changed only with respect to funds allotted for research and development. The

## FY 1976
### NASA Budget Appropriation
### Congressional Adjustments
### to Budget Request

(millions)

| Program | Budget Request | House Changes | Senate Changes | Conference Changes | Total Conference Recommendation |
|---|---|---|---|---|---|
| Research & Development | 2678.4 | −49.4 | +7.0 | −1.0 | 2677.4 |
| Construction of Facilities | 84.6 | −2.5 | −2.5 | −2.5 | 82.1 |
| Research & Prog Management | 776.0 | +0.5 | −0.5 | −0.5 | 775.5 |
| Totals | 3539.0 | −52.4 | +4.0 | −4.0 | 3535.0 |

(H.R. Comm Rpt 94−502; *CR*, 23 Sept 75, H9035−38)

committee appropriated to NASA $2 677 380 000 for R&D, $48.4 million more than the $2 628 980 000 proposed by the House [see 19 and 24 June] and $8.0 million less than the $2 685 380 000 proposed by the Senate [see 24 and 26 July].

The conference committee agreed that NASA could reprogram $7 000 000 from the R&D total for an upper atmospheric research, technology, and monitoring program. The committee also agreed that $1 000 000 of the total funding of $48 400 000 proposed for a Pioneer-Venus mission might be reprogrammed for further planning of the Large Space Telescope, though not beyond evaluation of phase B studies. Appropriations for construction of facilities and for research and program management had been previously agreed on and remained unchanged.

*23 September:* NASA Pilot William H. Dana made the last rocket-powered flight of the X-24B lifting body, from Flight Research Center. Although a four-chamber firing had been planned, only three chambers ignited, and a three-chamber flight plan was executed. The X-24B reached an altitude of 17 678 m and a speed of mach 1.2. Six more glide flights were planned.

Objectives of the flight were to observe stability and control with rudder bias at 10°, to test handling qualities with all damper gains at zero, to survey body pressure, to perform a qualification test of a high-density reusable surface insulator, and to make a lefthand fin tuft study. The X-24B was part of a joint NASA-Air Force program to test concepts for a manned vehicle that could reenter the atmosphere from space, and to provide advanced technology for a future hypersonic cruise aircraft. (FRC *X-Press*, 26 Sept 75, 2; NASA Hq X-24B Flt Rpt)

*25 September:* A NASA Atlas-Centaur vehicle launched *Intelsat-IVA F-1* at 8:17 pm EDT from Eastern Test Range into the planned synchronous transfer orbit. The communications satellite had been launched for the Communications Satellite Corp. under a cost-reimbursable contract. At 11:18 pm EDT 26 Sept., ComSatCorp fired the apogee kick motor to put the satellite in a geosynchronous orbit with a 35 821-km apogee, 35 294-km perigee, 23-hr 44-min period, and 0.1° inclination.

NASA objectives for the mission were to review design, performance, and flight readiness for the Federal Communications Commission and to assure compatibility of the Intelsat spacecraft with NASA launch vehicles and launch environmental conditions, and to launch *Intelsat-IVA F-1* into a transfer orbit that would enable the spacecraft apogee boost motor to inject the spacecraft into a synchronous orbit. Objectives were completed and the mission adjudged successful 29 Jan. 1976.

ComSatCorp objectives were to fire the apogee boost motor, position the satellite into the planned geostationary orbit, and operate and manage the system for the International Telecommunications Satellite Organization (INTELSAT).

*Intelsat-IVA F-1*, built by Hughes Aircraft Co., was 698.5 cm high and 238.2 cm in diameter, with a liftoff weight of 1515 kg and orbital weight of 825.5 kg. The spin-stabilized satellite consisted of two main elements: A rotating section contained the power subsystem, a cylindrical solar array and two nickel-cadmium batteries; the anhydrous hydrazine-powered positioning and orientation system; the solid-fueled apogee motor; and the despin-control system. The despun earth-oriented platform section contained the communications repeater, antennas, and associated elements of the telemetry and command subsystems. All spacecraft antennas were supported by a single tubular mast.

For 24 hr after apogee-motor firing, *Intelsat-IVA F−1* drifted eastward at a rate of 10° per day; then a velocity-correction maneuver slowed the spacecraft drift to 2.9° east per day. While the spacecraft was drifting, the Hawaii earth station checked out the spacecraft systems. Its final operating location, over the Atlantic Ocean at 335° east longitude, would be reached by mid-December. As primary Atlantic satellite in INTELSAT's global commercial communications satellite system, it would serve more than 40 earth stations, offering users 6250 two-way voice circuits and two TV channels.

*Intelsat-IVA F−1* was the first in the Intelsat IVA series; its communications capacity was two-thirds greater than satellites in the Intelsat IV series, seven of which were still operating. The Intelsat system began with the launch of *Intelsat-I F−1 (Early Bird 1)* on 6 April 1965; since then NASA had launched 16 successful comsats for ComSatCorp in the Intelsat II, III, and IV series.

Although Intelsat IVA satellites had been designed primarily for use over the Atlantic with its heavy communications traffic, INTELSAT planned to launch as many as five additional IVAs, some to provide service for the Pacific and Indian Ocean regions. Intelsat IVAs had a design lifetime of 7 yr and a capacity sufficient to meet international communications requirements through 1979, when the system would be augmented with the even larger capacity Intelsat V comsats.

INTELSAT, created in August 1964 by adoption of the Interim Agreements for the Establishment of a Global Commercial Comsat System, now comprised 91 member nations; a 24-member board of governors exercised overall responsibility for design, development, construction, establishment, operation, and maintenance of the Intelsat space segment. ComSatCorp, a privately owned carrier company operating under a U.S. congressional mandate, had provided technical and operational management services to INTELSAT and the global system under a cost-plus-fixed-fee management services contract.

NASA had been responsible for procuring the Atlas-Centaur launch vehicle, conducting preflight testing, and launching the spacecraft from the Eastern Test Range into a synchronous transfer orbit. ComSatCorp engineers in the INTELSAT Spacecraft Technical Control Center in Washington, D.C., then assumed responsibility for the mission.

Overall management for the NASA portion of the mission was under the direction of the Office of Applications. Lewis Research Center was responsible for Atlas-Centaur development and operation. Kennedy Space Center was responsible for vehicle checkout and for launch. (NASA MORs E−491−633−75−01, 25 Sept 75, 30 Jan 76; NASA Release 75−231; INTELSAT Releases, 21 Sept 75, 1 Oct 75; Singer, *Today*, 26 Sept 75; NASA PIO, interview, 26 Sept 75; GSFC *Wkly SSR*, 25 Sept−1 Oct 75)

*26 September:* Johnson Space Center's Glynn S. Lunney, U.S. Technical Director for the Apollo-Soyuz Test Project, had been named manager of the Space Shuttle Payload Integration and Development Program Office, JSC's *Roundup* reported. The office had been created to plan and develop JSC payload for the Space Shuttle. The Apollo Spacecraft Program Office had been abolished, and personnel from that office and the Payloads Coordination Office had been reassigned to the new program office. Besides managing Shuttle payloads and components—including the Interim Upper-Stage low-cost modular spacecraft, long-duration exposure facility, and Large Space Telescope—the new office would manage JSC participation in international space activities, including any future U.S.-U.S.S.R. activities. (JSC *Roundup*, 26 Sept 75, 1)

- X-ray satellites—including *Uhuru* (*Explorer 42* launched 12 Dec. 1970) and *Oso 7* (Orbiting Solar Observatory launched 29 Sept 1971)—had revealed a previously unidentified class of x-ray sources associated with very old star clusters, *Science* reported. In at least five documented cases, satellites had detected x-rays from globular star clusters coming from above and below the pinwheel disk of the Milky Way galaxy. These clusters, left behind when the rest of the primordial galactic matter fell into the disk of the galaxy over 10 billion yrs ago, still orbited the galactic center, bound to it by gravitational forces.

  Scientific evidence had indicated that the birth of new stars within clusters ended soon after the formation of the clusters themselves; the clusters had not been expected to contain any young binary star systems, those believed to make up the majority of galactic x-ray sources. This fact, along with the finding that sources were much more plentiful in globular clusters than in the galactic disk, indicated that some different process of x-ray production was at work.

  Scientists offered three explanations for the findings: existence of very large black holes, 1000 times as massive as the sun, within the clusters; a smaller black hole steadily fed with matter from the surrounding disk; and origin of the x-ray emissions in binary stars behaving like x-ray binaries in the galactic disk but with very different origins.

  Scientists were hoping that two more x-ray satellites, U.K.'s *Ariel 5* (launched 15 Oct. 1974) and *Ans 1* (Netherlands Astronomical Satellite launched 30 Aug. 1975), would contribute additional data to the study. (Metz, *Science*, 26 Sept 75, 1073-74)

- A vertical motion simulator to reproduce the critical maneuvers of aircraft during takeoff and landing was being built at Ames Research Center. Costing $3.5 million, the VMS would use an existing cab and computer system, but would have new hardware to provide large-scale vertical motions. It would move as much as 18 m vertically and 12 m horizontally and would accurately simulate flare and touchdown of such aircraft as the short takeoff and landing (STOL) and vertical takeoff and landing (VTOL). (FRC *X-Press*, 26 Sept 75, 4)

- Johnson Space Center's *Roundup* reported that Scott Aviation Co. would begin commercial production of a new lightweight breathing system for firefighters based on concepts and hardware developed by JSC engineers. Called the Scott Air-Pak 4.5, the new apparatus was the first major improvement in compressed-air breathing systems in more than 20 yrs. New features of the equipment included a 40% weight reduction; an improved pack frame and harness, with the unit's weight carried on the hips rather than the shoulders; and a redesigned face mask for better vision.

NASA's part in developing the apparatus had begun in 1970 when a national meeting of municipal officials identified the need for an improved system as their highest priority. NASA's effort had been carried out at JSC as an engineering applications project under the Technology Utilization Program. After a design had been chosen from a number of candidates, prototype units were successfully field tested under actual firefighting conditions by the Houston, New York City, and Los Angeles fire departments.

Scott Aviation was the first manufacturer to announce commercial production of an apparatus based on NASA effort. (JSC *Roundup*, 26 Sept 75, 4)

- Aetna Life and Casualty would become the third partner with Comsat General Corp. and International Business Machines Corp. in a venture to establish a domestic satellite system, the three companies jointly announced.

  In a letter to the Federal Communications Commission, the three companies had proposed a corporate structure of the jointly owned CML Satellite Corp. under which each of the three companies would own a part of the organization. IBM would purchase 42.5% and Aetna would purchase 15% of CML stock, with Comsat holding the remaining 42.5%. Aetna would also purchase a convertible note, bringing its investment to the same level as Comsat and IBM; the three partners would have equal representation on the board of directors. The arrangement would comply with FCC's requirement that no partner own less than 10% nor more than 49% of CML's stock [see 23 Jan.].

  After FCC approval, the three partners would each invest up to $55 million with additional funding obtained from outside financing.

  CML had been formed in 1972, with other stockholders, to enter the U.S. domestic communications satellite business; Comsat had owned all the stock of CML since July 1975 and, with IBM, had been seeking a third partner. Both Comsat and IBM had been providing the interim financing. (Aetna-Comsat-IBM Release, 26 Sept 75)

*28 September:* Plans had been completed for the first U.S. experiments to be flown on a Soviet spacecraft, Dr. David L. Winter, NASA Director of Life Sciences, announced. As a result of joint U.S.-U.S.S.R. working group meetings on space biology and medicine, 4 U.S. life-science experiments and 7 tissue investigations were to fly aboard a Soviet biological satellite scheduled for launch later in the year. The U.S. experiments—all passive and completely autonomous from spacecraft power, telemetry, and data recording—would be housed in 5 Soviet-built containers, each 17.8 cm long by 12.7 cm wide by 11.5 cm high. Maximum weight for each container, including experiments, would be 2.5 kg. Unique to this spacecraft was a Soviet-designed onboard centrifuge. U.S. experiments would be flown both on the centrifuge and on a stationary platform.

The 4 flight experiments included a plant tumor-growth experiment to study the effects of weightlessness on sensitive plant systems; a carrot-cell-culture experiment to assess the effects of weightlessness on plant systems and on normal development of embryonic tissues; a heavy-particle radiation experiment to measure high-charge and high-energy particle radiation aboard the spacecraft; and a killifish or mummichog (minnows) embryogenesis experiment to evaluate the effect of weightlessness on development of the vestibular system during embryonic development in a vertebrate animal.

Six of the tissue investigations would use rat tissue—and the seventh flies—supplied by the Soviets from their animal experiments. Soviet scientists would prepare the animal tissues for U.S. scientists to use in their investigations and would return the U.S. experiment packages upon completion of the flight.

Ames Research Center was managing 3 of the life-science experiments and all tissue investigations. Johnson Space Center would manage the fish-egg development experiment. (NASA Release 75-264)

- Transfer of the Launch Vehicle and Propulsion Program from the Office of Space Science to the Office of Manned Space Flight, and transfer of the NASA Directorate for Life Sciences from the Office of Manned Space Flight to the Office of Space Science became effective. Also, the name of the Office of Manned Space Flight had been changed to Office of Space Flight, and the name of the Launch Vehicle and Propulsion Program had been changed to Expendable Launch Vehicle Programs.

  In announcing the changes, NASA Administrator Dr. James C. Fletcher had said that the exchange of these functions would "provide a more logical alignment of responsibilities as we move into Space Shuttle operations. The new Office of Space Flight will be concerned primarily with launching... and conducting Shuttle operations, while Space Science will have responsibility for the science to be performed on these flights." (JSC *Roundup*, 26 Sept 75, 1)

*29 September:* Johnson Space Center announced that the first of two Space Shuttle training aircraft had been test flown at Grumman American Aviation's Bethpage, N.Y., plant. The modified Gulfstream II would be used to train crews for Orbiter descent and landing procedures. The cockpit layout and the aerodynamic modifications of the trainers provided motion and visual cues and handling qualities similar to those of the actual Orbiter. The two trainers would be delivered to Johnson Space Center early in 1976. (JSC *Roundup*, 10 Oct 75, 1)

*30 September:* Marshall Space Flight Center had resumed acoustic testing of a 6.4%-scale model of the Space Shuttle to provide information needed in designing the Shuttle and its launch facilities, MSFC announced. The tests would also produce data on effects of hot exhaust gases on the aft portion of the Shuttle immediately after ignition. Data from earlier testing of the model, which included the liquid-fueled engines and solid rocket motors, had alerted designers that special provisions would be needed to deflect hot gases out of the solid rocket boosters away from the facility fuel lines near the launch pad. A pad design based on this analysis had been tested and proven. During the tests, which had begun originally in August 1974, the model was held in a test stand while its engines were fired from 6 to 10 sec. (MSFC Release 75-205)

*During September:* NASA had invited nearly 5000 scientists worldwide to propose experiments for a possible 6-yr mission to Uranus, beginning with a November 1979 launch. The mission, which could include a swing by Neptune, was being studied for inclusion in NASA's FY 1977 budget request. A mission to Uranus and Neptune would mean that every planet in the solar system except Pluto would have been probed by a U.S. spacecraft by 1988. The spacecraft itself would be adapted directly from the Mariner probe now under development at Jet Propulsion Laboratory for a planned 1977 mission to Saturn; like Mariner-Saturn, the Uranus mission would use the gravity of Jupiter to hurl the spacecraft on toward the more remote planet. If the spacecraft remained healthy after it investigated Uranus, it would use the gravity of that planet to fly on to Neptune, more than 1.6 billion km deeper into space. (GSFC *Goddard News*, Sept 75, 2; UPI, *NYT*, 4 Sept 75, 20)

## October 1975

*1 October:* NASA and Rockwell International Corp. Space Div. had signed a $1.8-billion supplemental agreement for the follow-on development of the Space Shuttle Orbiter, NASA announced. The agreement formally incorporated into an existing contract awarded for Orbiter design, development, and test and evaluation, the construction of Orbiters 101 and 102, approach and landing tests, and six orbital flight tests. The supplemental agreement for the additional work brought the value of the Rockwell contract to more than $2.7 billion. (NASA Release 75–266; JSC Release 75–90)

- Marshall Space Flight Center had opened for competition a research and development program to develop and demonstrate cryogenic components for the Space Tug vehicle. Work would include design of a reusable cryogenic storage system, and fabrication and evaluation of a testing system. Proposals were to be submitted by 31 October. (SBD, 1 Oct 75, 159)

- NASA and the U.S. Coast Guard were jointly designing and developing a lightweight portable firefighting module to combat shipboard and dock fires, NASA announced. The module, a completely self-contained pumping system including pumps, hose, and firefighting suits, would pump sea water at a rate of up to 7300 liters per min (2000 gallons per min) for periods up to 3 hr. Marshall Space Flight Center was designing a prototype scheduled for construction early in 1976 and for testing 18 mos later. The Coast Guard had asked NASA to participate in the project because of its long experience with high-capacity lightweight fuel pumps developed for rocket engines. The project was being managed for NASA by MSFC's Technology Utilization Office. (NASA Release 75–267)

- NASA announced selection of Bendix Field Engineering Corp. and Raytheon Service Co. for competitive negotiations leading to award of a single contract to operate and maintain portions of NASA's worldwide Spaceflight Tracking and Data Network. Contractor services would include technical support, operation, and maintenance of the Goddard Space Flight Center's Network Operations Control Center, GSFC's Network Test and Training Facility, ten STDN stations, and the laser tracking subnet—one fixed station at Goddard Space Flight Center and 8 mobile stations at various sites around the world to support the San Andreas Fault Experiment, the Laser Geodynamic Satellite network, and the Geodynamic Experimental Ocean Satellite program—as well as operation of the magnetic-tape certification facility at GSFC. (NASA Release 75–268)

- Dr. Walter C. Williams' appointment as NASA Chief Engineer became effective. In his new position, Dr. Williams, who had been vice president and general manager of the Aerospace Corp. Vehicle Systems Div., would be responsible for reviewing the technical excellence of all NASA programs and for assuring their development on a sound engineering basis with proper programmatic controls.

From 1940 to 1958 Dr. Williams had been employed by the National Advisory Committee for Aeronautics and, after 1958, by NASA. His previous positions included chief of the High-Speed Flight Station, Associate Director of Manned Space Center, Associate Director of Mercury's Space Task Group at Langley Research Center, and Deputy Associate Administrator for Manned Flight Operations at NASA Headquarters. (MSFC *Marshall Star*, 10 Sept 75, 2)

- The appointment of Paul E. Cotton as Assistant Administrator for Center Operations became effective. Cotton would be responsible for special studies and analyses relating to institutional and organizational activities, and for implementing decisions resulting from these studies. Cotton joined NASA in 1958, serving in various management capacities until 1972, when he joined the Social Security Administration. He had returned to NASA in 1974 as special assistant to the Associate Administrator for Center Operations. (NASA anno, 1 Oct 75)

*3 October:* Johnson Space Center announced award of a $11.5-million contract to General Electric Co. Space Div. to provide acceptance-checked equipment for the Space Shuttle orbiter. The equipment would be used to conduct standard tests and to detect variances in spacecraft systems in assembly at the Rockwell International Corp. plant. GE would also provide checkout support during the approach and landing phases of the Shuttle missions. (JSC Release 75-92)

- Johnson Space Center announced signing of a $44 800 000 contract amendment with Lockheed Electronics Co., Inc., for electronic, scientific, and computing-center support services at JSC. The services would support JSC programs including Space Shuttle, earth resources, aircraft, health applications, and the Large Area Crop Inventory Experiment. The amendment brought the estimated value of the Lockheed contract to $173 100 000. (JSC Release 75-93)

*5 October:* Design and construction of a cost-saving municipal sewage-treatment plant, using a Jet Propulsion Laboratory-developed process, would begin late in the year in Orange County, Calif., JPL announced. The new system used a pyrolytic reactor to convert solid sewage materials to activated carbon, which was then used to treat incoming waste water. Besides reducing sewage-treatment costs by 25%, the new system was expected to virtually eliminate sewage solids, produce cleaner waste water for delivery to the ocean, remove heavy metals contained in the sewage, and eliminate odors. Gases generated by the sewage solids would be used as a source of power.

The plant—a pilot project funded by Federal, state, and municipal grants totaling approximately $2 million—would replace a small mobile unit installed by JPL in 1974. Project officials hoped that, following evaluation, the system could be scaled up 100-fold and adopted for urban waste treatment. (JPL Release 5 Oct 75)

- The Univ. of Chile would build and operate a ground station to receive earth-resources data directly from NASA's *Landsats 1* and *2* spacecraft launched 23 July 1972 and 22 Jan. 1975, NASA announced. A memorandum of understanding between NASA and the university provided for Chile's use of the Spaceflight Tracking and Data Network station at Santiago until the ground station becomes operational.

The station would aid South American investigators conducting research in cooperation with NASA and the United Nations. The Univ. of Chile station would be the tenth Landsat station worldwide. Three stations were located in the U.S. Brazil, Canada, and Italy each had stations in operation, and Iran and Zaire were constructing facilities for data acquisition. In addition, Canada was planning a second station. The Univ. of Chile station would be able to acquire data directly from the satellite as it passed over Bolivia, Uruguay, Argentina, Chile, and sections of Peru and Brazil. (NASA Release 75−265)

*6 October:* A massive new power source for earth—transmission of solar energy by microwave frequency—had been under study by the Jet Propulsion Laboratory at the Goldstone station near Barstow, Calif. The dish antenna equipped with a 450-kw klystron tube for microwave transmission, represented a space station; the tower a mile away with an array of solar receivers to which the experimental power converted to microwave energy had been beamed represented a ground station. This research had been designed to lead to a network of space stations that would collect solar energy 24 hr a day and beam it to ground stations; the Goldstone tests had been the first high-power field trials of the system. Tiny rectifying dipole antennas—called rectennas—collect and convert the radio-frequency beam; the 17 panels in the tower receiver contain 4590 of the rectennas. The system had demonstrated a maximum output of 30.4 kw converting the microwave beam to usable electricity, an efficiency of 82.5%. A space-to-earth system would be 6 times more effective than a surface system, the *Los Angeles Times* commented, and could probably supply 15% to 20% of the earth's energy needs in the next century. Development of a utility operating system delivering 10 000 mw of power would cost about $60 billion, as much as NASA has spent to date on space programs. A single equatorial space station carrying a 1-km-diameter antenna transmitting to a U.S. earth station 15 km long and 10 km wide could supply all the electrical needs of New York City or a third of the needs of the Los Angeles basin. (*LA Times*, 6 Oct 75, 1; *NYT*, 10 Oct 75, 2)

*7 October:* Dr. William R. Lucas, Director of Marshall Space Flight Center, announced the establishment of an advanced mineral-extraction task team within the Program Development Directorate. Working with the Dept. of Interior's Bureau of Mines, the team would use NASA expertise and technology to increase the nation's ability to extract energy and minerals at acceptable cost and with minimum effect on the environment. One of NASA's first tasks would be to improve the efficiency of the long-wall shearer—a coal-removing machine used by miners that could leave a foot or more of unrecoverable coal along the edges of a vein—by using sensors, remote controls, and TV monitors to guide the machine. (MSFC Release 75−215)

*8 October:* NASA announced award of a $1.5 million cost-plus-fixed-fee contract to Rockwell International Corp. for development of a solid-state bubble-memory data recorder to meet spacecraft data-storage requirements. The recorder, which would have no moving parts, would offer a capacity of 100 million bits, serial or parallel data input/output operation, user-selected data rates, and direct access to the memory. Rockwell would complete the work in two phases: A prototype would be completed by early 1977, and a flight-qualifiable model would be produced by early 1978. Langley Research Center would manage the contract for NASA. (NASA Release 75−272)

- *Cosmos 775*, believed to be third of a series in the Soviet Union's Statsionar communications-satellite program, was launched from Baykonur Cosmodrome into a very low near-circular earth orbit. Its predecessors, *Cosmos 637* and *Molniya I-S*, had been launched 26 March and 29 July 1974. The satellites were to supply communications, including telephone, telegraph, TV, and phototelegraph transmissions, between the Soviet Union and Europe and between the capital and extreme eastern portions of the U.S.S.R. Once the Statsionar communications system was established, the U.S.S.R. was expected to undertake implementation of a synchronous meteorological satellite system as early as the 1976−77 period. (*SBD*, 16 Oct 75, 232)

*10 October:* The first of a series of 5 conferences on design requirements for Space Shuttle and Spacelab payloads had been held at Marshall Space Flight Center. About 50 scientists and engineers from government, industry, and universities had met to define functional requirements for payload equipment to further NASA's Space Processing Applications (SPA) program, including analysis of selected design approaches and recommendations for getting the most use out of SPA payloads for the materials science and technology community. The meeting would hear presentations on the two SPA competitive engineering designs by General Electric and TRW Systems, Inc. The next 2 days would be used by working groups to discuss their specialties—biological preparations; chemical processes and fluid phenomena; glasses and ceramics; and solidification of metals, alloys, composites, and electronic materials. The 5 scheduled meetings should define payloads that are compatible with the Space Shuttle and Spacelab and that would meet the research needs of the industrial and scientific communities. (MSFC Release 75−218)

- NASA announced signing of an agreement with the Federal Aviation Administration to act as a third party for the purpose of receiving, processing, and analyzing safety reports filed under FAA's Aviation Safety Reporting Program. The third-party arrangement would stimulate the free and unrestricted flow of information, by assuring users that FAA's objective in the program was to provide the safest possible aviation system by identifying and correcting unsafe conditions before accidents occur. The agreement provided for specific procedures to protect the identity of persons involved in the reports submitted to NASA, except in criminal cases or accidents.

    NASA would set up an aviation safety reporting working group under its Research and Technology Advisory Council to advise NASA on the reporting system, to evaluate and review the program once it is under way, and to ensure the anonymity of those submitting reports. The group would include aviation and consumer groups and others involved in operational aspects of the national aviation system. (NASA Release 75−273)

*12 October:* Soviet cosmonauts Valery Kubasov and Gen. Aleksey Leonov arrived in Washington, D.C., to begin a 2-wk tour of the U.S. by visiting President Ford at the White House and touring the capital. The American crew of the Apollo-Soyuz Test Project and their wives would accompany the cosmonauts on the tour, which would also visit Chicago, Omaha, Salt Lake City, San Francisco, Reno, Lake Tahoe, Los Angeles, Atlanta, Nashville, and New York City. The cosmonauts were accompanied by their wives and three of their children, as well as the director of cosmonaut training, Gen. Vladimir Shatalov, plus a support staff and Soviet journalists. The Apollo-Soyuz crews had completed a 2-wk tour of the Soviet Union 4 Oct. (NASA Release 75−276; *SBD*, 14 Oct 75, 216)

*13 October:* At a news conference in Washington, D.C., with members of the Apollo-Soyuz Test Project crew today, astronaut Donald "Deke" Slayton said he is convinced that man eventually will fly to Mars, and that such a mission "is a natural for an international mission as opposed to any national mission." Maj. Gen. Aleksey Leonov of the visiting cosmonauts agreed that the ASTP mission was "only the beginning of our future cooperation in space" and that it would be a "shame" for the U.S. and the Soviet Union to conduct parallel space programs that could be conducted as joint missions. (*SBD*, 15 Oct 75, 224)

*14 October:* Manufacture of unique items in space is nearing reality with the availability of the Space Shuttle, scientists from TRW Systems Group told the 6th U.S.−European conference on partnership for space applications. Five practical processes in the space environment would be crystal growth, purification and separation, mixing, solidification, and processes in fluids. The gravity-free environment would permit the manufacture of items such as contamination-free glass for laser requirements, pure crystals for semiconductors, and highly pure vaccines. Also, if materials like copper-lead alloys that do not mix on earth could be combined in zero gravity, real breakthroughs in aircraft and automotive parts manufacture might be possible. (*SBD*, 14 Oct 75, 221)

- Jet Propulsion Laboratory scientists reported that a joint JPL−Univ. of Texas astronomy team had detected carbon monoxide in the atmosphere of Jupiter for the first time. Instrumentation at McDonald Observatory, the Univ. of Texas's 27-m telescope facility in Austin, made the discovery by using high-resolution spectra. Dr. Reinhard Beer, JPL leader of the team, reported that the near-infrared spectra had been produced by a Connestype Fourier spectrometer in 1974; spectral lines—the "signatures" of various molecules—had disclosed that carbon monoxide molecules were present to a depth of at least 50 km into the lower atmosphere. Carbon monoxide had been previously found in minor amounts in the atmospheres of earth, Venus, and Mars. Although the spectra had shown no indication of carbon dioxide in the Jovian atmosphere, Dr. Beer reported, future telescope searches from higher altitudes might reveal its presence there. (NASA Release 75−269)

*15 October:* A "rather exciting development" in the ability of manned spacecraft systems to navigate accurately should be available in 10 yr, General William J. Evans, Commander, Air Force Systems Command, told the 57th annual meeting of the American Defense

Preparedness Assn. in Los Angeles. The NAVSTAR Global Positioning System, the "most precise and accurate positioning, velocity, and navigation assistance air or surface vehicles have ever had," may replace as many as five separate systems now required for these functions. (AFSC Release 15 Oct 75)

*16 October:* NASA launched the first geostationary operational environmental satellite, GOES—A, at 6:40 pm EDT for the National Oceanic and Atmospheric Administration aboard the 106th successful Delta from Kennedy Space Center. Transfer orbit elements were: apogee, 36 795 km; perigee, 200 km; inclination, 23.7°; period, 650.9 min. Apogee boost motor fired at 10:51 am EDT 17 Oct. had put the spacecraft into a synchronous orbit at 36 200 km altitude, where it would drift to its scheduled position approximately 50°N. When the GOES—A had been moved to position and checked out, it would be designated *Goes 1* and turned over to NOAA for operational use.

GOES was the first operational version of the prototype synchronous meteorological satellites (*Sms 1*, launched 17 May 74 into geostationary orbit at 75°W, and *Sms 2*, launched 6 Feb. 75 and positioned at 115°W).

*Sms 1*, originally stationed over the eastern Atlantic in support of the Global Atmospheric Research Program, had provided the first near-continuous daylight coverage of a major hurricane (Carmen) in Sept. 1974. The day-night time-lapse motion pictures produced from *Sms 1* images had aided understanding of hurricanes and tropical storms, as well as the total weather pattern in the Atlantic. *Sms 2*, from its station over the equator, had viewed the western half of the U.S., including Hawaii. Other NOAA satellites in polar orbit (*Essa 8, Noaa 3,* and *Noaa 4*) had collected global data for transmission to ground stations in Alaska and eastern U.S. This information has been available worldwide to any individual or agency with equipment that could receive the transmissions.

*Goes 1* would provide NOAA scientists with images of a quarter of earth's surface at 30-min intervals, day and night. The satellite would also collect and relay nonvisual environmental data transmitted from thousands of manned and unmanned remote-sensing facilities on land and sea. Its radiation sensors would monitor solar activity and transmit its findings to NOAA's Environmental Research Laboratories in Boulder, Colo., for use in predicting solar storms that would affect radio communications on earth. *Goes 1* would also transmit weather maps and other material from NOAA's station at Wallops Island, Va., in a format suitable for use by ground stations at regional forecast centers. (NASA Release 75—270; KSC Release 251—75; MORs E—608—75—03, 22 Sept and 17 Oct 75; KSC *Spaceport News*, 31 Oct 75, 1; NOAA Release 75—188)

- The Ancient Mariner—a power package built as a backup for the unit flown past Venus on *Mariner 5* in 1967—was still going strong after more than 5 yr of operating in a space environment, the Boeing Space Center at Kent, Wash., said. Function of the package had been to transform raw power collected by solar panels into electric current acceptable to the spacecraft. NASA had asked Boeing to find out if electronic units like this could stand the rigors of long-distance space flights such as a 12-yr Grand Tour of the outer planets.

Boeing put the Mariner instrument in a test chamber—part of an extensive vacuum and space simulation facility—in which temperature had been fixed at 1.6°C and other conditions had been made to conform with those in space. The duplicate had performed perfectly, outlasting much of its ground-support equipment and all but one of the technicians originally assigned to the experiment. The test had proved that flight hardware could stand temperatures as much as 30°C cooler than those encountered on missions already completed, and that similar electronic packages had a life expectancy of more than 5 yr, far longer than the period for which they had been designed. (Newport News *Times-Herald*, 16 Oct 75, 15)

- Goddard Space Flight Center announced selection of PMI Facilities Management Corp. for a 2-yr $3 500 000 cost-plus-award-fee contract to provide processing, preparation, retrieval, reproduction, and distribution of space-science data for the National Space Science Data Center. PMI would also maintain, upgrade, develop, and implement plans for improved data-handling information systems, and would provide scientific analyses and programming, prepare reports, maintain files, order supplies and equipment, and operate photographic and microfilm equipment.

  The Data Center served as a NASA-wide depository for space-science data and a means of analyzing and disseminating this information beyond that provided by principal investigators. (GSFC Release G−24−75)

*17 October:* More than 1000 instrumented platforms—either floating buoys or balloons—were being used to provide information on tropical and polar weather and ocean and ice conditions, NASA announced. The platforms would transmit data from their sites to NASA's *Nimbus 6* meteorological research satellite, launched 12 June 1975, which would record and play back the data on command to ground stations around the world. The stations would then disseminate the information to investigators in Australia, Brazil, Canada, France, Norway, South Africa, and the U.S. Platforms would be located in such remote areas as the South Pole, Indian Ocean, Africa, Samoa, and the Arctic.

The National Center for Atmospheric Research was releasing more than 400 balloons to altitudes up to 14 km from American Samoa, Ghana, and Ascension I. to study the atmosphere in the tropics. About 50 low-level balloons were being released to an altitude of 750 m from the Seychelles island group, just north of the Malagasy Republic, by the Laboratoire de Meteorologie Dynamique in Paris to study the air-sea interface to determine its effect on monsoons in India. About 30 ice buoys were being placed in the Arctic and Antarctic, one only 45 m from the South Pole, as part of a major U.S. government project to understand ice packs, icebergs, and weather patterns near the poles. Scientists from the Norsk Polarinstitut in Norway were deploying four drifting buoys to help chart ice drift in the Greenland area. (NASA Release 75−271)

*20 October:* Kennedy Space Center announced a reorganization to prepare for the Space Shuttle era of space operations. John J. Neilon, Director of Unmanned Launch Operations since 1970, was named Deputy Director of Technical Support; George F. Page, Chief of Operations, Div. of Manned Spacecraft Operations, would move into Neilon's former post.

The Launch Operations Directorate would be reorganized into four major second-level groups: engineering, payloads, space transportation systems processing, and expendable vehicle operations. (KSC Release 253—75)

- The Soviet Union may establish a ground station to receive data directly from the U.S. Landsat earth resources satellites, under a cooperative program in which the U.S. would establish facilities to receive data from Soviet weather satellites, NASA Administrator James Fletcher announced in Moscow. Landsat could be of enormous value to the Soviet Union, which is judged to have vast untapped oil and mineral deposits, which depends heavily for power on its water resources, and which needs help in improving crop yields and forecasting. (SBD, 20 Oct 75, 255)

*20—22 October:* More than 100 astronomers from the United States and several foreign countries had gathered at Goddard Space Flight Center to exchange information on x-ray astronomy. The symposium had been arranged to discuss compact double-star systems that emit x-rays and that had been observed in other spectral bands—radio, optical, ultraviolet. Earth's galaxy alone contained 100 binary (double-star) x-ray objects, discovered by spacecraft observing their emissions beyond the radiation-absorbing blanket of earth's atmosphere.

Astronomers had considered the compact stars in these binary systems the most exciting objects of current research: the stars have masses close to that of earth's sun, but diameters of only about 10 km. Neutron stars, for example, had been found so dense that a teaspoonful would weigh a billion metric tons, and "black holes" would be even more dense. Five satellites currently in earth orbit had been observing celestial x-rays: *Copernicus, Oso 8,* and *Sas 3*, managed by GSFC, and the U.K.'s *Ariel 5* and the Netherlands' *Ans 1*, for which GSFC had coordination responsibility. In the past year spacecraft had discovered six new and extremely powerful x-ray sources, brightness in the sky, known as x-ray novas from their sudden brightening and subsequent fading after several months. One of these x-ray novas is as bright as earth's sun and more than 10 times brighter than the easily visible Crab Nebula, most famous of the supernovas (exploding stars). (NASA Release 75—279; JSC *Roundup*, 24 Oct 75, 1)

*21 October:* Goddard Space Flight Center had awarded a $45.8-million contract to RCA Astro-Electronics Division for design and production of eight third-generation Tiros-N spacecraft, the company announced. Scheduled for launch in 1978, the 635-kg spacecraft would forecast weather 2 days in advance with accuracy equal to that of present 1-day predictions. RCA earlier had received a $16.6 million NASA contract for long lead-time items for the Tiros-N series, bringing the total cost of the satellites to $62.4 million. (SBD, 21 Oct 75, 259)

- Jet Propulsion Laboratory and the Energy Research and Development Administration announced selection of 16 organizations to negotiate contracts for the development of low-cost solar cells for residential and commercial use. Twenty-two contracts, totaling approximately $12 million, would be awarded as part of the Low Cost Silicon Array Project managed by JPL for ERDA. Objective of the project was to create by 1985 an industrial capability of producing silicon solar cells or arrays at a market price of less than $500/kw. Cost to date had been $20 000/kw. The demand for solar cells was expected to increase from 100 kw annually in 1975 to 500 000 kw annually by 1985.

JPL negotiations included production of low-cost solar-grade silicon, economical production of silicon in large-area sheets suitable for use in solar cells, development of economical encapsulation materials and techniques for an array lifetime of 20 yr, and development of automated processes and facilities for low-cost production of arrays.

Future contracts would be negotiated to procure array modules for a series of ERDA-funded tests of photovoltaic systems, in cooperation with the Department of Defense to determine the technical feasibility of using photovoltaics for military applications, and in cooperation with Lewis Research Center for a similar civilian effort. (NASA Release 75—280)

*22—25 October*: Two unmanned Soviet probes, *Venera 9* and *10*, dropped landers to the surface of Venus after 4.5-mo journeys and relayed to earth through 297 million km of space the first photographs of the Venusian surface. The two main probes went into differing 2-day-period orbits about the planet. The *Venera 9* lander functioned for 75 min in the atmosphere and 53 min on the surface, and the *Venera 10* for 75 min in the atmosphere and 65 min on the surface.

The descent module of *Venera 9* separated from the main spacecraft 20 Oct., 2 days before it reached the Venus atmosphere, which it entered at 11:58 pm EDT 21 Oct., landing on the surface at 1:13 am EDT Oct. 22. When the atmosphere had slowed the descent sufficiently, a parachute system had been deployed at about 50 km above the surface, and the lander studied the cloud layers and atmosphere during the rest of the descent. The lander modules had been built to withstand the hostile environment at the Venus surface, with temperatures that reach 500°C and pressures up to 100 times that of earth's atmosphere. Before landing, the modules had been cooled to $-10°C$ to provide a reserve of cold temperature that would permit operation for about an hour before overheating. *Venera 10*, launched 14 June and following the same procedure, had separated from its orbiter 23 Oct. and had landed on Venus at 1:17 am EDT 25 Oct.

*Venera 9*, launched 8 June from the Baykonur cosmodrome, was the third Soviet spacecraft to make a soft landing on Venus: *Venera 7* and *8*, which landed on Venus in Dec. 1970 and July 1972, respectively, had transmitted technical data for 23 and 50 min, respectively, but did not carry cameras.

Pictures sent by the *Venera 9* lander showed a rugged surface with sharp-angled and flat-surfaced rocks 30 to 35 cm across and some boulders. Pictures sent by the *Venera 10* lander, which touched down 2200 km from the site of *Venera 9*, showed what seemed to be an older mountain formation with smooth rounded rocks. Soviet scientists had expressed amazement that both sites were bathed in light. The dense cloudcover that hides Venus from earth had been thought to keep the planet in perpetual shade. Dr. Arnold Selivanov, one of the TV system designers, said that they had "expected poor lighting conditions and built our instruments accordingly"; this is why they had received "good pictures...that can be used by newspapers without any preliminary processing."

The horizon of Venus discernible in a corner of the *Venera* photos did not appear concave (ends bent upward), as had been expected as a result of the theorized distortion resulting from the planet's dense atmosphere. The apparent geological youth of the rocks on Venus "testifies to recent catastrophic processes like volcanic eruptions or earthquakes—we should say Venusquakes—which took place recently," said Soviet mineralogist Dr. Aleksandr Basilevsky. "Venus apparently is internally active." This conclusion had affirmed American theory that Venus resembled earth in having a liquid core, a mantle, and a rocky crust.

Four Soviet scientists reviewing the *Venera 9* and *10* program in an article in *Pravda* had summarized the descent-module findings: Temperature and pressure at the landing sites, 460°C and 90 atm, respectively; surface illumination, about 10 000 lux; local wind velocity at landing site, 0.4 to 0.7 meter per sec (*Venera 9*) and 0.8 to 1.3 meter per sec (*Venera 10*); natural radioactive elements in Venus rock, 0.3% potassium, 0.0002% thorium, and 0.001% uranium. Orbiting module findings were: Temperature of Venus clouds at upper boundary, about −35°C; cloud temperatures on nocturnal side, about 10°C higher; atmospheric temperature decreased with altitude, but local temperature elevations were observable at the 66 to 55 km level; Venus ionosphere was thinner than the terrestrial and closer to the surface; and the electron concentration on the daytime side was significantly higher than on the nocturnal side but about 10 times smaller than in the terrestrial ionosphere. Much of the information relayed from Venus had still to be processed.

Cameras in the orbiting spacecraft of *Venera 9* and *10* had proceeded with photographing the dense cloudcover of Venus, their first pictures received 26 Oct. showing a panorama of more than 1500 km of the clouds circulating around the planet's equator. Scientists at the Deep Space Communications Center in Moscow said the pictures "confirm the presence of powerful streams of circulating flows in the planet's atmosphere, first detected on board the American probe *Mariner 10* in February 1974." There was no indication, the Baltimore *Sun* commented, how long the two Venus mother ships would continue in orbit. (*NYT*, 23 Oct 75, 1; 26 Oct 75, 1, sec. 4 p 7; FBIS 208, 210, 25 Oct 75; *W Star*, 23 Oct 75, A3; B *Sun*, 26 Oct 75, A11; 27 Oct 75, A4; *Defense/Sp Bus Daily*, 23 Oct 75, 272; 28 Oct 75, 289; 29 Oct 75, 301; *Pravda*, 21 Feb 76, 3−4)

*23 October*: Marshall Space Flight Center announced selection of Integrated Systems Support, Inc. to negotiate a $2.8-million cost-plus-award-fee contract for computer support and facility operations at MSFC's Slidell Computer Complex. The 1-yr contract had provisions for four additional 1-yr extensions. (MSFC Release 75−224)

*24 October*: Marshall Space Flight Center announced selection of nine experiments on processing materials in space to be carried aboard a Black Brant 5C sounding rocket scheduled for December launch. The launch would be first in a 5-yr series of space-processing flights in MSFC's Space Processing Applications Rocket (SPAR) project to find improved methods of materials processing on earth, and ultimately to produce materials in space that could not be produced on earth.

The nine experiments were: dendrite remelting and macrosegregation; liquid mixing; lead-antimony eutectic; foams from sputter-deposited metals; thoria-dispersed magnesium; dispersion-strengthened lead-silver alloys; particle-interface interactions; bubble behavior in melts; and contained polycrystalline solidification.

The low-cost rocket launches, each of which provided approximately 5.5 min of low gravity (0.001 of earth's) during the coast phase of the trajectory, would offer the only means of obtaining scientific data on space processing of materials in near-zero gravity until flights of the Space Shuttle in the early 1980s. (MSFC Release 75−225)

- Marshall Space Flight Center had issued requests for proposals for development of solar-energy cooling and heating systems for both residential and commercial applications. The RFPs had been issued in five areas of procurement: firms already producing and marketing solar heating and cooling subsystems; firms developing marketable subsystems and needing government assistance; firms with complete systems under development; firms designing innovative complete systems; and firms to compete for test and evaluation of the subsystems procured under the first two RFPs. The Solar Heating and Cooling Demonstration Act of 1974, which requires demonstration of the practical use of solar heating in the U.S. within 3 yr and of combined solar heating and cooling within 5 yr, made the Energy Research and Development Administration responsible for managing and coordinating the national program. ERDA assigned MSFC the responsibility for developing solar heating and cooling systems in an all-out effort to create a viable industrial and commercial capability of producing and distributing such systems. (NASA Release 75−284)

*25 October*: Travelers on the scenic Beartooth highway in Custer Natl. Forest, Mont., had access to the first solar-powered privy, the Baltimore *Sun* reported. The installation had advantages in addition to making the Rock Creek Vista area a national leader in progressive plumbing, according to the U.S. Forest Service: it had overcome two basic problems of the site, lack of water and lack of power. The oil-based flush system, which cost $29 878 with toilets and storage vaults, recycled a white oil. The solar panels over the two-sided toilet installation, which cost $9750, had almost no upkeep expense; they had offered unlimited power without the cost of installing electrical lines, solving the problem of power sources that plagued many mountaintop areas of the nation's forest system. (B *Sun*, 25 Oct 75, 11)

*26 October*: NASA announced approval of Rockwell International Corp. to begin final design and construction of two subscale models of highly maneuverable aircraft. The construction would be completed under an $11.3-million contract awarded earlier. The Rockwell design called for a 6.3-m-long 1260-kg model with a "coke bottle" fuselage and two large canards mounted on each end of the 2.5-m wing. The model would be powered by a General Electric Co. supersonic J85 supersonic jet engine.

The two unmanned models would be flown from NASA's Flight Research Center in a joint NASA−Air Force program to develop advanced technology in high-maneuverability aircraft. The models would be air-launched from large carrier aircraft and maneuvered by a pilot in a ground cockpit, using test techniques developed during NASA's Remotely Piloted Research Vehicle program. (NASA Release 75−279)

*27 October*: The Commerce Dept. had awarded Aeronutronic Ford, Inc., a contract for $5 581 824 to design and build a space environment monitor for the National Oceanic and Atmospheric Administration, which had commissioned it to make constant measurements of energetic solar particles when Tiros-N would be launched in 1978. Tiros-N, a joint mission of NOAA and NASA, would be first of a new series of polar-orbiting satellites to be operated by the National Environmental

Satellite Service as "a major improvement in our capability of monitoring environmental phenomena," according to Dr. George H. Ludwig, director of NESS's Office of Operations. Detecting changes in the space environment and learning how they affect conditions on earth would help scientists to predict and prepare for these effects. The NOAA space-environment monitor, the only set of sensors facing away from the earth, would provide much of the needed data. (DOC Release G 75–180)

*28 October: Heos 1*, Europe's longest-living satellite, reentered the earth's atmosphere as predicted after exceeding its nominal lifetime by nearly 6 years. Launched 5 Dec. 1968 by the European Space Agency, *Heos 1* was third of eight scientific satellites launched by ESRO/ESA between 1968 and 1975. Its full payload of 7 experiments gave excellent results 4 months longer than the nominal lifetime; the experiment to study the interplanetary magnetic field worked until reentry, providing invaluable data on the field for almost 7 of the 11 yr of a solar cycle. Experiments were: interplanetary magnetic field, high-energy galactic cosmic rays, and solar protons (Imperial College, London); solar wind (Universities of Florence, Rome, and Brussels); low-energy cosmic rays (Centre d'Etudes Nucleaires); high-energy electrons (CEN/University of Milan); barium cloud (Max Planck Institut). (ESA Release 29 Oct 75)

- NASA Administrator Dr. James C. Fletcher and Secretary of the Interior Thomas S. Kleppe announced the selection of Dr. William Nordberg, Goddard Space Flight Center scientist, and Dr. Carlos Brockmann, director of the Landsat project in Bolivia, to receive the 1975 William T. Pecora Award. Dr. Nordberg was selected for his meteorological contributions to the Landsat program and for his management of day-to-day research operations using data from *Landsat 1* and *2*, launched (as *Erts 1*) on 23 July 1972 and on 22 Jan. 1975. Dr. Brockmann was selected for his leadership of a multidisciplinary team of Bolivian scientists responsible for the evaluation and application of Landsat images. Within 1 year, the team had produced evaluation reports covering geology, volcanology, forestry, hydrology, and land use of Bolivia; the group also published an improved national atlas of Bolivia.

  Given annually by NASA and the Dept. of Interior in recognition of "outstanding contributions of individuals or groups toward the understanding of the Earth by means of remote sensing," the award would be presented at the joint U.S.–U.S.S.R. 4-day symposium on earth resources at Sioux Falls, S. Dak., 28–31 Oct. (NASA Releases 75–283, –285)

- Eight Soviet scientists would attend a meeting at the Earth Resources Orbiting Satellite (EROS) Data Center in S. Dak. this week to review previous U.S.–U.S.S.R. remote-sensing projects [see 20 Oct.] and to discuss future cooperation on remote sensing of geology and agriculture. The meeting was part of a joint effort by NASA and the Soviet Academy of Sciences to advance studies of the earth from space; the Soviets might set up a ground station to receive remote-sensing data directly from U.S. Landsats. (*SBD*, 28 Oct 75, 295)

*28–29 October:* Principal investigators and other scientists meeting at the Marshall Space Flight Center had presented preliminary findings on the space-processing experiments conducted during the U.S.–Soviet Apollo-Soyuz Test Project in July. First day's discussions would be on the U.S. and West German experiments on electrophoresis—separation of biological materials by means of an electrical field—considered a key to the development of drugs to fight strokes, heart attacks, clots, and blood diseases. Six of the seven materials-processing experiments using a multipurpose electric furnace would be discussed the second day; a Soviet experiment using the furnace would not be discussed. The furnace experiments had demonstrated use of a weightless environment to investigate crystal growth, convection, and solidification processes for future use in space as well as for applications to present earth technology. (MSFC Release 75—226)

*29 October*: NASA and the Energy Research and Development Administration dedicated a 100-kw wind turbine designed and built by Lewis Research Center at Plum Brook Station for ERDA. NASA Administrator Dr. James C. Fletcher and ERDA Administrator Robert C. Seamans, Jr., attended the ceremony conducted by LeRC Director Dr. Bruce T. Lundin. Objective of the wind turbine project was to test the performance and define the operating and economic characteristics of wind-energy systems for generating commercial electric power in the future. The installation will provide data for use in developing larger and more advanced systems that could be produced commercially by private industry. The Plum Brook installation was the largest wind-energy system in operation and the second largest ever built; the largest, a 1250-kw system operated between 1941 and 1945 in Vermont, had been abandoned because of economic difficulties and availability of less expensive fossil energy. The new generator, with its 38-m-wide propeller blades, was built on top of a 30-m-high open-truss steel tower and could supply the power needs of about 25 homes. NASA was participating in the ERDA program because of its capability in energy-related technology. (NASA Release 75—277)

- The Political & Security Committee of the U.N. General Assembly unanimously adopted a resolution seeking wider adherence to international agreements on the peaceful uses of outer space. The resolution called for continued U.N. work on space agreements, including one covering natural resources of the moon; on principles governing direct TV broadcasting by satellite; and on legal issues arising from remote sensing of earth and its resources from space. (*SBD*, 29 Oct 75, 300)

*30 October*: The Saturn V dynamic test stand at Marshall Space Flight Center was being modified under a $1 923 400 contract between the Army Corps of Engineers and the Universal Construction Co., to provide a Space Shuttle Mated Ground Vibration Test Facility, MSFC announced. The 123-m-high structure would be used to test the vehicle in launch and boost configuration, to determine the bending modes and dynamic response during launch and ascent conditions. (MSFC Release 75—230)

- NASA announced appointment of William H. Rock as Deputy Director of Flight Research Center, effective 9 Nov. Rock joined NASA in 1964 and had served at NASA Headquarters, at Goddard Space Flight Center, and since 1968 at Kennedy Space Center, where he was associated with various applications and manned flight programs. (NASA anno, 30 Oct 75)

*31 October*: Kennedy Space Center announced award of a $4 778 000 contract extension to Bendix Corp.'s Launch Support Div. for support services including operation and maintenance of Launch Complex 39, mobile launchers, mobile service structure, industrial complexes, technical shops, propellant systems, life-support facilities, and a components-cleaning laboratory. The 4-mo extension brought the total value of the contract to $266 662 753. Complex 39, launch site for the Apollo, Skylab, and Apollo-Soyuz Test Project missions, would be the launch and landing site for Space Shuttle. Modification of the facilities was under way. (KSC Release 260−75)

- Marshall Space Flight Center announced selection of Moog, Inc. for a $6 685 584 cost-plus-incentive-fee/award fee contract to design, produce, test, and maintain the Solid Rocket Booster Thrust Vector Control (TVC) electrohydraulic servoactuators. The devices move the solid rocket booster exhaust nozzle, providing directional control by gimballing the nozzle during propellant burn. (MSFC Release 75−232)
- Langley Research Center had awarded General Electric of Phila. a 2.5-yr $3.2 million contract to provide an orbital scatterometer for the SEASAT−A spacecraft scheduled for launch in 1978. The scatterometer would measure wind speed and direction. (*SBD*, 31 Oct 75, 317)

*During October:* Air Force scientists under contract to NASA had completed a study of the physiological effects on man resulting from transfers from one atmosphere to another during the 15−24 July Apollo-Soyuz Test Project. The difference between the Soyuz 7031-kg-per-sq-m atmosphere of 30% oxygen and 70% nitrogen and the Apollo's 3515.5-kg-per-sq-m 100% oxygen represented the transition between 3048 m and 8230 m in altitude. In 1973, scientists, using a hyperbaric chamber to simulate spacecraft environment, had found that volunteer subjects showed symptoms of decompression sickness (the bends) when moving from one atmosphere to the other. During the following year the scientists had found a way to lessen the chance of decompression sickness: The most satisfactory procedure was to increase the time allowed for denitrogenation (breathing 100% oxygen at ground-level pressure). This helped to remove excess nitrogen—cause of the bends—from the body. The Air Force study had helped NASA and Soviet ASTP planners schedule the actual transfers. (ASFC *Newsreview*, Oct 75, 8)

- A 50 000-kg towing tractor, first piece of operational ground-support equipment manufactured by International Harvester Co. for Kennedy Space Center's Space Shuttle operations, was accepted for delivery to KSC. The T−500S Paymover tractor would tow NASA's Boeing 747 carrier aircraft or the orbiter, or the mated 747-orbiter combination after use at the Flight Research Center, Edwards, Calif., during orbiter tests. (KSC Release 257−75)
- A new computer-controlled tracking device for telescopes, called VIP (video inertial pointing), had been developed at Ames Research Center to improve the aim of balloon-borne, airborne, and space telescopes. The two main functions of VIP were to help the astronomer steer the telescope, and to continuously update the automatic stabilization system that holds the telescope steady during viewing. Automatically stabilized telescopes heretofore had used gyroscopes to keep the telescope aimed at a point in space and counteract the rotation of the earth and other disturbances; gyroscopes had been found subject to drift, causing the telescope to wander off target. VIP, using a single video sensor, could generate the signals required for all

three control axes—roll, pitch, and yaw—and could use the visual information to point at nonvisible (infrared) sources. Previous systems using star trackers had needed from two to six conventional sensors to do the same job.

Development of VIP had made use of several state-of-the-art devices: the microprocessor, a complete digital computer on a single printed circuit card, was the computer inside the VIP electronics. Two types of video sensors used by VIP were the solid-state imaging device called a CCD (charge-coupled device) that incorporated a matrix of light-sensitive elements on the surface of a silicon chip measuring 0.01 mm on a side, and an advanced vidicon tube for night vision called an ISIT (intensified-silicon intensified-target) camera. By reducing the amount of hardware required for good performance, use of these new instruments had lowered the cost of the system. Ames would use VIP to provide accurate automatic pointing of the Spacelab Infrared Telescope Facility for the Shuttle that was under development at Ames and on contract. (Ames *Astrogram*, 9 Oct 75, 1)

- NASA had awarded a $45.8-million contract to Radio Corporation of America Astro-Electronics Div. to design and build eight Tiros-N third generation meteorological satellites. The 635-kg Tiros-N (Television InfraRed Operational Satellite) would have 4 times the payload capacity of the previous generation of metsats, and was expected to be able to forecast weather 2 days in advance with accuracy equal to the 1-day predictions. In addition to providing day and night imagery, the Tiros-N spacecraft would take atmospheric and sea-surface temperature readings; measure proton, electron, and alpha-particle activity surrounding the earth; and collect data from weather platforms on balloons and from ocean buoys. (GSFC *Goddard News*, Oct 75, 1)

- The Mass. Inst. of Technology Lincoln Laboratory, under contract to the Air Force, was developing an experimental ground electro-optical deep-space surveillance (GEODSS) system to scan nighttime skies for orbiting and geosynchronous satellites. The system was being installed at White Sands Missile Range, first of a proposed five-site GEODSS worldwide surveillance network. The network would augment the Air Force's Aerospace Defense Command's spacetrack coverage which maintained a day-by-day count of all satellites and debris circling the earth.

    GEODSS combined two sophisticated telescopes with electro-optics, a TV camera, and a digital computer, as well as system-related equipment. Mounted at the focal plane of the larger telescope was a TV camera to transmit the signal to the observatory control room for instant screening and analysis. (AFSC *Newsreview*, Oct 75, 14)

- The Air Force had successfully test-flown an F−111D turbofan jet engine with all-composite third-stage fan blades 40% lighter than conventional titanium fan blades. The tests—made at Edwards Air Force Base—marked the first military flight evaluation of a rotating structural engine component made of composite materials. Objectives of the flight-test program were to determine long-term environmental effects on the material and to verify potential operation and maintenance cost benefits. Use of composites could provide a 15 to 20% weight savings in jet engines, allowing increased payloads or extended range. (AFSC *Newsreview*, Oct 75, 1)

- Limitations imposed by the Strategic Arms Limitation Talks (SALT) agreements had made the development of the B-1 bomber essential, the Air Force Systems Command *Newsreview* reported Secretary of Defense James R. Schlesinger as telling Rep. Richard Bolling in an interview. Because the U.S. would be limited in megaton strike capability and approximately 60% of U.S. megatonnage was carried by bombers, it was essential to maintain a significant and highly effective bomber force. (AFSC *Newsreview*, Oct 75, 2)
- Two visiting research associates working under 2-yr associateships awarded jointly by NASA and the National Research Council completed their research at NASA. While at NASA Dr. Ditmar Kranzer of the Technical University of Vienna, Austria, had designed and fabricated charge-coupled devices on silicon and sapphire wafers. This type of electronic device would be used as part of an electronic circuit for imaging, storage, and signal processing. The device had a high speed capacity, a low power requirement, and a very high packing density, which would decrease the number of outside connections needed for the circuits and reduce the area required for data storage.

    Dr. Tsuneo Yoshikawa had developed steering laws for single-gimbal and double-gimbal control-moment gyro systems during his 2-yr tenure at NASA. These laws, which apply to any spacecraft requiring highly accurate attitude-control systems, had been formulated in an effort to minimize performance deterioration. (MSFC Release 75–213, –219)
- According to International Telecommunications Union figures, a satellite had been launched every 3 days in 1974. Of the total of 124 launched, 18 belonged wholly or partly to the U.S., and the Soviet Union accounted for 95. (GSFC *Goddard News*, Oct 75, 3)

## November 1975

*1 November*: A newly discovered dwarf galaxy called Snickers, 55 000 light years from the sun, and nearest neighbor to the Milky Way yet found, could lead to more accurate estimates of the size of earth's galaxy. Announcing his find in the *Astrophysical Journal Letters*, Dr. S. Christian Simonson of the U. of Md. said the new galaxy had been hidden from earth view by dense star fields and cosmic dust clouds. Spots of hydrogen gas detected on radioastronomy maps were the clue that led to the discovery, first to be made solely through radiotelescope observations of hydrogen gas. Until now, the Magellanic Clouds—twin galaxies about 205 000 light years away—had been considered the galaxies nearest to the Milky Way. Dr. Simonson's new galaxy had been christened Snickers by his colleagues, who said "it was like the Milky Way, only peanuts." (B *Sun*, 2 Nov 75, A23)

*2 November*: Efforts to charge batteries on the *Viking 2* lander failed for unknown reasons, but controllers at Jet Propulsion Laboratory were confident that time would allow them to solve the problem. Both *Viking 1* and *2* were launched with their lander batteries essentially uncharged to prolong their lives; the batteries aboard the *Viking 1* lander were charged without difficulty 2 wk ago, and the same procedure had been scheduled for Viking 2. The spacecraft carried a backup as well as a prime battery charger, but NASA had not attempted to use it in case the problem had originated outside the charger system. Use of the backup before the problem had been identified might have damaged it as well as the prime charger. *Viking 2*, originally scheduled to be launched first, had been pulled back for repair after its orbiter batteries had been accidentally drained; officials said there was no connection between prelaunch orbiter problems and the lander situation. (NASA Release 75–288; *Aero Daily*, 4 Dec 75, 10; *SBD*, 4 Nov 75, 16)

- Marshall Space Flight Center forwarded the experimental payload for the first launch of NASA's Space Processing Applications Rocket (SPAR) project to Goddard Space Flight Center, where the payload would be integrated with its launcher, a Black Brant sounding rocket. The integrated unit would undergo checkout before being sent to N. Mex. for firing by White Sands Missile Range early in Dec. The payload consisted of nine scientific experiments on processing materials in near-weightlessness. After the first launch, three flights per year carrying similar payloads had been planned through 1980. All payloads were to be recovered by parachute for ground analysis. The low-cost SPAR missions would provide data on space processing until Space Shuttle flights began in the early 1980s. (MSFC Release 75–233)

*4 November:* Inflation and procurement reforms had hiked NASA's charge for a Delta launch by about 20% for the average customer, *Aerospace Daily* said. Prospective users 2 yr ago had been quoted less than $10 million per launch; current cost had risen to $11–12 million, and those wanting 1978 launches had been quoted $13–14 million. The Delta had been the NASA launcher most in demand because of its size and price, most widely applicable among government and commercial users; the Kennedy Space Center, capable of one launch every 5 wk, had been booked solid for the foreseeable future. NASA had been renegotiating payment schedules, but no launch schedules had been changed except for technical reasons not related to cost. (*Aero Daily*, 4 Nov 75, 11)

- Largest of U.S. air carriers, United Airlines, and All Nippon Airways of Japan—one of the 10 largest airlines in the world—supported NASA's 10-yr program to reduce aircraft fuel use. Testifying before the Senate Aeronautical and Space Sciences Committee, executives of both companies said the airlines needed the improved technology but did not have the financial ability to conduct the research, estimated to cost $490 to $670 million. A spokesman for the Federal Energy Administration said more analysis was needed before Congress decided to fund the NASA program; the airlines should consider less expensive ways of conserving fuel, such as operating at higher load factors. (Transcript, 256, 271, 299)

- Soviet polar station Severny Polyus−22 had been scheduled to get a companion, a Moscow Tass broadcast reported. Drifting on an ice floe in the Arctic for the past 3 yr, the station's scientists had made about 200 scientific landings on the ice since last spring to observe the ice, the ocean, and the atmosphere. Another ice floe in the area of the "inaccessibility pole" was to receive station Severny Polyus−23, and the first deliveries of prefab houses and stocks of fuel and food had already been made by air. Scientists were to arrive at the new station in a few days. The high-latitude expedition, called Sever−27, was described as the biggest in the history of Arctic exploration, aimed at a comprehensive survey of the Arctic Ocean between the coasts of the U.S.S.R., Canada, and Greenland; about 50 scientists had participated. Work had also been done under an international program called Polex, to photograph the bottom of the Arctic Ocean and to study the lower surface of the polar ice cap. (FBIS No 214, 4 Nov 75)

*4−5 November:* A study called "Outlook for Aeronautics," on probable aeronautical progress during the next 25 yrs, had forecast development of a Concorde II supersonic transport and an entirely new second-generation SST, NASA witnesses told the House Subcommittee on Aviation and Transportation R&D. The hearing was first of a series called to get an early start on H.R. 11573, NASA's FY 1977 budget authorization. Earliest dates of introduction for the Concorde II and advanced SST had been set at 1985 and 1995 respectively. Recent NASA technology work on the advanced SST included an anticipated engine-noise breakthrough based on noise-suppressor research, as well as on changes in engines and aircraft configurations. (Transcript, Vol. II Part 1: 9, 67, 117)

*5 November:* "Clear and immediate benefits to society" that NASA could produce would justify a 25% increase in its budget, said Rep. Don Fuqua (D−Fla.), chairman of the House subcommittee on Space Science and Applications, urging the space agency to come up with plans for such a program. Noting that the Administration had been considering across-the-board cuts in the Federal budget, Rep. Fuqua said the President and the Office of Management and Budget should reconsider the NASA cuts, as returns from the space program should give it higher priority than nonproducing areas of the budget.

The subcommittee's three-volume report, "Future Space Programs," called for new space systems for educational and medical services, and for new earth-survey satellites to provide maritime, agricultural, geological, and demographic data. The report stressed that the agency should offer both short-term and long-term plans. The latter should include plans for lunar bases, orbital colonies, extraterrestrial communications, planetary and stellar exploration, satellite solar power, and disposal of nuclear waste.

Chairman Olin E. Teague (D–Tex.) of the House Committee on Science and Technology endorsed the subcommittee report, which he said contained "sensible recommendations," noting particularly the concept of space-based generation of electrical power. Growth of the nation, said Teague, "is fundamentally dependent upon our continuing its number one ranking in science and technology." (*W Post*, 6 Nov 75, A10; *NYT*, 6 Nov 75, 51; *SBD*, 6 Nov 75, 28–30)

- HASPA—the Navy's high-altitude superpressured powered aerostat—failed to reach its predicted altitude when the balloon-inflation mechanism malfunctioned. When fully extended, HASPA had measured 101.5 m long and 20.4 m in diameter. The launch at Kennedy Space Center had been planned as a 3- to 5-hr flight; during launch, the mylar container was inflated with the helium contained in the upper third and restrained there by a collar that would drop away to give the gas room to expand with altitude. The mechanism failed and the balloon inverted at 11.26 km, releasing the helium through a vent in the nose of the balloon, which impacted at KSC a half hour after launch. HASPA had been designed to function as a low-level satellite for a variety of payloads. (*Spaceport News*, 14 Nov 75; ETR PIO, interview, 5 Nov 75)

*6 November:* NASA had awarded a $2 548 265 contract amendment to Aeronutronic Ford Corp. of Houston for operation of the Mission Control Center at Johnson Space Center. The amendment to a contract originally awarded in 1963 called for additional labor and materials that would bring the value of the contract to $267 710 123. The work would employ 513 persons at the three AFC facilities in Houston, and in Penn. and Calif. (JSC Release 75–94)

- Workmen on the largest building in the world using solar energy for heating and cooling had doubted that the system would work, until they turned on a water hose, said Dr. Robert San Martin of the Energy Institute at N. Mex. State Univ. When the time came to put the collectors on the roof, the foreman told the crew to run water through them with a garden hose; the workmen forgot that the collectors had been in the sun and were quite hot. "When they ran the water through, violent steam came shooting out the other end," said Dr. Martin. "At that moment they were convinced."

  The one-story building, a facility for the N. Mex. Dept. of Agriculture, would use 330 units mounted on its roof as a solar collector. Fluid circulated through the collectors heated by the sun would provide 80% of the heating and cooling of the 2349-sq-m structure. Biggest test would be use of solar energy for cooling, relatively untried as yet. (*NYT*, 6 Nov 75, 34)

- Harnessing nuclear fusion reactions to generate electricity on earth moved a step nearer reality when scientists at the Mass. Inst. of Technology reported achieving a fivefold improvement in the confinement of hydrogen in a magnetic container where it could be heated and compressed to the point of fusion. A test reactor at MIT called Alcator had successfully raised the magnetic field containing the hot plasma to 75 000 gauss, about 150 000 times the strength of earth's magnetic field at the equator; this, said Dr. Robert C. Seamans, Jr., Administrator of the Energy Research and Development Administration, "exceeds by a factor of five anything previously achieved anywhere in the world."

The Soviet Union in the 1960s had developed toroidal (doughnut-shaped) machines called Tokamaks that produced better temperature, containment time, and particle density in combination than anything before; a similar machine built in France had achieved a figure of 2 trillion for the density-multiplied-by-containment-time, at a temperature of 20 million°C. MIT's Alcator had achieved 10 trillion, at a temperature of 10 million°C. ERDA officials and university scientists noted that the plasma confinement would have to be another 10 times as effective, at temperatures 10 times higher, than those of the Alcator experiments before fusion could be achieved artificially in the laboratory. Although the U.S. had begun a program to build the first fusion-power plant by the end of this century, the economy of this new power source would remain doubtful in spite of the availability of cheap fuel from the ocean: The costs of such plants had been estimated at billions of dollars each. (*NYT*, 6 Nov 75, 27; *CSM*, 6 Nov 75, 1; B *Sun*, 6 Nov 75, A1)

- The Aerospace Industries Association called on NASA to abide by government policy that relied on the private sector for goods and services, charging that increasing in-house activity by the government had been detrimental to the economics of the private sector. Karl G. Harr, president of AIA, told the House Subcommittee on Aviation & Transportation R&D during hearings on H.R. 11573 that a proper partnership between NASA and the aerospace industry should be maintained, and warned of the prospect that the government might move to "increase its share of that partnership." Involvement of government in postresearch activities such as prototype development "threatens to weaken our private technological base," Harr said. "We do not feel that NASA and other government agencies should both identify the needs and pursue the solutions on their own." Commending NASA for its 10-yr aircraft fuel-reduction program, Harr said the program was "an ideal opportunity to develop the government-industry team into its most efficient and cost-effective form." (Transcript, Vol II Part I, 149 ff)

*9 November:* All four batteries on the *Viking 2* lander had been fully charged, NASA's Jet Propulsion Laboratory reported. The problem had been solved by using a backup charger system carried on the orbiter after the primary system failed earlier [see 2 Nov.]. Recharging took about 24 hr per battery. The voltage had to be replenished before the spacecraft entered Mars orbit and before the lander separated from the orbiter. The difficulty with the lander's primary charger had been traced to failure of one of four resistors in the circuitry. To prevent unforeseen problems with the backup charger, Viking project officials had decided to keep at least one of the batteries charged throughout the remainder of cruise flight, to ensure activation of the switch shifting the lander from its solar-panel power to that from radioisotope thermoelectric generators. The two RTGs had not been used for primary power during flight to avoid thermal problems from the heat given off by the units. (*LA Times*, 6 Nov 75, 1; *Av Wk*, 10 Nov 75, 6; *SBD*, 12 Nov 75, 64; *Langley Researcher*, 14 Nov 75, 1)

- States from New England to California had joined the competition to be selected as the site of a proposed Federal Solar Energy Research Institute that might be spending up to $50 million annually within 3 yr after its establishment. Competition had begun even before issuance of formal criteria describing the center's mission and needs, expected later in November. A report on the center by a special committee of the National Research Council—an agency of the National Academy of Sciences and the National Academy of Engineering—had forecast a staff of about 800 professionals and 800 support people. The committee had recommended a single institute administered by a nonprofit corporation, like those that govern many other large national laboratories; geographical location had not been considered as important as "intellectual atmosphere, technical suppliers and availability of services for a substantial laboratory." Issuance of criteria by the Energy Research and Development Administration would call for proposals early next year, with site to be chosen in April or May 1976 by Dr. Robert C. Seamans, Jr., Administrator of ERDA.

   Among those already in competition were the state of N. Mex., which had put a proposal for the solar center in Feb. 1975 when ERDA was a month old; a consortium of universities and industries on Long Island, N.Y.; the governors of six New England states, sitting as the New England Regional Commission, who voted to "work as a regional force" to get the center for New England; and the state of N.Y., which had designated a coordinator to work on getting the center, including "assembling the necessary real estate" from various areas of the state. (*NYT*, 9 Nov 75, 33)

*10 November:* Marshall Space Flight Center had awarded contracts worth $500 000 each to Martin Marietta's Denver Division and to TRW Systems Group, to define the Spacelab payload called AMPS (atmospheric, magnetospheric, and plasmas in space). The definition studies would emphasize flexibility, low cost, and an evolutionary approach to environmental research.

One of the first payloads considered for Spacelab, AMPS had been conceived as a manned orbiting scientific laboratory to study the near-space environment of earth and the effects on it of changes in incident solar energy and of emissions from earth. Instrumentation would include a laser beam to define the composition of the various constituents of earth's atmosphere. Missions would last from 7 to 30 days, with extensive involvement of the scientist crews in conducting the experiments. (MSFC Release 75-239; *SBD*, 20 Nov 75, 108)

- The U.S. space program was an adventure whose potential benefits were undefinable and incalculable, NASA Administrator Dr. James C. Fletcher told the National Academy of Engineering. Speaking on the "Outlook for the Space Program," Dr. Fletcher noted that the public and the Congress were "now" oriented, and that concentration on immediate benefits might jeopardize "the vast potential" of the national space program. Admitting that "NASA's present actions seem to speak louder than its words," in that its money had been spent on current needs rather than on tomorrow's goals, Dr. Fletcher said this did not result from lack of vision but from "accommodation with current constraints." Projects to be considered for NASA's future should include satellite systems to beam solar energy to earth; a permanent manned space station in orbit, as a forerunner of the space colony that had been proposed; exploration of the planets expanded to interstellar missions; and making contact with another intelligent race—"the most significant achievement of this millennium." (Text; NASA Release 75—294)

- A 25-member delegation from the Soviet Union, headed by Professor K. D. Bushuyev, U.S.S.R. technical director for the Apollo-Soyuz Test Project in July, had begun work at the Johnson Space Center on a joint U.S.-U.S.S.R. report on the mission. Cosmonaut Aleksey S. Yeliseyev, who had been U.S.S.R. flight director for ASTP, was a member of the delegation, which had been scheduled to remain at JSC until November 21. Dr. Glynn S. Lunney, who had been U.S. technical director for ASTP, headed the NASA group working with the Soviet specialists. (JSC Release 75—95)

*11 November:* Development of a reusable single-stage Space Shuttle that could perform better than the current partially reusable design, and cost less, would be feasible before the end of this century "if pushed aggressively enough," Robert E. Smylie, Deputy Associate Administrator for Aeronautics and Space Technology for NASA, told the House Subcommittee on Space Science and Applications. Studies by the agency had shown that structural weight reductions of 25% and propulsion-system improvements of 5 to 6%, relative to current space flight technology, could be achieved in the late 1990s. In other areas of space technology, Smylie cited the concept of solar-electric propulsion that had been considered for planetary missions using the Space Transportation System, and described research in automated processing of microcircuits, robot vehicles for planetary exploration, vitreous materials resistant to laser radiation, and dry lubricant with application in space payloads. (Transcript, Vol I Part 1, 696, 726, 1130 ff)

- NASA had begun preliminary long-range planning for a fourth earth resources satellite, Landsat—D, Sen. Frank E. Moss (D—Utah) told the American Mining Congress, and "the odds are that it will be optimized for agricultural studies." Moss had predicted that the U.S. would eventually have a Landsat optimized for mineral exploration; "the people who will decide . . . if we need a specialized satellite will be the scientific and engineering community itself. They must guide NASA in the . . . proper utilization of the environment of space."

NASA had reported that both *Landsats 1* and *2* were operating satisfactorily but under reduced capability in areas without ground stations because of tape-recorder deterioration. To expedite flow of Landsat information to users, NASA had planned expansion by May 1977 to handling 200 Landsat scenes a day and delivering magnetic tapes from the data centers to the user within 24 to 48 hr after receipt of data at GSFC. (*CR*, 11 Nov 75, S19689; *SBD*, 17 Nov 75, 84)

*12 November:* NASA and the Energy Research and Development Administration were near agreement on a plan to build a coal gasification plant near NASA's Michoud assembly facility in La. to provide hydrogen gas for testing and operating the Space Shuttle, said Dr. Myron S. Malkin, director of the Space Shuttle program for NASA's Office of Space Flight. In testimony before the House Subcommittee on Space

Science and Applications, Dr. Malkin said ERDA would hire a private contractor to build a $65-million plant and would operate it for 3 yr to obtain data on coal gasification; after that, it would turn the plant over to NASA. The facility would produce enough gas daily to provide 30 tons of liquid hydrogen for the Shuttle. (Transcript, Vol I Part 1, 858, 910)
- NASA had decided to eliminate all expendable launch vehicles except Scout after the Space Shuttle became operational, NASA Associate Administrator John F. Yardley told the House Subcommittee on Space Science and Applications, but the Air Force wanted to play it safe. AF officials wanted extra vehicles on hand after Shuttle in case problems developed, he said. NASA had planned about 20 launches of expendable vehicles a year through 1980—4 Scouts, 10 Deltas, 3 or 4 Atlas-Centaurs, and 1 or 2 Titan-Centaurs—with 17 launches in FY 1977 and 27 in FY 1978. (Transcript, Vol I Part 1, 899)
- Marshall Space Flight Center had selected Moog Inc. Controls Div. to negotiate a cost-plus-incentive-fee/award-fee contract totaling $6 685 584 for the design, development, production, and acceptance testing of a thrust-vector control electrical-hydraulic servoactuator on the Space Shuttle solid rocket booster. The servoactuator would control the SRB's exhaust nozzle. (*Marshall Star*, 12 Nov 75, 4)

*13 November:* The earth's ozone layer, a shield of heavy molecules of oxygen that prevents some harmful ultraviolet radiation from reaching the earth's surface, might be in less danger from aerosol sprays than from the increasing worldwide use of chemical fertilizers. Michael B. McElroy, professor of atmospheric sciences at Harvard University, had predicted that a projected increase in the use of nitrogen fertilizers over the next 25 yr could reduce the ozone layer 20 to 25%. Theoretically, each 1% reduction in the ozone layer could lead to thousands—as many as 15 000—additional cases of skin cancer in the U.S. alone.

All life needs nitrogen to make proteins, McElroy pointed out. The earth's atmosphere would have lost all its nitrogen long ago except for the process by which bacteria "denitrify" decayed matter in the soil; most nitrogen returned to the atmosphere in the same inert pure form in which it was taken and converted to use by earth's plant life, which then formed a source of nitrogen for other plants and animals; however, some nitrogen returned in a form that acted as a catalyst to destroy ozone. Man had recently been extracting nitrogen from the air and putting it into plant life in the form of fertilizer on a scale rivaling nature; by the year 2000, said McElroy, the amount of nitrogen extracted from the air would be double that removed by natural causes. With a sharp increase in the natural denitrification process, more nitrogen catalysts destructive to ozone would be released. Serious effects should be evident within the next 40 yr, McElroy said. Although aerosol sprays could be dispensed with, nitrogen fertilizers were a necessity for increased world food production. Increased incidence of skin cancer concerned light-skinned humans, who were unlikely to get much sympathy from darker-skinned inhabitants of areas where the fertilizers—and increased food production—are urgently needed. (*WSJ*, 13 Nov 75, 13)
- First major inflight tests of the *Viking 1* lander scheduled to touch down on Mars in July 1976 had shown the craft to be in excellent health,

flight controllers at Jet Propulsion Laboratory reported. JPL engineers completed thorough checks of the nuclear power, propulsion, guidance, attitude control, computer, tape recorder, telemetry, heat control, and radar landing systems, as well as four of the scientific instruments, including two cameras to photograph the Martian surface in closeup. The second Viking lander, scheduled to reach Mars in Sept. 1976, would undergo similar checkout the following week. (Pasadena *Star-News*, 14 Nov 75, 1)

- Calculation of crop acreage from satellite images had proven "generally adequate" in NASA tests involving wheat acreage in nine Great Plains states during the past year, NASA Associate Administrator Charles W. Mathews told the House Subcommittee on Space Science and Applications. Reporting on the 9.9% error rate in the LACIE (large area crop inventory experiment) test using Landsat data, Mathews said that the U.S. had demonstrated the feasibility of an operational global forecasting system for wheat. Mathews mentioned other Landsat data applications: The Geological Survey and NOAA had developed computer techniques for processing multispectral Landsat data that made oil slicks clearly visible. Other computer techniques using the data had detected subtle differences in surface soils and rocks that revealed mineralization and potential ore deposits. (Transcript, Vol I Part 1, 931, 938, 939)

- Estimated cost of the Spacelab being built for the first Space Shuttle flight had risen from $400 million to $500 million during the last year, and the program had fallen 3 to 6 mo behind schedule, NASA's Spacelab program director Douglas Lord told the House Subcommittee on Space Science and Applications. The European Space Agency, which was developing Spacelab, was confident of delivering it in time to fly on the Shuttle. Lord and NASA Associate Administrator John F. Yardley said the increased development cost would not affect NASA's cost in purchasing the modules, because NASA's agreement with ESA precluded the Europeans from recouping development costs. NASA had been committed to buying one set of Spacelab hardware at a cost "reasonable to NASA," although pricing proposals would not be received for 6 to 8 mo. The two officials said European cooperation on the program had been "exceptional." (Transcript, Vol I Part 1, 889)

*14 November:* Pan American World Airways announced the industry's first nonstop flights from New York and Los Angeles to Tokyo, beginning early in 1976. The announcement came a day after a new Boeing 747SP jetliner had made a demonstration flight New York to Tokyo in 13 hrs 33 min, fastest commercial service to Japan from the U.S. Current commercial time on that route, which required a refueling stop, was more than 17 hr. The Boeing 747SP (special performance) was 100 seats smaller than the 747-21, but had the same fuel capacity and the same four engines as the larger model. The SP had been capable of cruising faster and higher than any other commercial jetliner and had a range of 1200 km beyond that of the standard 747. (B *Sun*, 15 Nov 75, A9)

*15 November:* Observation of the sun's magnetic fields might make it possible to predict weather 4 days in advance with greater reliability, said scientists at the Institute for Plasma Research at Stanford University. Investigators at the National Oceanic and Atmospheric Administration in Boulder, Colo., and at Stanford said that magnetic forces from the sun played a role in formation of pressure troughs on earth, which bring rainy weather; observations from space, corroborated from earth measurements, indicate influence on weather from solar flares, sunspots, and high velocities of the solar wind.

The weather effect involved two phenomena, the sun's magnetic-sector structure and a concept called the vorticity-area index. The index is a measure of the size of low-pressure troughs in square kilometers, calculated by computer from twice-daily standard weather-service maps. As solar magnetic sectors and boundaries were measured regularly at Mt. Wilson Observatory, the Astrophysical Observatory in the Crimea, and the Stanford Solar Observatory, scientists could compare the solar magnetic structure with the vorticity-area index to see whether the sun structure affected the nature and size of low-pressure troughs that bring rain and stormy weather. (*NYT*, 16 Nov 75, 40)

- A nova that exploded about 10 000 yr ago and attracted the attention of primitive farmers in the Near East may have been a stimulus to the rise of man's first known civilization, according to the *New York Times*. Reporter Boyce Rensberger, writing on the theory published in the *Explorers Journal* by amateur scholar George Michanowsky, described clues found by Michanowsky in cuneiform texts regarding a gigantic star in the constellation Vela. No particularly bright star was to be found in Vela in modern times, Michanowsky pointed out, but astronomers had agreed that a supernova of uncertain date had probably been the forerunner of the present Gum nebula, a huge glowing cloud of hot gases centered on a relatively cold pulsar. Occurrence of the supernova had been placed anywhere from 6000 to 15 000 yr ago. In the times after the starburst, the primitive peoples—especially Sumerians—developed astronomy, mathematics, and writing; Michanowsky had suggested that the supernova had stimulated great attention to stellar phenomena and led to a more careful study of the heavens, to formal astronomy and mathematics, and to the need to make written records. (*NYT*, 15 Nov 75, 29)

*17 November:* The U.S.S.R. launched its 76th space mission for 1975, an unmanned *Soyuz 20*, from Baykonur Cosmodrome near Tyuratam, and established a new procedure by identifying the unmanned flight as a spacecraft designed for manned flight. Reuters quoted Soviet scientist Dr. Konstantin Feoktistov as saying the purpose of the flight was to develop a transport system either to supply a space station with expendables, or to serve as an emergency rescue ship. One possibility would be to automatically dock a Soyuz with the manned station after one of the cosmonauts had removed the original Soyuz, or perhaps a new Soyuz would be used as a new return vehicle: one of these procedures would be required unless the Salyut space station had been designed with two docking positions.

The Soviet director of July's Apollo-Soyuz Test Project, Konstantin Bushuyev, in Houston for a conference, said *Soyuz 20* had not been designed to transfer expendables to the current *Salyut 4* space station; Christopher S. Wren, reporting from Moscow for the *New York Times*, pointed out that it was not known how provisions could be transferred from the Soyuz to the space station, since neither craft had been reported to have such an automatic capability. (*NYT*, 20 Nov 75; 21 Nov 75, 25; *SBD*, 18 Nov 75, 90; 21 Nov 75, 114)

- Dr. Detlev W. Bronk, president emeritus of Rockefeller Univ. and former president of Johns Hopkins Univ., the Natl. Academy of Sciences, and the American Assn. for the Advancement of Science, died at the age of 78. A scientist who had been adviser to Presidents Truman, Eisenhower, and Kennedy, Dr. Bronk had been credited with formulating the modern theory of the science of biophysics—application of physics to the life systems of plants and animals. He had been chairman of the National Research Council from 1946 to 1950, and president of the National Science Foundation for three terms from 1950 to 1960. In 1964 he had received the Medal of Freedom, highest civilian award to be granted by a U.S. President. (B *Sun*, 18 Nov 75, A13; *NYT*, 18 Nov 75, 38M)
- Alexander P. Vinogradov, leading geochemist of the U.S.S.R. and vice president of the Soviet Academy of Sciences, had died at Moscow at the age of 80, according to Tass, the Soviet news agency. Vinogradov had identified 40 chemical elements in earth's soil zones and had developed a new field of investigation called cosmochemistry—the chemical analysis of lunar and Venus composition. He had remained active as the head of Moscow's Institute of Geochemistry and Analytical Chemistry. (Tass, in FBIS, 17 Nov 75; *NYT*, 19 Nov 75, 38; *Aero Daily*, 20 Nov 75, 112)

*18 November:* NASA expected to add 3800 contractor personnel in FY 1976, mostly for the Space Shuttle, NASA Comptroller William E. Lilly told a joint session of the House Subcommittees on Space Science and Applications and on Aviation and Transportation R&D. However, overall employment on the space program had been forecast down 68%, or 278 000 workers, from its peak of 409 000 in 1965. NASA would lose 17 more Federal positions in FY 1976 after losing 521 Federal employees in FY 1975. NASA is paying more for fewer employees, Lilly pointed out, because agency manpower expenses go up about $6 million a year, offset by only $1 million in replacing retired workers with new workers in lower grades. Total Federal employment at NASA had hit a high of 33 924 in 1966; NASA Federal employees now numbered 24 316. (Transcript, Vol I Part 1, 995, 1024, 1031)
- Boeing Co., which had a 44-mo $3-million subcontract from Rockwell to detect electrical problems with the Space Shuttle, reported that 35 design or hardware changes had been necessary in main-engine electrical circuits of the Shuttle so far. Boeing was conducting a sneak circuit analysis of the first Shuttle orbiter, to be used for approach and landing tests; later studies were scheduled for the second orbiter, the solid rocket booster, and the external fuel tank, as well as critical ground-support equipment. (*SBD*, 18 Nov 75, 92)
- At least five of NASA's 25 000 employees had made use in past years of hunting facilities owned by Rockwell International, and the agency and Sen. William Proxmire (D–Wis.) had begun investigating the matter. The investigation had not been confined to entertainment at the Rockwell facility in Md., but included "receipt of other gratuities such as Redskin (football) tickets and use of hospitality suites," the senator said. Rockwell had been prime contractor for the $6.4-billion Space Shuttle program that accounted for more than a third of NASA's budget. Sen. Proxmire, vice chairman of the Joint Committee on Defense Production, had directed the committee staff to assist NASA in determining the full extent of "unethical or illegal lobbying by NASA

contractors." NASA had placed five inspectors on the case. (*NYT*, 19 Nov 75, 1)

- INTELSAT had awarded a 15-mo $60 000 contract to EIC Inc. of Newton, Mass., to develop improved hydrogen electrodes and separators for use in nickel-hydrogen cells in communication-satellite batteries. NiH batteries would provide improved reliability and weight-to-power ratios over nickel-cadmium batteries now in use. (INTELSAT Release 75−19)

*19 November:* The Space Shuttle orbiter might travel by ground rather than air from Rockwell International's assembly facility at Palmdale, Calif., to Edwards Air Force Base for its approach and landing tests, a Rockwell official said as the orbiter made its first appearance for the press. Original plan had been to send the orbiter piggyback on NASA's Boeing 747 scheduled to launch the unpowered craft during its tests at Edwards. Travel on the ground would take 6 hr to go about 32 km, using an off-the-road vehicle inherited from the Apollo program, and would cost less. (NASA Release 75−293; *Aero Daily*, 20 Nov 75, 107)

*19 November:* *Soyuz 20*, the unmanned Soviet spacecraft launched from Baykonur 17 Nov., docked automatically with *Salyut 4* using onboard radio devices and computer installations for a program of "testing unit constructions and onboard systems...in joint flight," Tass announced. *Salyut 4*, the orbiting space station, had been visited by the crew of *Soyuz 17* for almost 30 days and by the crew of *Soyuz 18* for more than 60 days. The *Soyuz 20* mission had been analyzed as transfer of fuel to *Salyut 4*; in orbit since 26 Dec. 1974, the space station was thought to be running low on maneuvering fuel. Reporters had speculated that a new two-man crew might be sent to *Salyut 4* later this year for a stay of 3 mo or longer that would break the 84-day record set by U.S. astronauts in 1974. The value of the automatic docking for resupply would be the economy of not sending men with each flight, although the Soviets had never indicated the cost of their launches.

Tass reported that the "space complex Salyut-4−Soyuz-20" was in an orbit with apogee 367 km, perigee 343 km, 51.6° inclination, and 91.4 min period of revolution. (*W Post*, 21 Nov 75, B14; *NYT*, 21 Nov 75, 25; FBIS No. 225, 19 Nov 75, No. 226, 20 Nov 75; *Aero Daily*, 20 Nov 75, 108)

*19 November−23 December*: Atmosphere Explorer−E, third in a series of new deep-dipper spacecraft designed to investigate the ionosphere, was launched from Eastern Test Range aboard a Delta vehicle at 9:07 pm EST (0206GMT) into an almost perfect orbit: apogee, 3026 km; perigee, 157 km; inclination, 19.7°; period, 118.1 min. In orbit the satellite became *Explorer 55*.

Major difference between the original and new generation of atmosphere explorers was the unusually large onboard propulsion system that enabled the later spacecraft to move themselves up or down in space by hundreds of km or go from highly elliptical to circular orbits. The name "deep-dippers" meant that the spacecraft could drop into the atmosphere and pull out again, taking measurements only while in the upper reaches of the thermosphere. This region, important because it absorbed most ultraviolet light and heat received from the sun and because movement of that heat determined the structure of the upper atmosphere, had been difficult to study because the altitude was too high for all aircraft and most balloons and too low for other satellites. The dipping orbit prevented accumulation of excessive heat in the metal skins of the AE spacecraft.

*Explorer 55* would also carry a backscatter ultraviolet spectrophotometer to obtain a vertical profile and density measurements of the ozone layer between 20°N and S; the measurements would be compared with those from *Nimbus 4*, launched in 1970, to evaluate the long-term accuracy of the Nimbus instrumentation. A mission operations report 23 Dec. said *Explorer 55* had completed more than 400 orbits and all subsystems were performing well. (NASA Release 75-290, 70-29, 72-212; KSC Release 266-75; *SBD*, 21 Nov 75, 114; KSC *Spaceport News*, 31 Oct 75, 3; MORs 13 Nov 75, 3 Dec 75, 23 Dec 75)

*20 November*: *Pioneer 11* had been scheduled to take its first look at Saturn, according to controllers at Ames Research Center. Upon its encounter with Jupiter on 2 Dec. 1974, Pioneer's trajectory was altered by the gravitational effects of Jupiter to one that would take Pioneer to Saturn on or about 1 Sept. 1979. The spacecraft had returned the only pictures ever taken of Jupiter's polar regions and probed the planet's intense inner radiation belts for the first time, attaining the highest speed ever attained by a manmade object (171 000 km per hr).

Pioneer was still more than 1287 million km from Saturn, but in a position to observe the ringed planet at a phase angle 4 times larger than the largest angle at which Saturn could be seen from earth. The total amount of light and the polarization of sunlight reflected from a planet's atmosphere at different angles could reveal the composition of the atmosphere; scientists would watch Saturn throughout Pioneer's approach to note changes in intensity and polarization of light reflected from the planet with changes in phase angle. The viewing angle would also permit study of the light-reflecting characteristics of Saturn's rings in large areas of shadow from the planet, an effect not visible from earth. Pioneer was headed upward out of the plane containing the sun's planets, and in 1977 would attain a distance above this plane of about 161 million km, allowing it to measure phenomena coming from higher latitudes on the sun than previously observable. (ARC Release 75-53; MOR S-811-73-07, 19 Nov 75)

- Jet Propulsion Laboratory had awarded a $20-million contract to Lockheed Missiles and Space Co. for design and development of Seasat-A, an ocean survey satellite that would circle the earth 14 times a day and view 95% of the surface every 36 hrs. Scheduled for launch in 1978, Seasat would record radar imagery of waves and ice fields, ocean topography, tides, and currents, and would measure wave height, length, and direction; force and direction of sea-surface winds; and temperatures of the sea surface and the air/sea interface. (*W Star*, 21 Nov 75, A2; *Av Wk*, 8 Dec 75, 6; *SBD*, 21 Nov 75, 118)

- A new satellite-fed printing plant that would speed delivery of the *Wall Street Journal* to its readers in the southeastern United States had been introduced to Orlando, Fla., area business leaders and dedicated by the Journal's publishers, Dow Jones and Co. The $2-million plant could print 70 000 copies of the newspaper per hour, and would serve readers in Ga. and Fla. at first, with distribution into five other southeastern states coming later.

The paper would be printed in Orlando on the date of publication in Chicopee, Mass., by the use of *Westar 1* facilities provided by

American Satellite Corp. Fullsize facsimiles of pages would be put under a high-intensity scanner that would convert the images to electronic impulses beamed to *Westar 1*. The satellite would relay the impulses to the Orlando plant where they would be received on pagesize photo film; sending and receipt of data for a full page had been clocked at 3.5 min. The film would be used to make photolithographic plates for the Orlando press.

The Orlando plant would be the tenth regional printing plant for the Journal. The eastern edition previously delivered to 60 000 subscribers in Fla. and Ga. had been airmailed from Silver Spring, Md., arriving a day late. (Orlando *Sentinel-Star*, 20 Nov 75, 20A)

- The Air Force's Space and Missile Systems Organization awarded a $36.3-million contract to McDonnell Douglas Astronautics Co. for building and testing in earth orbit a space laser communications system. The 5-yr contract would produce three synchronous satellites to relay high-data-rate messages by laser to aircraft, ground stations, or other satellites, with a capability of transmitting a billion pieces of data per sec—about 20 times the data volume possible with present commercial communications-satellite systems. The award includes development of the space and ground-station equipment. McDonnell would work on high-accuracy pointing of the laser beams, within 10 millionths of a degree. (*St. Louis P-D*, 20 Nov 75, 13; *LA Times*, 9 Dec 75, 15C; *SBD*, 20 Nov 75, 107)

*21 November*: Goddard Space Flight Center awarded Boeing Co. a fixed-price incentive contract for $8 236 000 to provide modules for the first two Atmospheric Explorer missions: AEM−A (heat capacity satellite) and AEM−B (the stratospheric aerosol gas experiment—SAGE). AEM−A, scheduled for launch in 1978, would carry a heat-capacity mapping radiometer that NASA would acquire independently, to gather thermal-inertia data on rock types, location of mineral resources, soil moisture, and vegetation-cover temperature measurements. AEM−B's SAGE would measure spatial distribution of stratospheric aerosol and ozone on a global basis. It had been scheduled for launch in 1979. (NASA Release 75−299; *WSJ*, 24 Nov 75, 28)

*23 November*: Dr. John E. Naugle became Associate Administrator of NASA, a position in which he had acted since the departure of Dr. Rocco A. Petrone in April. Beginning his service with NASA in 1959, Dr. Naugle had been appointed Associate Administrator for Space Science and Applications in 1967, and Deputy Associate Administrator of NASA in 1974. As Associate Administrator, he would be responsible for directing the research, development, and operational activities conducted by the Offices of Applications, Aeronautics and Space Technology, Space Science, and Tracking and Data Acquisition. (NASA anno, 23 Nov 75; SBD, 24 Nov 75, 122)

*25 November:* The U.S.S.R. launched *Cosmos 782*, a biological satellite carrying 4 U.S. experiments as well as materials from Russia, Czechoslovakia, France, Hungary, Poland, and Romania, to be recovered in Soviet territory in about 3 wks. Launched from Plesetsk into an orbit with 405 km apogee, 227 km perigee, 62.8° inclination, and 90.5 min period, the satellite provided an opportunity for U.S. scientists to fly biology and weightlessness experiments that would otherwise have had to await flights of the Space Shuttle in the 1980s. Last U.S. biology research spacecraft, *Biosatellite 3*, had been launched in 1969.

*Cosmos* 782 was the first satellite equipped to simulate gravity in space, to compare effects on biological processes with those of weightlessness. A specially constructed centrifuge, consisting of a small disk that would rotate at a constant speed, would create a centrifugal force equal to gravity on earth for packages placed at the right spot on the disk. Comparing the specimens exposed to lack of gravity during orbital flight with similar specimens feeling the pull of gravity would prove any effects had resulted from weightlessness. Results of the experiments should be available in about 90 days, according to the Tass news agency.

Ames Research Center managed the U.S. participation in the *Cosmos* 782 flight. It had had no advance notice of the launch date, place of launch, or mode of orbital operation, although life limitations of the biology materials had dictated a launch near this date. NASA had been invited to participate in a second Soviet biosat mission in 1977, and had planned to fly Russian experiments on its Space Shuttle in the early 1980s. (*NYT*, 27 Nov 75, 62M; NASA Release 75–292; *SBD*, 28 Nov 75, 145; *Aero Daily*, 28 Nov 75, 148)

- A metering truss made of graphite—one of the few materials that expands with cold and contracts with heat—and epoxy, which expands with heat and contracts with cold in the usual way, had been designed to hold stable the mirrors of NASA's Large Space Telescope without being distorted by temperature changes. Built by Boeing Aerospace Co. under a $185 000 contract with Marshall Space Flight Center, the structure had been subjected to a 2-day vacuum-chamber test at Boeing's Space Center in Seattle under temperatures ranging from −84°C to +21°C without suffering lens-limiting thermal distortion. (*Huntsville Times*, 25 Nov 75; Boeing Release A−0459)

- The Soviet Academy of Sciences chose Anatoly Alexandrov, a 72-yr-old nuclear physicist, as its new president, the Tass news agency reported. Peter Osnos commented in the *Washington Post* that the main speaker at the Academy meeting—Mikhail Suslov, the Communist Party's chief ideologist, who is not even a member of the Academy—made the only speech on Alexandrov's behalf. This, said Osnos, showed the extent to which the choice was determined by Kremlin leadership and the Party's growing influence over the once-autonomous scientific establishment. Academy members traditionally had voted in secret ballots and had maintained their own organizational structure in running the country's important research facilities; Alexandrov was chosen on a secret ballot, but apparently there were no other candidates. Alexandrov, head of the atomic research institute in Moscow, was one of the few Academy members also a member of the Party central committee. (*W Post*, 26 Nov 75, A8)

*26 November:* Under a decision of the Federal Communications Commission, the Communications Satellite Corporation (ComSat) had been making too much money and would have to reduce its rates on international communications services by up to 35%. The order followed a 10-yr FCC investigation of the entire range of ComSat services. Essentially a wholesaler of communications, ComSat had been selling satellite circuits to telephone and telegraph firms that retailed the services to consumers. The FCC order could substantially reduce overseas telephone and telegraph charges; an accompanying order required the international firms that had been dealing with ComSat to pass the reduced rates on to their customers. FCC ruled that ComSat was

entitled to a return ranging from 10.8 to 11.8%, but ComSat's rate of return in 1974 was about 34.6%. (*W Star*, 27 Nov 75, A2)
- The People's Republic of China successfully launched an earth-orbiting satellite into an orbit with 483 km apogee, 173 km perigee, 63° inclination, and 91-min period, the New China News Agency announced. (FBIS No. 230, 26 Nov 75; *W Star*, 27 Nov 75, A2)
- The U.N. General Assembly adopted a resolution [see 29 Oct.] calling for wider observance of international agreements on peaceful uses of outer space, and providing for continued U.S. work on new international accords relating to space. Among these were a moon treaty, principles governing direct television broadcasting by satellite, and legal issues involved in remote sensing of earth from space. The resolution also called for continuing review of ways to help developing countries apply space technology. (*SBD*, 26 Nov 75, 140)

*28 November:* The National Oceanic and Atmospheric Administration (NOAA) awarded a $145 000 contract to RCA's Astro-Electronics Div. for a first-phase design study of the Tiros-N data-acquisition and control system (DACS). DACS would provide operational control of Tiros-N in orbit, and would collect and transmit telemetry and weather information to NOAA facilities in Md. RCA's Astro-Electronics Div., which had been developing the new generation of weather satellites for NASA, had scheduled the first flight for 1978. As part of the design study, RCA would provide descriptions of materials and procedures needed to implement the DACS, a network of ground stations including the Md. center and the command and data-acquisition stations (CDA) in Alaska and Va. (*Aero Daily*, 28 Nov 75, 149)

*29 November:* Radio astronomers in both the U.S. and U.S.S.R. had intensified their search for possible civilizations out in space, Walter Sullivan wrote in the *New York Times*. Little publicity had been given the search to prevent raising public expectation for "what is likely to be a prolonged effort." Costs were minimal, because the observations had made use of equipment also used for other research. Dr. Hans Mark, Director of Ames Research Center, had told students at the Polytechnic Inst. of N.Y. that the search for other civilizations would become "one of the major scientific enterprises," and cited his belief that "communicative civilizations" had been a natural consequence of biological evolution. Dr. Mark pointed out that the U.S. giant antenna at Goldstone, Calif., used for spacecraft communications, could pick up emissions only from the vicinity of the nearest stars. One American search is using the largest antenna on earth—the dish at Arecibo, P.R.—to scan the five nearest galaxies at 1420 mhz, the frequency of radio waves given off by hydrogen atoms, long considered a logical frequency for civilizations trying to make contact. The Arecibo search had been conducted by two Cornell professors, Dr. Frank B. Drake and Dr. Carl Sagan. Another American search had been using the National Radio Astronomy Observatory at Green Bank, W. Va., to scan 700 stars (resembling earth's sun) within 80 light years of the solar system for signals at the 1420-mhz wavelength. No obvious signals had been detected, but the recordings had not been subjected to detailed analysis, according to Dr. Patrick Palmer of the Univ. of Chicago and Dr. Benjamin M. Zuckerman of the Univ. of Md.

The Soviet program, scanning the entire portion of the celestial sphere visible from the Soviet Union, had been using two networks with a total of eight stations spread across the vast width of the Soviet

Union. The November issue of *Icarus*, international journal of solar-system studies, had carried a prospectus of the Soviet CETI (communication with extraterrestrial intelligence) program divided into two phases. CETI 1, scheduled from 1975 to 1985, included the whole-sky scan by eight stations plus monitoring by two space stations and a survey of nearby galaxies like the one under way at Arecibo. Only the eight-station project had been initiated, as far as U.S. scientists knew. CETI 2, from 1980 to 1990, had been scheduled to continue satellite monitoring but would also use semirotatable antennas such as the RATAN−600 antenna 600 m wide, known to be located high in the Caucasus; the system consisted of numerous plates that had to be aimed individually, calling for hand alignment since no computer had been designed to perform this role. The Soviets also envisioned a monitoring station at the point beyond the moon where the gravitational fields of earth and moon would balance each other; a station at that point would be protected from the earth's own radio emissions. Identical fluctuations of signal recorded simultaneously at widely separated points would be considered to have extraterrestrial origin rather than to result from a local manmade effect. (*NYT*, 29 Nov 75, C13)

*During November:* An x-ray telescope assembled at Marshall Space Flight Center for the study of remnants of an exploded star had been shipped from England, where it had undergone payload integration, to the Woomera Rocket Range in Australia for launch. The project, called Skylark for the British sounding rocket on which the telescope would fly, was a joint British-American undertaking to study a supernova remnant known as Puppis A for information on the evolution of stars and perhaps on the formation of neutron stars. The flight would provide at least 3 min 20 sec for the 346.5-kg payload to make its observations above 120-km altitude. (NASA Release 75−287; MSFC Release 75−238)

- States involved in NASA's land-use satellite applications had included Ala., Miss., and Tenn., which had organized statewide survey programs, and Mo., which had just begun such a program. The Southern Growth Policies Board, a 15-state group monitoring the South's development, had investigated application of Landsat capabilities to identify prime farm land, potential commercial sites, and land values. Georgia, working through Ga. Tech., Marshall Space Flight Center, and the U.S. Dept of Agriculture, had begun aerial surveys of Georgia's peach orchards to study the problem of premature loss of the fruit trees. MSFC aircraft provided multispectral photos and thermal data to be processed through Ga. Tech. computers, detecting stages of decline not visible from the ground. (MSFC Release 75−247, 75−251)

- The first successful air drop of data-collection platforms for use by the *Nimbus 6* polar-orbiting weather satellite would lead to another first: successful day-to-day tracking of ice-pack movement in the Arctic Sea north of Alaska's oil-rich Prudhoe Bay area. Knowledge about the interaction between the winter sea ice and the continental shelf would be important to government and to oil companies; government would use the information in allocating drilling areas and overseeing their use by private industry. Oil companies would need to know whether to locate drilling rigs on the ice or sink them into the sea bed; whether to bury pipelines under the ocean floor, lay them along the bottom, or run them over the ice to loading areas. The exterior sphere of the platforms, built to be dropped on the ice from small aircraft, had been constructed of a tough material to foil the hunger and curiosity of polar bears, which reportedly would attempt to chew on anything appearing different from the usual surroundings.

The tracking and data-relay experiment aboard *Nimbus 6* had successfully completed its share of a transmission test that sent sensor, telemetry, and ranging signals from its near-polar 1110-km orbit to NASA's *Ats 6* in geostationary orbit 35 900 km above India. The signals, relayed to NASA's Madrid receiving station, were immediately retransmitted to GSFC. Information gathered in the test would be used to design the Tracking and Data Relay Satellite System (TDRSS) planned for late 1979. The system would use two geostationary satellites to relay tracking data, commands, and communications between a central ground terminal and a number of spacecraft in low earth orbit; this would increase low-orbit spacecraft access to ground stations from 15 to 85%, reducing NASA requirements for ground-station networks. General Electric's Space Div., reporting on the success of the test, said the quality of all data received was excellent. (*Marshall Star*, 26 Nov 75, 4; *San Diego Union*, 9 Nov 75, 1; *Aero Daily*, 26 Nov 75, 140; *SBD*, 26 Nov 75, 143)

- The Bangui Anomaly—a massive magnetic disturbance in the earth along the equator in Africa, suggesting rich deposits of heavy metals like iron and uranium—had been accidentally discovered by *Ogo 1*, launched in 1964 with magnetometers that were still measuring the earth's magnetic field in 1970 when the Geological Survey decided to use them for other purposes.

  Attempting to measure the ionosphere for traces of a jetstream above the equator, Survey scientists noticed a "kink" in the data and had checked their findings by scanning the region with magnetometers in aircraft. Results showed a magnetic difference so great that it could be caused only by an ore body larger than the Mesabi range in Minn., largest in the U.S. Scientists had deemed the find so important that NASA had suggested putting a satellite in orbit to do nothing but chart the earth's magnetic field.

  The anomaly was named for the capital of the Central African Republic, an extremely poor nation almost as big as Texas with a population of about 2 million. The deposits had been localized in hilly regions away from the Ubangi River, so that a railroad or highway would be needed to carry ores to the river for shipping. State Dept. sources said that interest in the region among foreign mineral-extraction companies had risen. (O'Toole, *LA Times*, 15 Nov 75, 1)

- Energy from the Mideast had been forecast for cooling buildings in the Persian Gulf countries—but without using a drop of oil, according to the *Christian Science Monitor*. Solar energy would provide power to sweeten the waters of the Dead Sea, heat water for householders in Cyprus and Israel, and desalinize water for military units in Saudi Arabia. First desalination plant of its kind to use solar energy would be built at Aqaba by the Jordanian Royal Scientific Society working with Dornier of West Germany; the latter company would also build the small desalting plants for the Saudis. A world conference on solar energy had been scheduled for the new Dhahran campus of the Saudi Arabia University of Petroleum and Minerals, at the end of November. (*CSM*, 18 Nov 75, 2)

- A simple device familiar for centuries—the flywheel—had been the subject of a national symposium as a possible solution for some of America's critical energy problems. The first conference on flywheel technology, held at Berkeley, Calif., under sponsorship of the Energy Research and Development Administration and the Lawrence Livermore Laboratory, had been told that modern materials and design had made possible a superflywheel that could transform the energy picture. The principle upon which flywheels performed—storage of mechanical energy by a spinning wheel—had been enlarged by the use of fiber composites developed for use in spacecraft nosecones that had greater strength-to-weight ratios, being 10 to 20 times stronger than steel and much lighter. As the amount of energy stored by the flywheel varied as the square of the rotation speed, the limit had been on the tensile strength of the material used; the fiber wheel could be spun faster to store more energy per kg than the conventional metal one, and would also cost less. A superflywheel might spin at 100 000 or even 200 000 rpm. Both the Soviet Union and American industry had displayed interest in developing the flywheel; use of the technology in automobiles would reduce pollution and would represent a safety improvement. The flywheels also could act in backup systems for homes using solar energy and windmills for their energy needs. Further investigation of the materials under stress-rupture tests was being conducted at the Lawrence laboratory and elsewhere. (*NYT*, 30 Nov 75, L84)
- National laws and national governmental institutions were inadequate to deal with problems arising from possible depletion of ozone in the stratosphere by manmade fluorocarbons, said a study by the Library of Congress. The Senate Committee on Aeronautical and Space Sciences said the study would be "most useful" in its continuing investigation of pollution in the upper atmosphere. The study called for an international system of data gathering and institutional control, at the same time emphasizing the lack of conclusive evidence that fluorocarbons had reduced atmospheric ozone—although it found "very strong grounds" for believing that probable cause of harm existed. (*SBD*, 4 Nov 75, 12)

## December 1975

*1 December:* National Science Foundation held in Chicago the first of 7 meetings scheduled nationwide to obtain public views on a Federal program for greater public involvement in science policy issues. Subsequent meetings would be held during Dec. in Atlanta, Dallas, Denver, San Francisco, Washington, D.C., and Boston. Oral and written statements presented at the meetings would be included in an NSF report to Congress. (*SBD*, 3 Dec 75, 173)

*2 December:* Space programs that create new commercial activity, new wealth, and new jobs were essential to the growth of both the U.S. and Europe, Sen. Frank E. Moss (D–Utah) told the Assembly of Western European Union. Reviewing policies that had governed agreements on space with Western Europe, Moss warned that continued cooperation in space would hinge on Europe's paying its full share of the expenses. Concentration on programs with immediate application to earth problems should not mean abandonment of space science, or loss of "the vision and imagination that have been so important to our achievements to date," the senator said. (*CR*, 10 Dec 75, S 21650; *SBD*, 5 Dec 75, 185)

- *Defense/Space Business Daily* reported that "a NASA spokesman" had said the proposed Mariner Jupiter-Uranus mission scheduled for 1979 had not been eliminated from the agency's FY 1977 budget and was "still planned." The Federal budget squeeze announced for FY 1977 had put the $177-million project in serious question. NASA reportedly might try to use one of its Mariner Jupiter–Saturn 1977 spacecraft as an alternative to permit a flight to Uranus. (*SBD*, 3 Dec 75, 175)

*3 December:* Dr. Maxime A. Faget, director of engineering and development at Johnson Space Center, received the gold medal of the American Society of Mechanical Engineers at its annual winter meeting honors assembly in Houston. Faget had been responsible for design and testing of the Mercury, Gemini, and Apollo spacecraft and the Skylab space station, as well as of the Space Shuttle scheduled for flight in the late 1970s. Previous JSC recipients of the ASME gold medal had been JSC Director Dr. Christopher C. Kraft, Jr., in 1973 and former JSC Director Robert R. Gilruth in 1970. (JSC Release 75–98)

- Dr. Ernst Stuhlinger, associate director for science at Marshall Space Flight Center, would retire after 30 yrs of Federal service on 28 Dec., NASA announced. Born and educated in Germany, Dr. Stuhlinger became a physicist and had been involved in rocketry and space work since 1943 when he joined the rocket development team at Peenemunde, Germany. He had come to the U.S. after World War II and worked for the U.S. Army before transferring to NASA when MSFC was wet up in 1960. Early planning for lunar exploration and the Apollo telescope mount had been carried out under his direction, as well as early planning on the High Energy Astronomy Observatory and initial phases of the Space Telescope project. His work had included electric propulsion studies and scientific payloads for the Space Shuttle. (MSFC Release 75–256)

- The Peoples Republic of China announced it had successfully recovered an artificial earth satellite for the first time, becoming the third country after the U.S. and the Soviet Union to develop a technique for returning a satellite from orbit. The satellite, which had been launched 26 Nov., was China's fourth; the third had been orbited in July and the first two in 1970 and 1971. Edward K. Wu, Hong Kong correspondent of the Baltimore *Sun*, noted that Chinese scientists and engineers apparently had solved technological problems including development of a heat-resistant alloy to withstand reentry temperatures, automatic techniques for remote control, and a trigger system to fire the satellite back to earth. (B *Sun*, 4 Dec 75, A4)
- The automated docking of the U.S.S.R.'s pilotless *Soyuz 20* with the unmanned *Salyut 4* space station [see 19 Nov.] had been a preliminary to Soviet experiments aimed at building a permanent space base, the *Christian Science Monitor* said. Recalling the failure of the automatic docking system in August 1974, *CSM* quoted Maj. Gen. Vladimir Shatalov—head of the cosmonaut group—as saying that the Soyuz was being developed as a universal spacecraft for carrying crews, fuel, and provisions to scientific stations and for assembling complex structures in orbit; craft of this kind would "undoubtedly become assembly sites for large space stations to be set up in orbit." European observers had said the Soviet plan would be to launch the central part of the space platform first, followed by separate laboratory modules that would plug into docking ports to draw on the power and facilities of the mother station. The orbital stations would be supplied with consumables by a modification of the existing Soyuz, and this experience would be applied later to shuttles and space tugs. (*CSM*, 3 Dec 75, 7)
- Evidence that aerosols from spray cans had damaged earth's ozone layer had been under study by an independent scientific panel that would make its report early in 1976, the *Wall Street Journal* said, but Federal action against fluorocarbons was not considered likely before that time, and foreign governments were considered unlikely to act before the U.S. did. Fluorocarbon manufacturers and users had insisted that the ozone-depletion theory had not been proved, and warned of damage to an industry that had made products worth $400 million to $450 million a year, with manufacturing facilities worth $300 million. U.S. producers had accounted for about half the world's production of fluorocarbons, which was approaching about 0.9 billion kg a year.

  Two Univ. of Chicago chemists—Mario J. Molina and F. Sherwood Roland—had advanced a theory 18 mo ago that fluorocarbons used in spray containers and in refrigerators and air conditioners were getting into the upper atmosphere, encountering ultraviolet light (more intense at high altitudes) that split the fluorocarbon molecules to release highly active chlorine. The chlorine had depleted the atmospheric ozone by converting it to ordinary oxygen. The theory had been partly confirmed in the laboratory by National Bureau of Standards chemists who showed that ultraviolet lamps would split fluorocarbons to produce either one or two chlorine atoms, depending on the ultraviolet wavelength used. However, no one had discovered a way to measure the ozone to tell whether it had been affected by chlorines from fluorocarbon dissociation. Another clue had been to search for chlorine products such as hydrochloric acid in the stratosphere; tests conducted by the Jet Propulsion Laboratory using U-2 flights and balloons had confirmed the presence in the stratosphere of hydrochloric acid that had not originated from the earth. Granting the effect of aerosols, opinions had been divided upon the significance of ozone depletion: possible consequences of additional ultraviolet radiation impinging upon earth included increased incidence of skin cancer, harmful effects on animal and plant life, and alteration of temperature patterns in the stratosphere with resulting alteration of weather patterns on earth. Even without proof of damage, public apprehension of risk had increased pressure for a ban on the suspect chemicals, especially in aerosols. (Tannenbaum in *WSJ*, 3 Dec 75, 1)

- The Aerospace Corp. had donated its $2-million San Fernando Observatory to Calif. State Univ. at Northridge as an educational center for solar research, Aerospace president Dr. Ivan Getting announced. The nonprofit company, which had used the observatory in support of flights in the Apollo and Skylab projects, was giving away the 7-yr-old facility because its planned solar research had been completed. (*SBD*, 3 Dec 75, 173)

*4 December:* NASA's winter phase of meteorological rocket and satellite comparisons had been completed at Wallops Flight Center, where 31 single and two-stage rockets had been launched since 18 Nov. Data obtained from the launches had been compared with measurements received from the *Nimbus 6* satellite passing over the Wallops facility. The Super-Loki single-stage rocket had carried three types of payload: inflatable spheres of mylar to obtain atmospheric density and wind measurements; datasondes to obtain temperature and wind data; and ozone-measuring devices. The Nike-Cajun, a two-stage vehicle, carried acoustic grenade payloads to measure wind and temperature by recording time and location of detonation on arrays of sensitive microphones at the launch site that could measure time lapse and direction of sound arrival. The rocket and satellite data comparison had been the latest of several conducted from the Wallops facility since 1970. (WFC Release 75—14)

- *Landsat 2,* launched 22 Jan. 75, had successfully completed more than 10 mos of orbital operations during which it had acquired more than 53 000 frames of imagery. Nearly 300 000 messages from data-collection platforms had been received, processed, and sent to users; imagery from more than 2600 sample sites for the Large Area Crop Inventory Experiment (LACIE) had been processed and shipped to Johnson Space Center. About 62% of domestic investigations and 43% of foreign investigations had been supplied enough *Landsat 2* data to meet the major objectives. Primary mission objectives had been declared satisfied, as well as all secondary objectives that were not time-dependent. (MOR E−641−75−02, 4 Dec 75)

- The Soviet Union had landed Arctic explorers on an ice floe 600 km from Chukotka to set up Severny Polyus−23, a drifting observatory to study weather and ice conditions in the Arctic Sea, Tass announced. Air search last spring had discovered a flat berg 7 km long and 3 km wide about 500 km from Wrangel Island north of the eastern tip of

Siberia. N. Blinov, head of the Soviet high-latitude expedition Sever-27, called this a rare find because the ice appeared to be 20 m thick. Research stations on such floating islands made it possible to trace circulation of Arctic ice over many years; the Severny Polyus–22 station operating in the Arctic for 3 yr had moved in a clockwise anticyclone circular drift and "repeatedly found itself in both the eastern and western hemispheres." The new station, larger than its predecessor, was expected to last longer. (Tass, in FBIS No. 240, 4 Dec 75)

*5 December:* Launch of the Dual Air Density (DAD) spacecraft from the Western Test Range failed when a catastrophic event during third-stage burn caused loss of vehicle control and failure to achieve orbit. The destruct signal was transmitted 341 sec after ignition.

The four-stage Scout-D launch was to have placed into orbit two satellites—a small rigid sphere of aluminum and a large inflatable sphere of mylar—similarly instrumented with a mass spectrometer and associated electronics. The satellites were to have been in $90°$ polar orbits intersecting periodically, to measure atmospheric densities at two differing altitudes but at the same latitudes at the same local-hour angle of the sun. The measurements would show effects on upper atmosphere readings of changes in solar heat input and the relationships between upper and lower atmospheres.

The launch, scheduled for 3 Dec., had been delayed because of a malfunction in the ignition system of the Scout launch vehicle. (NASA Release 75–300; MOR S–863–75–05, 6 Jan 76; HQ memo, Goozh, 9 Dec 75; *SBD*, 5 Dec 75, 186)

- Marshall Space Flight Center issued a request for quotations from industry for development of a flat conductor-cable wiring system suitable for use in residences and commercial establishments. Surface mounting, a feature of the flat cable system originated by NASA to reduce the size and weight of wiring harness in space capsules, would eliminate the need for routing wires in walls, ceilings, and floors because it could be covered by paneling, tile, or even wallpaper. Use of the flat wiring would reduce costs of wiring new construction and permit savings in renovating, remodeling, and rewiring of existing buildings. The successful proposer would develop a surface-mounted baseboard system using a cable 63.5 mm wide; total thickness of the baseboard—channel, airspace, and cover—would be about 11 mm, thickness of the thinnest lumber baseboard strips in use. (MSFC Release 75–257)

*8 December:* The Senate passed H.R. 8674, the Metric Conversion Act of 1975, after substituting the provisions of its own bill S. 100. The legislation would establish a 17-member United States Metric Board, appointed by the President, to carry out a program of planning and public education to implement the adoption of the metric system on a voluntary basis. Board members would serve staggered terms of up to 6 yr, and would report to the President annually on the conversion process with recommendations on necessary legislation or executive action. (*CR*, 8 Dec 75, S21368)

*9 December:* The Communications Satellite Corporation filed with the U.S. Court of Appeals for the District of Columbia a petition for review of the Federal Communications Commission decision of 4 Dec. on the allowed rate base, rate of return, and rate structure for ComSat's international satellite services. ComSat intended to apply to the

FCC—and to the court—for a stay of the FCC order that new tariffs be filed by 5 Jan. 1976. The corporation had prepared a letter to its stockholders on the FCC decision. (ComSat Release 75–61)

*10 December:* Marshall Space Flight Center announced it would distribute the first NASA Standard Parts List (NSPL) to all NASA centers and contractors in Jan. The NSPL was to include electronic parts selected from experience in flights, tests, failure analyses, and recent plant surveys, to ensure reliability. The standardization had been undertaken to solve problems of small-quantity buying, high cost, long delivery schedules, market fluctuations, part failures, and extensive rework. The cost of correcting problems resulting from failure of parts had been estimated as 100 times the cost of preventing the use of poor-quality parts to begin with. The NASA Standard Parts Lead Office, located at MSFC because of its experience in parts management developed during the Saturn program, would establish requirements for parts, coordinate the effort with all NASA Centers, and issue and maintain the list. (NASA Release 75–303; MSFC Release 75–259)

- The Air Force had formally entered into an agreement with NASA on development of an experimental hypersonic research aircraft capable of mach 6 flight. The two agencies had agreed after a joint 1974–75 study that "the combined objective and the national interest" would be best served by a joint hypersonic flight-test program. The agreement provided that the report of the joint study would be the baseline document for program development; that technology development and experimental tests would be completed in 1976 so preliminary design study could begin in FY 1977; that personnel from both agencies would participate in all phases to the maximum extent practicable; and that an ad hoc steering committee would be established to prepare a project plan for use in vehicle procurement. (Memo of understanding, 10 Dec 75)

- A long-duration free-flying manned space station in earth orbit appeared to be the most economical way of providing a continuing manned presence in space, according to "Manned Orbital Systems Concepts," a study conducted for Marshall Space Flight Center by McDonnell Douglas Astronautics Co. A station that could significantly extend the time in space planned for the Space Shuttle would offer advantages for doing a given amount of work with fewer flights and reducing the number of turnarounds, checkouts, and similar operations. A permanent facility could offer maintenance and repair and even reconfiguration of payloads. The study envisioned development of four-man facilities that would serve as building blocks for growth into larger stations for 12 or 24 workers as needed; the four-man configuration would consist of a subsystems and habitability module carried into orbit by one Shuttle and joined with a logistics and payload module carried on a second mission. Several payload modules could be docked in tandem with the core vehicle, or in radial locations through use of a multiple-port docking adapter. The core vehicle would be left on station for a nominal lifetime of 5 yr, logistic and payload modules being replaced at 90-day intervals. Crew exchange and resupply would occur every 90 days. The study concluded that problems anticipated by the world in the 1990s should be solved in the 1980s, and that an extended-duration manned orbital facility could contribute significantly to those solutions. (MSFC Release 75–26)

- INTELSAT—the International Telecommunications Satellite Organization—had awarded a 20-mo fixed-price contract for $220 000 to the Messerschmitt—Boelkow—Blohm Co. of Munich for design, fabrication, and test of a prototype lightweight deployable solar array for use in synchronous satellites. The new array would provide 30 to 33 watts per kg at the end of a 7- to 10-yr mission, compared to the current level of 10 to 15 watts per kg. (INTELSAT Release 75—20)

*11 December:* A Black Brant VC sounding rocket, first of a series scheduled over the next 5 yrs in a Space Processing Applications Rocket (SPAR) project managed by Marshall Space Flight Center, was launched to an altitude of 225 km from the Army's White Sands missile range and traveled downrange about 80 km, providing about 5 min of near weightlessness during its coast phase. The rocket carried nine space-processing experiments in a 143-kg payload assembled at MSFC. The next flight had been scheduled for the spring of 1976. (MSFC Release 11 Dec 75)

- NASA issued a request for proposals from industry on a space-station systems analysis study. The proposals would be due 26 Jan. 1976. Two $700 000 contracts would be issued in April for parallel 18-mo concept studies, one to be managed by Marshall Space Flight Center, the other by Johnson Space Center. Contractors would study low-orbit and synchronous-orbit facilities, to begin module construction in the mid-1980s. The space station would be used as a test facility and construction base to support the manufacture and assembly in space of various kinds of space structures—e.g., generation of electric power by large solar collectors and transmission to earth by microwave antennas. Other uses would be the retrieval and repair of automated spacecraft, or storage and transfer of fuel and other expendables. Elements of the station would be compatible with the Shuttle cargo bay, 18.3 m long by 4.6 m in diameter. (NASA Release 75—310; MSFC Release 75—262; JSC Release 75—99)

- Kennedy Space Center awarded a $1 279 000 contract to George A. Fuller Co., a division of Northrop Corp. of Chicago, for fabrication and erection of a structure for mating and demating the Space Shuttle Orbiter and its Boeing 747 carrier aircraft at NASA's Flight Research Center in Calif. KSC had been given responsibility for Space Shuttle launch operations and for Orbiter ground operations and support equipment both at KSC in Fla. and at FRC, which was to be Orbiter landing site for the first four missions. The contractor was to complete the structure—including hoist system, access platforms, and utility systems—within 328 days after notice to proceed. (KSC Release 290—75)

- Fluorocarbon levels in the stratosphere over N. Mex. had more than doubled since 1968, reported Dr. David G. Murcray, professor of physics at Univ. of Denver. The average annual increase of about 14% approximated the annual increase in fluorocarbon use. Dr. Murcray's experiments, funded by the National Science Foundation and the Manufacturing Chemists Assn., used a balloon-borne high-resolution infrared spectrometer to measure the wavelengths at which fluorocarbons absorbed light. "The measurements of the fluorocarbons made using this technique are of particular interest," said Dr. Murcray, "since I had made similar measurements in 1968." The 1968 measurements had been the only ones of the amounts of fluorocarbons present in the stratosphere made before 1974. In 1968, fluorocarbon 11 was measured at 20 parts per trillion; in 1975, at about 50. Comparable figures for fluorocarbon 12 were 50 to 60 parts per trillion in 1968 and 140 to 160 parts per trillion in 1975. (NSF Release PR75−101)

- NASA announced selection of Boeing Commercial Airplane Co. of Seattle for negotiations leading to award of a $9.8-million contract to develop a computer software system that could substantially reduce design time and costs of aircraft and space vehicle design and ensure improved vehicle performance. The system, called IPAD (integrated programs for aerospace-vehicle design), would process installation of engineering information on computers at major U.S. aerospace companies, acting as the key communications and calculations integrator for many designers and increasing their effectiveness by speeding up computations and data management required in design. It would serve large engineering staffs from the design concept through detailed design, and would help to organize and assemble design data in support of manufacturing processes. (NASA Release 75−312)
- The Soviet Union launched *Intercosmos 14* into an orbit with 1707-km apogee, 345-km perigee, 105.3-min period, and 74° inclination. The satellite had been equipped to study low-frequency electromagnetic fluctuations in the magnetosphere, the structure of the ionosphere, and the intensity of micrometeorite fluxes. Tass reported that flags of nine socialist countries had waved over "one of the Soviet cosmodromes" where the launching took place; mentioned as participating in this particular mission were scientists from Bulgaria, Czechoslovakia, and Hungary, as well as the U.S.S.R. Satellites of the Intercosmos series were considered "closest to purely earth problems" in that they had been connected with specific orders placed by geophysicists, physicians, meteorologists, and radiocommunications experts. Academician Boris Petrov, head of the Soviet Intercosmos, said the launch ended the first decade of cooperation between scientists and engineers of the socialist countries, who had made significant discoveries regarding sun and earth climate, influence of solar activity on magnetic weather, and dispersal of radio waves. Collectives of experts in all nine member countries of Intercosmos would take part "as usual" in processing information received from the satellite, working at the scientific flight headquarters near Norilsk, beyond the polar circle, where parallel ground observations would be conducted. (Tass, in FBIS 240 and 243, 12 Dec 75)

*12 December:* *Satcom 1*, first of three domestic communications satellites to be placed in geostationary orbit around the equator for RCA Corp. to transmit long-distance telephone calls and service data, had been launched at 8:56 pm EST from Cape Canaveral on a Delta vehicle into a parking orbit with 35 980-km apogee and 182-km perigee. A kick motor was to be fired at 5:41 pm EST 15 Dec. to put the satellite into a synchronous orbit. The RCA Satcom was the second domestic communications satellite to be launched by communications companies; the Western Union Corporation's Westar system became operational earlier this year. Each Satcom would receive and transmit data on 24 channels. Westar—a two-satellite system—had only 12 message channels per satellite. Each Satcom channel was built to transmit 1000 telephone calls at a time, or one color TV channel, or 64 million bits per sec of computer data. *Satcom 1* would be stationed at 119°W, and the two subsequent Satcoms would be located at 99°W and 129°W over the equator to handle television, voice channels, and high-speed data between the contiguous U.S., Hawaii, and Alaska. (NASA Release 75-302; *NYT*, 13 Dec 75, 37; *SBD*, 16 Dec 75, 242; KSC Releases 289-75, 295-75)

- NASA marked the 10th year in space of *Pioneer 6*, launched 16 Dec. 1965 to make the first detailed measurements of the interplanetary medium, some spanning more than 805 million km. *Pioneer 6* had measured the sun's corona, returned data on solar storms from the invisible side of the sun, and measured the tail of Comet Kohoutek, during what was believed to be the longest operating life yet attained by an interplanetary spacecraft; it had also helped to chart the solar wind, solar cosmic rays, and the solar magnetic field, all of which extend far beyond the orbit of Jupiter. *Pioneer 6*, and three sister craft—*Pioneers 7, 8,* and *9*—all years beyond their 6-mo design lives, had constituted a network of solar weather stations that circle the sun in locations millions of kilometers apart. Two younger craft—*Pioneer 10* and *11*—had flown past Jupiter in 1972 and 1973 and were headed for the outer reaches of the solar system. (NASA Release 75-311; ARC Release 75-65)

*14 December:* The Air Force launched a "spy satellite" from Kennedy Space Center at 12:15 am EST into a stationary orbit at about 36 000 km above earth. Although no announcement had been made of the success or failure of the launch, observers next day concluded that the launch had been successful. The satellite was one of a series designed to monitor missile launchings in China and the Soviet Union to give a 30-min warning of a land-launched attack or about 10-min warning of a missile attack from submarines. It had been launched by a Titan IIIC, 24th of this type launched from Cape Canaveral, and was thought to carry both infrared sensing equipment and still and TV cameras to permit day and night monitoring. (*NYT*, 16 Dec 75, 14; *C Trib*, 15 Dec 75, 5)

*15 December:* Giant designs stretching for miles on the Nazca plains of Peru had possibly been laid out by observers in hot-air balloons, *Time* magazine reported, as members of the International Explorers Society constructed and flew such a balloon over the area using materials and techniques available to the ancient Indians. Documents discovered in Portugal had revealed that in 1709 a missionary had demonstrated a model of a hot-air balloon reportedly used by the Indians. Textiles recovered from desert graves provided evidence that the Nazcas had the materials to make a balloon envelope, and a picture of an ancient ceramic appeared to represent a hot-air bag. The IES members built a balloon about 27 m high with fabric resembling the Nazca material, with lines and fastenings from local fibers and a gondola woven from reeds growing in Peru's Lake Titicaca. Named *Condor I*, the device flew about 4 km in 18 min and reached an altitude of more than 300 m. (*Time*, 15 Dec 75, 50)

- Kennedy Space Center had awarded a $265 700 contract to Charter Industries of Raleigh, N.C., for rental of 14 geodesic domes to house displays at a bicentennial exposition of science and technology to be sponsored by the government next summer at the spaceport. KSC would be host for the exposition, which would consist of exhibits from government and industry showing applications of science and technology to improve living conditions during the next 200 years. (NASA Release 75-313)

*16 December:* Astronaut Donald K. (Deke) Slayton had been appointed Deputy Director of Flight Operations for Approach and Landing Test at the Johnson Space Center, announced Center Director Dr. Christopher C. Kraft, Jr. In the newly created position, Slayton would be responsible for planning and implementing the approach and landing test project for the Space Shuttle program. He most recently had been docking module pilot of the U.S. crew for the Apollo-Soyuz Test Project. (NASA Release 75–314; JSC Release 75–100)

- Langley Research Center awarded a $1.5-million 30-mo contract to Rockwell International's Autonetics Group for a highly reliable solid-state satellite data recorder using bubble-domain technology. The new recorder, with a memory capacity of 100 million bits, would achieve a tenfold improvement in reliability over mechanically driven recorders now in use. The bubble-domain technology would make the recorder about a third the size and half the weight of present recorders and would require only half the power for operation. Dr. R. L. Stermer of LaRC said a NASA decision to define hardware using the bubble-domain technology for use in the late 1970s would be greatly influenced by success of the new recorder. (Rockwell Release AG–9)

*17 December:* The Peoples Republic of China announced the launch of its fifth earth satellite, third launched this year. Reuters, quoted in the *New York Times*, said observers believed it had been a step toward Peking's first manned space flight. The fourth China spacecraft, launched 26 Nov., had been recovered, leading observers to believe the Chinese engineers had perfected soft-landing techniques. Weights of the three craft launched this year had not been disclosed, to hide the capacity of the type of rocket used for launch, nor had the current announcement disclosed the satellite's orbit or height. (*NYT*, 18 Dec 75, 26; New China News Agency, in FBIS 243, 17 Dec 75)

- Evidence that the Arabian peninsula is rotating toward Asia exists in photographs taken during the Apollo-Soyuz Test Project, according to Dr. Farouk el-Baz, director of the Smithsonian Center for Earth and Planetary Studies at the National Air and Space Museum. Dr. el-Baz, who coordinated the earth-observation study of ASTP, said the photographs provided geologic evidence of the continental drift theory in color shots of a vast complex of fault lines extending north from the Red Sea through the Gulf of Aqaba, the Dead Sea, and the Sea of Galilee. The fan-shaped pattern revealed the movement of the Arabian landmass to be a rotation rather than simply an eastward drift, Dr. el-Baz said. Until the ASTP pictures became available, the fault line had not been known to extend beyond the Golan Heights region, probably a pivotal spot geologically as well as politically. (*NYT*, 18 Dec 75, 24)

- Kennedy Space Center awarded a $74 998 contract to the Univ. of Fla. Institute of Food and Agricultural Sciences for continuation of a freeze-prediction study. A 1973 contract had provided for study of the application of satellite thermal and infrared imagery to development of a freeze-prediction model; results were so encouraging that a second contract had been awarded in 1974. Also involved in this effort had been meteorologists from the Lakeland office of the National Oceanic and Atmospheric Administration and its Environmental Services Center at Auburn, Ala. Freeze-warning information had previously been provided by NOAA's Ruskin, Fla., facility on the basis of data from 300 recording thermometers operated by a Federal-state agricultural weather system and other reporting stations; the KSC contracts had aimed at developing a computerized system using thermal scanning and infrared imagery from spacecraft and aircraft for more accurate predictions. The study used data from all agencies, including spacecraft readings and actual ground- and leaf-temperature measurements provided by Univ. of Fla. personnel at specific times in selected test areas, combined at KSC for analysis in its earth resources data-analysis facility. The analyzed data had then gone to the university and to NOAA for predictive model development. (KSC Release 294—75)

- NASA had awarded Bendix Field Engineering Corp. a $104-million contract to operate and maintain portions of the worldwide Spaceflight Tracking and Data Network that serves the needs of NASA's earth-orbital space programs, manned and unmanned. The 2-yr cost-plus-award-fee contract would permit negotiation of up to three additional 1-yr extensions, and would be supervised by Goddard Space Flight Center. The contractor would provide technical support for STDN stations and facilities as well as for the Laser Tracking Subnet and for the Magnetic Tape Certification Facility and the NASCOM facilities located at GSFC. (NASA Release 75—315)

*18 December:* Allegations that the Soviet Union had been testing a laser system to blind U.S. spy satellites were discussed in an article in the *Christian Science Monitor*. A satellite watching Russia from a spot in the sky over the Indian Ocean had set off false alarms of a missile attack 18 Oct. when its sensitive instruments detected heat such as might result from the firing of missiles. The heat had originated from a fire caused by breaks in a natural gas pipeline, as later satellite photographs confirmed, but between the alarm and the photographs U.S. intelligence had speculated that the Soviets had trained lasers on the satellite to prevent detection, in violation of the U.S.-Soviet agreement on arms. The gas fires had set off other warnings after 18 Oct., apparently because the ruptures and fires occurred at different times and places in Russia in both October and November. *CSM* quoted an article in *Aviation Week* magazine questioning whether lasers could be used to prevent monitoring of military activity on the ground by satellites, noting that officials at the U.S. Arms Control Agency had discounted the reported incidents. Both the U.S. and the Soviet Union were known to be spending large amounts on laser research, and the December issue of the authoritative *Jane's Weapon Systems* had discussed the implications of laser research for military uses in both countries. (*CSM*, 18 Dec 75, 3; *Miami Herald*, 24 Dec 75, 8A; Ft. Myers *News-Press*, 28 Dec 75, 1)

- The Soviet Union for the first time had discovered atoms of "natrium"—sodium, as it is known in the U.S.—at an altitude of nearly 100 km, in the lower ionosphere. The experiment was difficult, said Tass, because at that altitude only 1 atom of natrium exists per 1000 million molecules of nitrogen and oxygen. A laser unit had established a relationship between natrium content in the upper atmosphere and meteor showers; the announcement said the discovery would be significant because of the effect of natrium on long-distance television and radio communications. (Tass, in FBIS No. 248, 18 Dec 75)

*22 December:* A new domestic communications-satellite partnership named Satellite Business Systems filed applications with the Federal Communications Commission seeking authorization to establish an all-digital system to serve large industrial, government, and other users beginning in 1979. The new partnership had been formed by subsidiaries of Aetna Life & Casualty, Comsat General, and IBM, each of which intended to own one-third. The proposed system, using higher frequencies at 12 and 14 gigahertz and advanced technology, would permit customers in geographically dispersed locations to combine voice, data, and image communications into a single integrated private-line network; small automated ground stations would be located at customers' premises and connected where possible to existing terminals to minimize terrestrial communications costs. The system would use two satellites in geostationary orbit at about 36 000 km altitude, one serving as a backup, to provide coverage of the 48 contiguous states. (Comsat Release 75–63)

- The Soviet Union launched *Prognoz 4*, a 905-kg automatic station to study corpuscular and electromagnetic emissions of the sun, flows of solar plasma, and near-earth magnetic fields, to determine solar influence on the earth's magnetosphere. The satellite's orbit would have a 40 836-km apogee, 451-km perigee, 12 hr 16 min period, and 62.8° inclination. (Tass, in FBIS No 247, 22 Dec 75)

- The Soviet Union launched a communications satellite called *Raduga* (*Rainbow*) into a stationary circular orbit at 35 000 km, with a 23 hr 54 min period and a 0.3° inclination. The satellite had been instrumented to provide continuous round-the-clock telephone and telegraph radio communications with simultaneous transmission of color and black-and-white central TV programs to the network of Orbita stations. (Tass, on FBIS No. 248, 23 Dec 75)

*23 December:* *Pioneer 11*, about a third of the way on its course toward Saturn, had successfully completed its riskiest course-change maneuver in response to signals from earth, 458 million km distant. Controllers at Ames Research Center ceased communications with the Pioneer for several hours, giving the spacecraft time to command itself into a change of position, fire its thrusters, and reposition itself with antenna pointing back to earth. The maneuver increased spacecraft velocity by 108 km per hr, to ensure that Pioneer would have a choice of approach as it neared Saturn: between the rings and the planet, or under the rings and upward outside them. Recent tests had shown the onboard camera working well; *Pioneer 11* had been scheduled to take the first closeup pictures of Saturn and its rings upon arrival at the planet in 1979. If its course had been set inside the rings, *Pioneer 11* should be able to look closely at Saturn's sixth moon, Titan, which is larger than the planet Mercury. (NASA Release 75–319; ARC Release 75–69)

- The Pentagon, checking out reports of Soviet laser devices that had blinded U.S. reconnaissance satellites, had found the reports to be wrong, said new Secretary of Defense Donald H. Rumsfeld at his first press conference since taking over from James R. Schlesinger a month ago. Rumsfeld declined to say how it had been determined that intelligence reports were wrong, but said: "After examination, any implication that the activity that was reported... did in fact exist was not correct." He said it was clear there was no violation of any arms control agreement, nor any evidence to support the laser report, which he sought to portray as press speculation. (*W Star*, 23 Dec 75, 6; *NY News*, 23 Dec 75, 2)

- The National Oceanic and Atmospheric Administration had developed a new remote-sensing technique to "see" nitrogen oxide in the stratosphere, permitting scientists for the first time to calculate the actual rate at which the nitrogen compounds destroy stratospheric ozone. Much of the work on the ozone cycle had been theoretical, based on computer models and laboratory tests; seasonal variations in the stratosphere at high latitudes, revealed by spectrometers on jet aircraft, had not been predicted by theory, which meant that the computer models had not provided close simulation. The new technique, using a spectrometer to measure the light spectrum and isolate that portion caused by nitrogen dioxides, would permit routine checks on daily, seasonal, and latitudinal variations in the content of the stratosphere. The method would be applied by small automated instruments run by minicomputers at NOAA stations in Alaska, Samoa, Hawaii, and the South Pole, and at the Air Force base in Greenland. (NOAA Release 75–220)

*24 December*: The National Research Council, in its program of conducting technical studies on behalf of the National Academy of Sciences and the National Academy of Engineering, had mailed out announcements to more than 5000 individuals and organizations inviting statements on the role of nuclear power in the context of alternative energy systems. Hearings in five major cities had been scheduled for January and early February 1976 to obtain the broadest possible perspective for a comprehensive study. The Energy Research and Development Administration had asked the academies to conduct a 2-yr study, and the announcement represented the first time in the 112-yr history of the National Academy of Sciences that such an extensive effort had been made to solicit the views of the informed public on a projected study. The committee conducting the study would attempt to set forth the current and probable states of energy technology through the year 2010, various policy options, and consequences of each. (NRC Release 24 Dec 75)

- The Air Force had awarded a $28.5-million contract to Lockheed Aircraft Corp. for design, development, and testing of a strengthened wing for the C–5A transport airplane that the *New York Times* called a symbol of Pentagon "cost overruns." Aim of the original C–5A contract with Lockheed had been to produce a transport capable of carrying a 99 700-kg payload and having a service life of 30 000 flying hours. Even before it had gone into full operational use, the C–5A wing had been shown in static ground tests to be too weak to carry the prescribed load and had developed fatigue cracks in the metal structure. The cracking had been attributed to high stress levels inherent in the design, ineffective fasteners on the wing panels, and difficulty in mating the stiff contours of the wing and joints during manufacture. The Air Force had paid $4.5 billion on delivery of 81 C–5As; it had already spent more than $198 million through June 1975 correcting deficiencies for which Lockheed was responsible under the original contract, renegotiated when Lockheed was in financial difficulty in 1971. In addition to $1.3 billion for fixing the wings, the AF planned to spend another $200 million to fix other defects. Cost of the repairs would nearly double the original AF estimate of $3.4 billion for the C−5A program. However, without the repairs, the C−5A fleet would have to be grounded beginning in 1979 when the flying-time safety limit had been reached. Because of the defects, service life had recently been re-estimated at 8750 hr and payload limit at 78 900 kg. (*NYT*, 15 Dec 75, 1; *WSJ*, 24 Dec 75, 3)

- Citing "new scientific evidence" supporting ozone-depletion theories, the Natural Resources Defense Council and 10 states had petitioned the Consumer Product Safety Commission to ban the use of fluorocarbon propellants in aerosol products. The commission had denied a previous petition from the resources council; the Food and Drug Administration—which had responsibility for regulating all food, drug, and cosmetic aerosol products—had also rejected a petition from the group. Copetitioners with the resources council were the Environmental Defense Fund and the Minnesota Public Interest Research Group; the 10 states were Colo., Fla., Mass., Mich., Minn., N.H., N.Y., Ore., Vt., and Wis. Both N.Y. and Ore. had already passed laws banning the use of fluorocarbon sprays, and more than a dozen other states were said to be considering this action. The Aerosol Education Bureau—sponsored by industry to oppose a ban of fluorocarbon-containing products—issued a statement that restrictions would be "inappropriate." (*WSJ*, 24 Dec 75, 5)

*26 December:* The Christmas Day inaugural flight announced for the U.S.S.R.'s supersonic jetliner had been delayed a day by weather conditions variously reported as runway ice or fog at the destination, but its first flight carrying freight and mail from Moscow to Alma Ata—capital of the central Asian republic of Kazakhstan—took only 4 hr for the 8000-km round trip. The Tupolev−144 airliner, rival of the British-French Concorde, could fly at an altitude of up to 46.3 km, higher than conventional airliners. Although the flights had been billed as the first commercial service by a supersonic jet, and had been scheduled to get ahead of the Concorde passenger service set for January between Paris and Rio and between London and Bahrain, Tu−144 would not carry passengers until the second half of 1976 at the earliest, and then on internal routes, according to Deputy Aviation Minister Nikolai Boikov. (*NYT*, 27 Dec 75, 2; *Miami Herald*, 27 Dec 75, 20A)

*27 December:* In a *Miami Herald* article presenting "the case for the Concorde," S. Fred Singer—who had served in 1970−71 as chairman of an interagency task group to evaluate environmental effects of supersonic transports—pointed out that the effect of Concorde pollution would be completely swamped by other factors that had been shown to decrease atmospheric ozone—nitrogen fertilizers, carbon tetrachloride, and fluorocarbons. Banning the Concorde from the U.S. would not prevent it from depositing pollutants elsewhere; in any case, military planes flying in the stratosphere and subsonic planes pushing to higher altitudes to avoid traffic would produce the same pollution effect as the Concordes. Singer, now a professor of environmental sciences at U. of Va., called for more study of the effects of Freon and fertilizers and for modification of the engines of future SSTs. (*Miami Herald*, 27 Dec 75, 7A)

*28 December:* Lutz T. Kayser, an aerospace engineer from Stuttgart, Germany, had established a business called Orbital Transport and

Raketen A.G. to build and launch rockets on a commercial basis, starting in 1980: if successful, it would be the world's first private space-launch venture. NASA had been charging at least $25 million to put a $12-million satellite into space; Kayser had designed a "space truck" to put satellites in orbit for half the NASA price. His rockets could not be used for military purposes, he said, as he had no plans to sell equipment, but would do the launches himself and sell only booster services. As the launches could not take place in crowded West Germany, they would have to be done from either a ship at sea or from a base in some sparsely populated country. After ground tests of Kayser's rocket engines last summer, German scientists had been convinced that his spacecraft would fly.

Kayser said that if the Germans would not support him, he would seek backers elsewhere, in South America or Asia. He had hoped more than half the backing would come from private investors, and the rest from banks. The West German government had spent about $2 billion to encourage German and European space projects that could offer an alternative to the NASA monopoly, but such projects had not succeeded.

The first "space truck"—now under construction for suborbital test next year—would be 39.62 m high and consist of inexpensive conventionally produced materials. To make mass production possible, all stages would use identical engines; the stages would be placed in concentric rings that would drop off as the craft climbed. (*NYT*, 28 Dec 75, 18)

*30 December:* Pulsars had been discovered to be not only superdense but also superfluid, Columbia University physicists reported in *Astrophysical Journal*. Pulsars, collapsed remnants of exploded stars formed almost entirely of neutrons, had been shown to be so dense that a spoonful would weigh millions of metric tons. Theorists had suspected that such objects, even with a thin rigid crust of iron, would be superfluid—a state produced experimentally on earth in small helium specimens cooled almost to absolute zero, where both friction and viscosity disappear. Material inside such an object, stirred into motion, would swirl indefinitely.

The Columbia Univ. experimenters had attempted to measure the surface temperature of a young pulsar, such as the one in the Crab Nebula. If the temperature were less than 8.8 million°C, the object would probably be superfluid; a fast-spinning neutron star would cool rapidly because of its reduced heat capacity. Temperatures of such objects could be detected by x-ray wavelengths, but the x-rays produced could not penetrate earth's atmosphere. Space observation was necessary, and rockets launched by NASA from Hawaii and N. Mex. tested the Crab nebula pulsar, using the moon as a curtain. As the moon passed across the whole nebula several times, cutting off its various x-ray emissions, the physicists had been able to identify the x-rays indicating temperature. Since the Crab nebula pulsar's temperature was shown to be less than 5 million °C, the physicists concluded that the pulsar must be superfluid. (*NYT*, 30 Dec 75, 40)

- Predictions of a trend toward harsher winters for North America had been put in question by satellite monitoring of snowcover over the northern hemisphere during the past 9 yr, according to the National Oceanic and Atmospheric Administration. NOAA's National

Environmental Satellite Service had published a document, "Monthly Winter Snowline Variation in the Northern Hemisphere from Satellite Records, 1966–1975," containing the most complete record available of hemisphere snowcover. Analysis of the charts had revealed no significant change in North American snowcover during the entire period. Predictions of harsher weather had raised the possibility of southward movement of the polar ice cap, which could be caused by a drop of only a few degrees in average winter temperature. The document concluded that "Lack of systematic increase in . . . snow cover tends to contradict the evidence . . . that the current climate is changing adversely." (NOAA Release 75–217)

- President Ford had signed legislation establishing a 17-member U.S. Metric Board to plan for voluntary conversion of the nation to the metric system. Dr. Ernest Ambler, acting head of the National Bureau of Standards, said the signing of the bill would encourage U.S. industry and commerce to export products made to metric dimensions, which "should stimulate additional international trade." (SBD, 30 Dec 75, 299)

*During December:* Goddard Space Flight Center and the Fla. Dept. of Natural Resources had approved the first long-range, space-assisted program to develop an "early warning system" for the onset of red tide, an ocean-borne organism that had left tons of dead fish rotting on beaches around the world. Dr. Warren A. Hovis of GSFC had designed an ocean-color scanner capable of detecting subtle variations in the color of coastal waters that might indicate changes in concentration and species of marine phytoplankton, especially the red-tide dinoflagellate *Gymnodinium breve*. Red tide had been a major problem because of oxygen depletion in ocean water from fish decomposition, and eye and respiratory infections caused by toxic particles produced in the organism. Shellfish beds affected had been closed for months by authorities. An outbreak of red tide in 1971 had cost the state of Fla. more than $20 million in lost tourist business. The ocean scanner, developed for use in the Nimbus-G pollution-sensing spacecraft scheduled for launch in 1978, could be used in photographic overflights by aircraft and Landsat spacecraft, in connection with on-site sampling of ocean waters by research ships, to detect the organism in time to take countermeasures and warn the public about affected areas. (*Spaceflight*, Vol 17 No. 12, Dec 75, 425)

- The boards of directors of the All-Women Transcontinental Air Race and the Ninety-Nines, Inc., had decided to terminate the Powder Puff Derby—world's largest and longest speed race for light aircraft—after the 1976 bicentennial race, because of the nation's current financial and energy problems, the National Aeronautic Association announced. Sponsored by NAA since its beginning in 1947, the derby had a safety record unparalleled in air race history: no fatalities. The NAA newsletter noted that the derby had inspired countless women to fly, and that recent races had more than 100 aircraft competing. The final derby had been scheduled for 12–19 July 1976 from Sacramento, Calif., to Wilmington, Del. (*NAA News*, Dec 75)

- Space scientists and astronomers had been wrestling with a "cultural crisis" in putting names to features of distant worlds that had never been clearly seen before, as closeup views of several planets became available and more were on the way. Working through the International Astronomical Union, the scientists had been forced to look beyond the names of scientists and explorers that had been used almost exclusively for objects on the moon and on Mars. A meeting on the problem in Moscow in July 1974 had agreed that objects on Venus would be called after women famous in mythology and history. New maps of the recently photographed surface of Mercury would bear the names of contributors to the arts and other humanities; the scientists had sought advice from scholars in the humanities to get the widest possible representation from past and present cultures.

The Moscow meeting on new approaches to names for planetary features had been preceded by intense work on naming newly discovered configurations on the moon and on Mars. A complete scheme of names for more than 500 large formations on the far side of the moon—visible only to spacecraft because one side of the moon had always been turned toward earth—had been approved in 1970. Soviet scientists had rejected suggestions of names from the humanities for features on the moon and on Mars; they had urged a fresh start on Mercury, a new planet, with formations named not only for authors, as the Americans had originally suggested, but also for persons in other branches of the humanities. (*NYT*, 25 Dec 75, C10)

- Space communities that could supply enough energy to the earth to end its energy crisis permanently had been described in an article by Dr. Gerard K. O'Neill in *Aerospace* magazine. Recalling the conclusions of Princeton Univ. conferences on space colonization in 1974 and 1975, and his own testimony before the Subcommittee on Space Sciences and Applications of the House Committee on Science and Technology 23 July 1975, Dr. O'Neill said that within 13 yrs after beginning construction in space, the amount of usable electricity delivered to the point where power would enter the distribution system would exceed the peak capacity of the Alaska pipeline. An investment of about $96 billion over a 6-yr period would serve to establish the beachhead colony, and even with program costs of up to $300 billion the payback after 9 yr would exceed the total investment and interest, he estimated. (*Aerospace*, Dec 75, 3)

- Establishment of manufacturing facilities in space might be feasible within this century, Dr. Gerard K. O'Neill wrote in the December issue of *Science* magazine. Construction of satellite solar power stations (SSPSs) from lunar materials, for example, could be shown more economical than building such stations on earth and lifting them into orbit. The space manufacturing facility (SMF) that had been envisioned by Dr. O'Neill and others would be a self-sustaining habitat for a large number of people whose energy needs could be met by solar power used directly for agriculture, as process heat for industry when concentrated by mirrors, or indirectly as electricity. Dr. O'Neill had assembled tables and figures from the literature to accompany his argument for the SSPS as a means of meeting increased demands for electricity on earth in a time of shortage of fuels and concern about use of nuclear energy. (*Science*, Vol 190, 4218)

# Appendix A

## SATELLITES, SPACE PROBES, AND MANNED SPACE FLIGHTS, 1975

World space activity increased in 1975. Total launches, 125, increased from 106 in 1974. Of this total, the U.S. had 28, described in the introduction to Appendix B. The U.S.S.R. had 89 launches with 112 payloads, 85 of these in the Cosmos series; Cosmos payloads included such specialized spacecraft as *Cosmos 775*, part of the Statsionar series for TV relay, and the *Cosmos 782* biosat carrying 4 U.S. experiments. The Soviets also launched 10 spacecraft in the Molniya comsat series, 4 in the Meteor series, 2 Intercosmos spacecraft carrying experiments from other countries, the 2 Venus landers (*Venera 9* and *10*), the magnetospheric investigator *Prognoz 4*, a color-TV comsat called *Raduga (Rainbow)*, and 5 Soyuz spacecraft including the April 5 "anomaly" and the *Soyuz 19* that participated in ASTP. The U.S.S.R. also launched *Aryabhata* for India in April, and *Sret 2* for France as a piggyback payload with *Molniya 1–30* in June.

The number and variety of launches by other countries also increased in 1975. In Feb., France launched *Starlette*, a passive geodetic satellite, first for the Diamant booster developed by CNES; in May it launched *Castor* and *Pollux*, first 2 satellites launched by France on a single booster; in Sept., the industry-managed satellite *Aura* went into orbit on the last Diamant booster. The U.S.S.R. had launched France's *Sret 2* in June, and a French-West German consortium sponsored the launch of *Symphonie 2* by NASA in Aug. In Feb., Japan launched *Taiyo* to observe solar radiation and thermospheric structure; in Sept., it launched *Kiku* to measure temperatures in space. The People's Republic of China in July launched its third satellite, first in 4 yr (previous launches were in 1970 and 1971); in Nov., it launched a fourth satellite and shortly announced its successful recovery, becoming the third country to retrieve a payload from orbit. In Dec., the PRC launched its fifth satellite, third for 1975, in what observers considered a step toward its first manned space flight.

The following table includes all payloads that have (a) orbited; (b) as probes, ascended to at least the 6500-km altitude that traditionally had distinguished probes from sounding rockets; or (c) conveyed one or more human beings into space, whether or not orbit was attained. The 1975 table lists for the first time in this series the identities of Soviet launch vehicles; for details on these vehicles, see *Soviet Space Programs 1971–75*, U.S. Senate Comm. on Aeronautical & Sp. Sciences, 30 Aug. 76, Vol. I, 39–61. For details on U.S. launch vehicles, see *Aeronautics & Space Report of the President, 1975 Activities*, GPO stock number 033–000–00649–1, App. D, 110.

Sources of these data include the United Nations Public Registry of Space Flights; the *Satellite Situation Report* compiled by Goddard Space Flight Center's Operations Control Center; press releases of NASA, Department of Defense, National Oceanic and Atmospheric Administration, and other government agencies, as well as the Communications Satellite Corporation. Soviet data derive from statements in the Soviet press, translations from the Tass news agency, and international news-service reports, as well as announcements and briefings by Soviet officials on matters affecting the Apollo-Soyuz Test Project. Data on satellites of other nations also come from the announcements of their respective governments and international news services.

# SATELLITES, SPACE PROBES, AND MANNED SPACE FLIGHTS, 1975

| Launch Date | Spacecraft, Country, Int'l Designation, Vehicle, Launch Site | Payload Data | Apogee (km) | Perigee (km) | Period (min) | Inclination (degrees) | Remarks |
|---|---|---|---|---|---|---|---|
| 10 Jan. | Soyuz 17 U.S.S.R. 1975-1A A-2 Baykonur-Tyuratam | Total weight: Unavailable. Objective: Rendezvous, docking, joint experiments with Salyut 4. Description: Unavailable. | 348 | 336 | 91.4 | 51.6 | Crew (Lt. Col. Aleksey A. Gubarev, Georgy M. Grechko) docked with Salyut 4 on 12 Jan., undocked & reentered 9 Feb. |
| 17 Jan. | Cosmos 702 U.S.S.R. 1975-2A A-2 Baykonur-Tyuratam | Total weight: Unavailable. Objective: "Investigation of upper atmosphere and outer space." Description: Unavailable. | 309 | 203 | 89.6 | 71.3 | Reentered 29 Jan. |
| 21 Jan. | Cosmos 703 U.S.S.R. 1975-3A B-1 Plesetsk | Total weight: Unavailable. Objective: "Investigation of upper atmosphere and outer space." Description: Unavailable. | 1514 | 195 | 102.0 | 82.0 | Probable military satellite. Reentered 20 Nov. |
| 22 Jan. | Landsat 2 (ERTS-B) U.S. 1975-4A Thorad-Delta WTR | Total weight: 953 kg. Objective: Improve remote-sensing techniques; demonstrate practical application of earth-resources data. Description: Like Nimbus spacecraft, three sections: toroid base, hexagonal housing, 2 solar panels; 3-axis stabilized; 3.04 m high, 1.52 dia.; carries multispectral scanner, return-beam vidicon, 2 video-tape recorders, and data-collection subsystem. | 915 | 901 | 103.2 | 99.1 | Still in orbit; provided contiguous images of earth every 18 days, initially to more than 40 countries. |
| 23 Jan. | Cosmos 704 U.S.S.R. 1975-5A A-2 Plesetsk | Total weight: Unavailable. Objective: "Investigation of upper atmosphere and outer space." Description: Unavailable. | 337 | 166 | 89.5 | 72.8 | Reentered 6 Feb. |
| 28 Jan. | Cosmos 705 U.S.S.R. 1975-6A B-1 Plesetsk | Total weight: Unavailable. Objective: "Investigation of upper atmosphere and outer space." Description: Unavailable. | 498 | 269 | 92.2 | 71.0 | Probable military satellite. Reentered 18 Nov. |

# Appendix A

| Launch Date | Spacecraft, Country, Int'l Designation, Vehicle, Launch Site | Payload Data | Apogee (km) | Perigee (km) | Period (min) | Inclination (degrees) | Remarks |
|---|---|---|---|---|---|---|---|
| 30 Jan. | Cosmos 706 U.S.S.R. 1975-7A A-2-3 Plesetsk | Total weight: Unavailable. Objective: "Investigation of upper atmosphere and outer space." Description: Unavailable. | 39 821 | 621 | 719.6 | 62.9 | Probable early-warning satellite; still in orbit. |
| 5 Feb. | Cosmos 707 U.S.S.R. 1975-8A C-1 Plesetsk | Total weight: Unavailable. Objective: "Investigation of upper atmosphere and outer space." Description: Unavailable. | 546 | 500 | 95.1 | 74.0 | Probable electronic ferret; still in orbit. |
| 6 Feb. | Molniya 2 U.S.S.R. 1975-9A A-2-3 Plesetsk | Total weight: Unavailable. Objective: Operation of long-range telephone and telegraph, radiocommunications system; transmission of TV programs to stations in the Orbita network. Description: Unavailable. | 39 879 | 504 | 718.4 | 62.8 | Twelfth in series; still in orbit. |
| 6 Feb. | Starlette 2 France 1975-10A Diamant BP4-1 Kourou, French Guiana | Total weight: 47 kg. Objective: Passive geodetic satellite. Description: Polyhedral spheroid 0.26m in dia., carrying 60 flat laser-reflectors for triangulation experiments. | 1137 | 804 | 104.5 | 49.8 | Still in orbit. |
| 6 Feb. | Sms 2 U.S. 1975-11A Thorad-Delta ETR | Total weight: 272 kg in orbit. Objective: Provide day-night cloudcover data; transmit environmental information to NOAA from thousands of data-collection platforms worldwide. Description: Spin-stabilized solar-cell-covered cylinder 190.5 cm in dia., 254 cm long without apogee motor; carries space-environment monitor, visible infrared radiometer, meteorological data-collection system. | 36 685 | 35 680 | 1456.4 | 1.1 | Still in orbit. |
| 12 Feb. | Cosmos 708 U.S.S.R. 1975-12A C-1 Plesetsk | Total weight: Unavailable. Objective: "Investigation of upper atmosphere and outer space." Description: Unavailable. | 1411 | 1368 | 113.6 | 69.2 | Probable navigation satellite; still in orbit. |

| Date | Name/Designation | Description | | | | Remarks |
|---|---|---|---|---|---|---|
| 12 Feb. | Cosmos 709<br>U.S.S.R.<br>1975-13A<br>A-2<br>Plesetsk | Total weight: Unavailable.<br>Objective: "Investigation of upper atmosphere and outer space."<br>Description: Unavailable. | 310 | 183 | 89.4 | 62.9 | Reentered 25 Feb. |
| 24 Feb. | Taiyo (Srats)<br>Japan<br>1975-14A<br>M-3C-2<br>Kagoshima | Total weight: 86 kg.<br>Objective: Study intereffects of solar radiation and earth's thermosphere.<br>Description: Unavailable. | 3132 | 250 | 120.32 | 31.5 | Still in orbit; 3rd Japanese scientific satellite. |
| 26 Feb. | Cosmos 710<br>U.S.S.R.<br>1975-15A<br>A-2<br>Baykonur-Tyuratam | Total weight: Unavailable.<br>Objective: "Investigation of upper atmosphere and outer space."<br>Description: Unavailable. | 332 | 173 | 89.6 | 65.0 | Reentered 12 Mar. |
| 28 Feb. | Cosmos 711<br>U.S.S.R.<br>1975-16A<br>C-1<br>Plesetsk<br>and | Total weight: Unavailable.<br>Objective: "Investigation of upper atmosphere and outer space."<br>Description: Unavailable. | 1494 | 1463 | 115.4 | 73.9 | Eight spacecraft on single carrier; all still in orbit; probable comsats. |
| | Cosmos 712<br>U.S.S.R.<br>1975-16B<br>C-1<br>Plesetsk<br>and | Total weight: Unavailable.<br>Objective: "Investigation of upper atmosphere and outer space."<br>Description: Unavailable. | 1492 | 1412 | 114.9 | 73.9 | |
| | Cosmos 713<br>U.S.S.R.<br>1975-16C<br>C-1<br>Plesetsk<br>and | Total weight: Unavailable.<br>Objective: "Investigation of upper atmosphere and outer space."<br>Description: Unavailable. | 1488 | 1398 | 114.7 | 73.9 | |
| | Cosmos 714<br>U.S.S.R.<br>1975-16D<br>C-1<br>Plesetsk<br>and | Total weight: Unavailable.<br>Objective: "Investigation of upper atmosphere and outer space."<br>Description: Unavailable. | 1492 | 1446 | 115.2 | 73.9 | |

# Appendix A ASTRONAUTICS AND AERONAUTICS, 1975

| Launch Date | Spacecraft, Country, Int'l Designation, Vehicle, Launch Site | Payload Data | Apogee (km) | Perigee (km) | Period (min) | Inclination (degrees) | Remarks |
|---|---|---|---|---|---|---|---|
| | *Cosmos 715* U.S.S.R. 1975-16E C-1 Plesetsk | Total weight: Unavailable. Objective: "Investigation of upper atmosphere and outer space." Description: Unavailable. | 1506 | 1471 | 115.7 | 73.9 | |
| | and | | | | | | |
| | *Cosmos 716* U.S.S.R. 1975-16F C-1 Plesetsk | Total weight: Unavailable. Objective: "Investigation of upper atmosphere and outer space." Description: Unavailable. | 1516 | 1480 | 115.9 | 73.9 | |
| | and | | | | | | |
| | *Cosmos 717* U.S.S.R. 1975-16G C-1 Plesetsk | Total weight: Unavailable. Objective: "Investigation of upper atmosphere and outer space." Description: Unavailable. | 1538 | 1481 | 116.1 | 73.9 | |
| | and | | | | | | |
| | *Cosmos 718* U.S.S.R. 1975-16H C-1 Plesetsk | Total weight: Unavailable. Objective: "Investigation of upper atmosphere and outer space." Description: Unavailable. | 1491 | 1430 | 115.1 | 73.9 | |
| 10 March | DOD spacecraft U.S. 1975-17A Titan III-Agena WTR | Total weight: Unavailable. Objective: "To develop space flight techniques and technology." Description: Unavailable. | 39 338 | 295 | 702 | 63.5 | Still in orbit. |
| 12 March | *Cosmos 719* U.S.S.R. 1975-18A A-2 Baykonur-Tyuratam | Total weight: Unavailable. Objective: "Investigation of upper atmosphere and outer space." Description: Unavailable. | 306 | 174 | 89.3 | 65.0 | Reentered 25 Mar. |

| Date | Satellite | Description | | | | Notes |
|---|---|---|---|---|---|---|
| 21 March | Cosmos 720<br>U.S.S.R.<br>1975-19A<br>A-2<br>Plesetsk | Total weight: Unavailable.<br>Objective: "Investigation of upper atmosphere and outer space."<br>Description: Unavailable. | 270 | 210 | 89.3 | 62.8 | Reentered 1 April. |
| 26 March | Cosmos 721<br>U.S.S.R.<br>1975-20A<br>A-2<br>Plesetsk | Total weight: Unavailable.<br>Objective: "Investigation of upper atmosphere and outer space."<br>Description: Unavailable. | 228 | 205 | 88.8 | 81.3 | Reentered 7 April. |
| 27 March | Cosmos 722<br>U.S.S.R.<br>1975-21A<br>A-2<br>Baykonur-Tyuratam | Total weight: Unavailable.<br>Objective: "Investigation of upper atmosphere and outer space."<br>Description: Unavailable. | 327 | 172 | 89.5 | 71.4 | Reentered 9 April. |
| 27 March | Intercosmos 13<br>U.S.S.R.<br>1975-22A<br>C-1<br>Plesetsk | Total weight: Unavailable.<br>Objective: Study dynamic processes in earth's magnetosphere and ionosphere, and low-frequency electromagnetic waves.<br>Description: Unavailable. | 1687 | 284 | 104.8 | 82.9 | Still in orbit. |
| 1 April | Meteor 1<br>U.S.S.R.<br>1975-23A<br>A-1<br>Plesetsk | Total weight: Unavailable.<br>Objective: Acquire meteorological information for use in operational weather system.<br>Description: Unavailable. | 892 | 865 | 102.5 | 81.2 | Twenty-first in Meteor series; still in orbit. |
| 2 April | Cosmos 723<br>U.S.S.R.<br>1975-24A<br>F-1-m<br>Baykonur-Tyuratam | Total weight: Unavailable.<br>Objective: "Investigation of upper atmosphere and outer space."<br>Description: Unavailable. | 264 | 251 | 89.7 | 65.0 | Probable ocean-surveillance satellite; still in orbit. |
| 5 April | Soyuz (anomaly)<br>U.S.S.R.<br>A-2<br>Baykonur-Tyuratam | Total weight: Unavailable.<br>Objective: Continuing joint experiments with Salyut 4 station in orbit.<br>Description: Unavailable. | | 1600 km from launch site | | | Malfunction of 3rd stage of RD-107C carrier rocket aborted mission at T+5; crew (Col. Vasily Lazarev, Oleg Makarov) recovered safely after soft landing in Siberia. |
| 7 April | Cosmos 724<br>U.S.S.R.<br>1975-25A<br>F-1-m<br>Baykonur-Tyuratam | Total weight: Unavailable.<br>Objective: "Investigation of upper atmosphere and outer space."<br>Description: Unavailable. | 264 | 251 | 89.7 | 65.0 | Probable ocean-surveillance satellite; still in orbit. |

| Launch Date | Spacecraft, Country, Int'l Designation, Vehicle, Launch Site | Payload Data | Apogee (km) | Perigee (km) | Period (min) | Inclination (degrees) | Remarks |
|---|---|---|---|---|---|---|---|
| 8 April | Cosmos 725 U.S.S.R. 1975-26A B-1 Plesetsk | Total weight: Unavailable. Objective: "Investigation of upper atmosphere and outer space." Description: Unavailable. | 482 | 271 | 92.1 | 71.0 | Probable military satellite; reentered 6 Jan. 76. |
| 9 April | Geos 3 U.S. 1975-27A Thorad-Delta 1410 WTR | Total weight: 340 kg. Objective: Use radar altimeter for precise sea-surface mapping and comparison with other geodetic measurement systems. Description: Gravity-gradient-stabilized octahedron 132 cm wide, topped by truncated pyramid for height of 81 cm.; 8 trusses on center tube support 16 solar panels; carries radar altimeter, C-band transponders, S-band instrumentation, laser reflectors, and doppler transmitters. | 844 | 837 | 101.9 | 115.0 | Still in orbit. |
| 11 April | Cosmos 726 U.S.S.R. 1975-28A C-1 Plesetsk | Total weight: Unavailable. Objective: "Investigation of upper atmosphere and outer space." Description: Unavailable. | 994 | 958 | 104.6 | 83.0 | Probable navigational satellite; still in orbit. |
| 14 April | Molniya 3 U.S.S.R. 1975-29A A-2-R Plesetsk | Total weight: Unavailable. Objective: Operate long-range telephone and telegraph radiocommunications system; transmit TV programs to stations in Orbita network. Description: Unavailable. | 40 651 | 599 | 736.0 | 62.9 | Second of this type; still in orbit. |
| 16 April | Cosmos 727 U.S.S.R. 1975-30A A-2 Baykonur-Tyuratam | Total weight: Unavailable. Objective: "Investigation of upper atmosphere and outer space." Description: Unavailable. | 334 | 171 | 89.6 | 65.0 | Reentered 28 April. |
| 18 April | Cosmos 728 U.S.S.R. 1975-31A A-2 Plesetsk | Total weight: Unavailable. Objective: "Investigation of upper atmosphere and outer space." Description: Unavailable. | 321 | 200 | 89.7 | 72.8 | Reentered 29 April. |

| Date | Satellite | Description | (col) | (col) | (col) | Notes |
|---|---|---|---|---|---|---|
| 18 April | DOD spacecraft U.S. 1975-32A Titan IIIB-Agena WTR | Total weight: Unavailable. Objective: "Develop space flight techniques and technology." Description: Unavailable. | 408 | 126 | 90.0 | 110.5 Reentered 5 June. |
| 19 April | Aryabhata India 1975-33A C-1 Kapustin Yar, U.S.S.R. | Total weight: 360 kg. Objective: Measure celestial x-rays and ionospheric parameters; look for neutrons and gamma rays from the sun. Description: Spin-stabilized polyhedron, 26 sides covered with solar cells; 3 scientific experiments; telemetry transmitters. | 611 | 568 | 96.5 | 50.7 First scientific satellite of India; launched by U.S.S.R.; still in orbit; its 3 scientific experiments were turned off after 60th orbit. |
| 22 April | Cosmos 729 U.S.S.R. 1975-34A C-1 Plesetsk | Total weight: Unavailable. Objective: "Investigation of upper atmosphere and outer space." Description: Unavailable. | 1009 | 977 | 105.0 | 83.0 Probable navigational satellite; still in orbit. |
| 24 April | Cosmos 730 U.S.S.R. 1975-35A A-2 Plesetsk | Total weight: Unavailable. Objective: "Investigation of upper atmosphere and outer space." Description: Unavailable. | 265 | 168 | 88.8 | 81.3 Reentered 6 May. |
| 29 April | Molniya 1 U.S.S.R. 1975-36A A-2-e Plesetsk | Total weight: Unavailable. Objective: Operate long-range telephone and telegraph radiocommunications system; transmit TV programs to stations in Orbita network. Description: Unavailable. | 40 850 | 437 | 736.8 | 62.8 Twenty-ninth of this series; still in orbit. |
| 7 May | Explorer 53 (Sas-C) U.S. 1975-37A Scout San Marco | Total weight: 195 kg. Objective: Search for radiation sources in x-ray and other spectral regions inside and outside earth's galaxy. Description: Spin-stabilized 145.2-cm-high spacecraft in 2 sections (housekeeping and experiments) with 4 foldable solar panels 470.3 cm tip to tip; experiments include galactic monitor, galactic absorption, Scorpio monitor. | 516 | 509 | 94.9 | 3.0 Third in series of small astronomy satellites; launched by Italian crew from platform in Indian Ocean; still in orbit. |
| 7 May | Anik 3 (Telesat C) Canada 1975-38A Thorad-Delta ETR | Total weight: 544 kg. Objective: Provide TV, voice, and data communications throughout Canada. Description: Spin-stabilized cylinder 1.8 m dia., 3.4 m high, carries 1.5-m stationary antenna; provides 10 color TV channels, up to 9600 telephone circuits. | 35 919 | 231 | 634.3 | 24.8 Domestic comsat launched by U.S. for Canada; still in orbit. |

Appendix A  ASTRONAUTICS AND AERONAUTICS, 1975

| Launch Date | Spacecraft, Country, Int'l Designation, Vehicle, Launch Site | Payload Data | Apogee (km) | Perigee (km) | Period (min) | Inclination (degrees) | Remarks |
|---|---|---|---|---|---|---|---|
| 17 May | Castor/Pollux France 1975-39A, -39B Diamant B-P4 Kourou, French Guiana | Total weight: -A (Pollux), 36 kg; -B (Castor), 76 kg. Objective: -A, space test of hydrazine propulsion unit; -B, space test of Cactus accelerometer. Description: Unavailable. | -A, 1283 -B, 1269 | -A, 270 -B, 273 | -A, 100.4 -B, 100.3 | -A, 30.0 -B, 30.0 | Two payloads on single carrier; 1975-39A (Pollux) reentered 5 Aug.; 1975-39B (Castor) still in orbit. |
| 20 May | DOD spacecraft U.S. 1975-40A Titan IIIC WTR | Total weight: Unavailable. Objective: "Develop space flight techniques and technology." Description: Unavailable. | Orbital elements not available. | | | | Reentered 26 May. |
| 21 May | Cosmos 731 U.S.S.R. 1975-41A A-2 Baykonur-Tyuratam | Total weight: Unavailable. Objective: "Investigation of upper atmosphere and outer space." Description: Unavailable. | 280 | 201 | 89.3 | 65.0 | Reentered 2 June. |
| 22 May | Intelsat IV F1 U.S. 1975-42A Atlas-Centaur ETR | Total weight: 700 kg in orbit. Objective: Domestic comsat. Description: Spin-stabilized solar-cell-covered cylinder 5.28 m high, 2.38 m dia.; 12 communications transponders, 6 antennas (2 global-transmit, 2 global-receive, 2 steerable spot-beam). | 36 184 | 35 704 | 144.2 | 0.5 | Reimbursable launch for Comsat General Corp.; last of Intelsat IV series; still in orbit. |
| 24 May | DOD spacecraft U.S. 1975-43A Thor-Burner 2A ETR | Total weight: Unavailable. Objective: "Develop space flight techniques and technology." Description: Unavailable. | 893 | 807 | 101.9 | 98.9 | Still in orbit. |
| 24 May | Soyuz 18 U.S.S.R. 1975-44A A-2 Baykonur-Tyuratam | Total weight: Unavailable. Objective: Joint experiments with Salyut 4 orbiting space station. Description: Unavailable. | 349 | 338 | 91.4 | 51.6 | Crew (Col. Pyotr Klimuk, Vitaly Sevastyanov) docked with Salyut 4 25 May; reentered 26 July; 63-day mission, new Soviet record. |

| | | | | | | |
|---|---|---|---|---|---|---|
| 28 May | Cosmos 732<br>U.S.S.R.<br>1975-45A<br>C-1<br>Plesetsk<br>and | Total weight: Unavailable.<br>Objective: "Investigation of upper atmosphere and outer space."<br>Description: Unavailable. | 1472 | 1405 | 114.6 | 74.0 | Probable comsats; 8 spacecraft carried by single booster; all still in orbit. |
| | Cosmos 733<br>U.S.S.R.<br>1975-45B<br>C-1<br>Plesetsk<br>and | Total weight: Unavailable.<br>Objective: "Investigation of upper atmosphere and outer space."<br>Description: Unavailable. | 1554 | 1473 | 116.2 | 74.0 | |
| | Cosmos 734<br>U.S.S.R.<br>1975-45C<br>C-1<br>Plesetsk<br>and | Total weight: Unavailable.<br>Objective: "Investigation of upper atmosphere and outer space."<br>Description: Unavailable. | 1473 | 1445 | 115.0 | 74.0 | |
| | Cosmos 735<br>U.S.S.R.<br>1975-45D<br>C-1<br>Plesetsk<br>and | Total weight: Unavailable.<br>Objective: "Investigation of upper atmosphere and outer space."<br>Description: Unavailable. | 1475 | 1463 | 115.2 | 74.0 | |
| | Cosmos 736<br>U.S.S.R.<br>1975-45E<br>C-1<br>Plesetsk<br>and | Total weight: Unavailable.<br>Objective: "Investigation of upper atmosphere and outer space."<br>Description: Unavailable. | 1488 | 1471 | 115.5 | 74.0 | |
| | Cosmos 737<br>U.S.S.R.<br>1975-45F<br>C-1<br>Plesetsk<br>and | Total weight: Unavailable.<br>Objective: "Investigation of upper atmosphere and outer space."<br>Description: Unavailable. | 1531 | 1471 | 115.9 | 74.0 | |

| Launch Date | Spacecraft, Country, Int'l Designation, Vehicle, Launch Site | Payload Data | Apogee (km) | Perigee (km) | Period (min) | Inclination (degrees) | Remarks |
|---|---|---|---|---|---|---|---|
| | Cosmos 738 U.S.S.R. 1975-45G C-1 Plesetsk and | Total weight: Unavailable. Objective: "Investigation of upper atmosphere and outer space." Description: Unavailable. | 1511 | 1470 | 115.7 | 74.0 | |
| | Cosmos 739 U.S.S.R. 1975-45H C-1 Plesetsk | Total weight: Unavailable. Objective: "Investigation of upper atmosphere and outer space." Description: Unavailable. | 1473 | 1425 | 114.8 | 74.0 | |
| 28 May | Cosmos 740 U.S.S.R. 1975-46A A-2 Baykonur-Tyuratam | Total weight: Unavailable. Objective: "Investigation of upper atmosphere and outer space." Description: Unavailable. | 326 | 172 | 89.5 | 65.0 | Reentered 10 June. |
| 30 May | Cosmos 741 U.S.S.R. 1975-47A A-2 Plesetsk | Total weight: Unavailable. Objective: "Investigation of upper atmosphere and outer space." Description: Unavailable. | 226 | 205 | 88.8 | 81.4 | Reentered 11 June. |
| 3 June | Cosmos 742 U.S.S.R. 1975-48A A-2 Plesetsk | Total weight: Unavailable. Objective: "Investigation of upper atmosphere and outer space." Description: Unavailable. | 333 | 145 | 89.3 | 62.8 | Reentered 15 June. |
| 5 June | Molniya 1/ Sret 2 U.S.S.R./France 1975-49A, -49B A-2-e Plesetsk | Total weight: —A, 29.6 kg; —B, 30 kg. Objective: A, Operate long-range telephone and telegraph radiocommunications system; transmit TV programs to stations in Orbita network. B, Study behavior of passive cryogenic radiation system and the aging of plastic films. Description: Unavailable. | (A) 39 852 (B) 40 856 | 510 434 | 717.9 736.8 | 62.8 62.8 | Two payloads on single booster, still in orbit; 30th of Molniya 1 launches. |

| Date | Name/Designation | Objective/Description | Col4 | Col5 | Col6 | Remarks |
|---|---|---|---|---|---|---|
| 8 June | Venera 9 U.S.S.R. 1975-50A D-1-e Baykonur-Tyuratam | Total weight: Unavailable. Objective: Continuation of scientific investigation of Venus and space surrounding it, and physical character of interplanetary space. Description: Unavailable. | | Trans-Venus trajectory; circum-Venerean orbit | | Landed on Venus 22 Oct. |
| 8 June | DOD spacecraft U.S. 1975-51A, -51C Titan III WTR | Total weight: Unavailable. Objective: "Develop space flight techniques and technology." Description: Unavailable. | (A) 262 (C) 1399 | 155 1389 | 88.6 113.6 | 96.3 95.0 (A) Reentered 5 Nov. (C) Still in orbit. |
| 12 June | Nimbus 6 U.S. 1975-52A Delta WTR | Total weight: 832 kg. Objective: Experimental monitoring of environmental conditions and demonstrating capabilities of meteorological instrumentation. Description: Butterfly-shaped spacecraft 3 m high, 3.3 m wide, with toroid base, hexagonal housing for stabilization and control system, 2 solar panels 1 m by 2.5 m; 3-axis-stabilized; nine experiments. | 1101 | 1093 | 107.3 | 100.0 Last of Nimbus series; still in orbit. |
| 12 June | Cosmos 743 U.S.S.R. 1975-53A A-2 Plesetsk | Total weight: Unavailable. Objective: "Investigation of upper atmosphere and outer space." Description: Unavailable. | 290 | 168 | 89.1 | 62.8 Reentered 25 Jun. |
| 14 June | Venera 10 U.S.S.R. 1975-54A D-1-e Baykonur-Tyuratam | Total weight: Unavailable. Objective: Continue scientific investigation of Venus and space around it, and of physical characteristics of interplanetary space. Description: Unavailable. | | Trans-Venerean trajectory, circum-Venerean orbit | | Landed on Venus 25 Oct. |
| 18 June | DOD spacecraft U.S. 1975-55A Atlas-Agena D ETR | Total weight: Unavailable. Objective: "Develop space flight techniques and technology." Description: Unavailable. | 40 800 | 30 200 | 1422.0 | 9.0 AF early warning satellite; still in orbit. |
| 20 June | Cosmos 744 U.S.S.R. 1975-56A A-1 Plesetsk | Total weight: Unavailable. Objective: "Investigation of upper atmosphere and outer space." Description: Unavailable. | 634 | 600 | 97.1 | 81.3 Probable electronic ferret; still in orbit. |

# Appendix A

| Launch Date | Spacecraft, Country, Int'l Designation, Vehicle, Launch Site | Payload Data | Apogee (km) | Perigee (km) | Period (min) | Inclination (degrees) | Remarks |
|---|---|---|---|---|---|---|---|
| 21 June | Oso 8 U.S. 1975-57A Delta ETR | Total weight: Unavailable. Objective: Study interface of solar corona and chromosphere in UV to better understand transport of energy. Description: Spin-stabilized 324.6-cm-high 2-section spacecraft: 152.4-cm dia. rotating wheel carrying nonpointing experiments, and rectangular "sail" on top consisting of solar array and pointing experiments. | 560 | 543 | 95.7 | 51.8 | Eighth of OSO series; all 8 experiments functioning; still in orbit. |
| 24 June | Cosmos 745 U.S.S.R. 1975-58A B-1 Plesetsk | Total weight: Unavailable. Objective: "Investigation of upper atmosphere and outer space." Description: Unavailable. | 512 | 262 | 92.3 | 71.0 | Probable military spacecraft; reentered 12 Mar. 76. |
| 25 June | Cosmos 746 U.S.S.R. 1975-59A A-2 Plesetsk | Total weight: Unavailable. Objective: "Investigation of upper atmosphere and outer space." Description: Unavailable. | 323 | 180 | 89.5 | 62.8 | Reentered 8 July. |
| 27 June | Cosmos 747 U.S.S.R. 1975-60A A-2 Plesetsk | Total weight: Unavailable. Objective: "Investigation of upper atmosphere and outer space." Description: Unavailable. | 285 | 191 | 89.3 | 62.8 | Reentered 9 July. |
| 3 July | Cosmos 748 U.S.S.R. 1975-61A A-2 Plesetsk | Total weight: Unavailable. Objective: "Investigation of upper atmosphere and outer space." Description: Unavailable. | 270 | 170 | 88.9 | 62.8 | Reentered 16 July. |
| 4 July | Cosmos 749 U.S.S.R. 1975-62A C-1 Plesetsk | Total weight: Unavailable. Objective: "Investigation of upper atmosphere and outer space." Description: Unavailable. | 549 | 508 | 95.2 | 74.0 | Probable electronic ferret; still in orbit. |

## ASTRONAUTICS AND AERONAUTICS, 1975 — Appendix A

| Date | Name | Description | | | | Remarks |
|---|---|---|---|---|---|---|
| 8 July | Molniya 2 U.S.S.R. 1975-63A A-2-e Plesetsk | Total weight: Unavailable. Objective: Operate long-range telephone and telegraph radiocommunications system in U.S.S.R.; transmit central TV programs to stations in Orbita network; international cooperation. Description: Unavailable. | 40 858 | 433 | 736.9 | 62.9 | Thirteenth of Molniya 2 series; still in orbit. |
| 11 July | Meteor 2 U.S.S.R. 1975-64A A-1 Plesetsk | Total weight: Unavailable. Objective: Acquire meteorological data for use in operational weather service. Description: Unavailable. | 889 | 856 | 102.4 | 81.3 | First of series; still in orbit. |
| 15 July | Soyuz 19 (ASTP) U.S.S.R. 1975-65A A-2 Baykonur-Tyuratam | Total weight: 6800 kg. Objective: Experimental flight in Soyuz-Apollo program. Description: Soyuz capsule used since 1967, modified with compatible docking system and air composition and pressure to match that of Apollo. | 230 | 219 | 89.0 | 51.8 | Reentered 21 July; crew (Aleksey A. Leonov, Valery N. Kubasov) took part in 1st docking in space of vehicles from 2 different countries, joint engineering and scientific experiments. |
| 15 July | Apollo (ASTP) U.S. 1975-66A Saturn I-B ETR | Total weight: 14 856 kg (combined modules). Objective: Rendezvous, docking, and crew transfer with Soviet spacecraft; demonstrate interaction of crews and of U.S. and U.S.S.R. control centers. Description: Command module as used in lunar landings, modified with controls and equipment for docking module cylinder 1.5 m in dia. and 3 m long with hatches at both ends. | 230 | 219 | 89.0 | 51.8 | Reentered 24 July; crew (Thomas P. Stafford, Vance D. Brand, Donald K. Slayton) controlled docking in space with Soyuz 19; took part in joint engineering and scientific experiments. |
| 17 July | Cosmos 750 U.S.S.R. 1975-67A B-1 Plesetsk | Total weight: Unavailable. Objective: "Investigation of upper atmosphere and outer space." Description: Unavailable. | 802 | 270 | 95.4 | 71.0 | Reentered 29 Sept.; probable military satellite. |
| 23 July | Cosmos 751 U.S.S.R. 1975-68A A-2 Plesetsk | Total weight: Unavailable. Objective: "Investigation of upper atmosphere and outer space." Description: Unavailable. | 315 | 196 | 89.6 | 62.8 | Reentered 4 Aug. |
| 24 July | Cosmos 752 U.S.S.R. 1975-69A C-1 Plesetsk | Total weight: Unavailable. Objective: "Investigation of upper atmosphere and outer space." Description: Unavailable. | 513 | 468 | 94.6 | 65.9 | Probable military satellite; still in orbit. |

# Appendix A

| Launch Date | Spacecraft, Country, Int'l Designation, Vehicle, Launch Site | Payload Data | Apogee (km) | Perigee (km) | Period (min) | Inclination (degrees) | Remarks |
|---|---|---|---|---|---|---|---|
| 26 July | *China 3* Peoples Republic of China 1975–70A Vehicle unknown Shuang Cheng-tze | Total weight: Unavailable. Objective: Unavailable. Description: Unavailable. | 455 | 182 | 90.9 | 69.0 | Reentered 14 Sept. |
| 31 July | *Cosmos 753* U.S.S.R. 1975–71A A-2 Plesetsk | Total weight: Unavailable. Objective: "Investigation of upper atmosphere and outer space." Description: Unavailable. | 314 | 166 | 89.3 | 62.8 | Reentered 13 Aug. |
| 9 Aug. | *Cos-B* Eur. Space Agency 1975–72A Delta WTR | Total weight: 277.5 kg. Objective: Celestial observation satellite, for gamma-ray astronomy. Description: Solar-cell-covered cylinder, 4 monopole antennas on bottom. | 101 568 | 343 | 2277.1 | 90.2 | Celestial observation satellite, first for ESA, 8th designed by former ESRO; still in orbit. |
| 13 Aug. | *Cosmos 754* U.S.S.R. 1975–73A A-2 Baykonur-Tyuratam | Total weight: Unavailable. Objective: "Investigation of upper atmosphere and outer space." Description: Unavailable. | 325 | 204 | 89.8 | 71.4 | Reentered 26 Aug. |
| 14 Aug. | *Cosmos 755* U.S.S.R. 1975–74A C-1 Plesetsk | Total weight: Unavailable. Objective: "Investigation of upper atmosphere and outer space." Description: Unavailable. | 1010 | 972 | 104.9 | 82.9 | Probable navigation satellite; still in orbit. |
| 20 Aug. | *Viking 1* (Viking B) U.S. 1975–75A Titan III-Centaur ETR | Total weight: 3400 kg (orbiter and lander). Objective: Investigation of Mars from orbit and on surface. Description: Octagonal body, conical heatshield, 4 solar panels. | Trans-Mars trajectory, heliocentric orbit | | | | Soft-landed on Mars 20 Jul. 76; first on-site analysis of surface material of another planet. |

| Date | Satellite/Designation | Description | Value1 | Value2 | Value3 | Value4 | Remarks |
|---|---|---|---|---|---|---|---|
| 22 Aug. | Cosmos 756<br>U.S.S.R.<br>1975-76A<br>A-1<br>Plesetsk | Total weight: Unavailable.<br>Objective: "Investigation of upper atmosphere and outer space."<br>Description: Unavailable. | 631 | 620 | 97.2 | 81.2 | Probable electronic ferret; still in orbit. |
| 27 Aug. | Symphonie 2<br>Fed. Rep. Ger./France<br>1975-77A<br>Thor-Delta<br>ETR | Total weight: 221 kg (empty).<br>Objective: Experimental telecommunications satellite for FRG and France.<br>Description: Octagon 0.5 m long, 1.85 m in dia., 3 solar paddles. | 37 974 | 413 | 678.3 | 13.2 | Second of two experimental comsats launched by NASA for French-West German consortium; still in orbit. |
| | Cosmos 757<br>U.S.S.R.<br>1975-78A<br>A-2<br>Plesetsk | Total weight: Unavailable.<br>Objective: "Investigation of upper atmosphere and outer space."<br>Description: Unavailable. | 314 | 180 | 89.5 | 62.8 | Reentered 9 Sept. |
| 2 Sept. | Molniya 1<br>U.S.S.R.<br>1975-79A<br>A-2-e<br>Plesetsk | Total weight: Unavailable.<br>Objective: Operate long-range telecommunications; transmit Soviet TV programs to stations in Orbita network.<br>Description: Unavailable. | 40 665 | 622 | 736.8 | 62.9 | Thirty-first of series; still in orbit. |
| 5 Sept. | Cosmos 758<br>U.S.S.R.<br>1975-80A<br>A-2<br>Plesetsk | Total weight: Unavailable.<br>Objective: "Investigation of upper atmosphere and outer space."<br>Description: Unavailable. | 308 | 172 | 89.3 | 67.1 | Exploded into 84 pieces 6 Sept.; probably first of a series of fourth-generation observation craft; payload reentered 25 Sept. |
| 9 Sept. | Molniya 2<br>U.S.S.R.<br>1975-81A<br>A-2-e<br>Plesetsk | Total weight: Unavailable.<br>Objective: Operate long-range telecommunications; transmit Soviet TV programs to stations in Orbita network.<br>Description: Unavailable. | 40 846 | 423 | 736.4 | 62.8 | Fourteenth of series; still in orbit. |
| | Kiku (ETS-1)<br>Japan<br>1975-82A<br>N rocket<br>Kagoshima | Total weight: 85 kg.<br>Objective: Engineering test to confirm launch technology, and test of extendable antennas to measure environment, altitude, etc.<br>Description: Polyhedron (cylindrical) 0.9 m long, 0.85 m in dia. | 1104 | 977 | 106.0 | 47.0 | First launching by new Japanese N rocket; still in orbit. |
| | Viking 2<br>(Viking A)<br>1975-83A<br>Titan III-Centaur<br>ETR | Total weight: 3521.7 kg (orbiter and lander).<br>Objective: Scientific investigation of Mars from orbit and on surface.<br>Description: Octagon, with conical heat shield and 4 solar panels, 3 m high and 3 m long. | Trans-Mars trajectory; heliocentric orbit | | | | Soft landing on Mars successful 3 Sept. 76. |

# Appendix A

| Launch Date | Spacecraft, Country, Int'l Designation, Vehicle, Launch Site | Payload Data | Apogee (km) | Perigee (km) | Period (min) | Inclination (degrees) | Remarks |
|---|---|---|---|---|---|---|---|
| 12 Sept. | Cosmos 759 U.S.S.R. 1975-84A A-2 Plesetsk | Total weight: Unavailable. Objective: "Investigation of upper atmosphere and outer space." Description: Unavailable. | 273 | 227 | 89.5 | 62.8 | Reentered 23 Sept. |
| 16 Sept. | Cosmos 760 U.S.S.R. 1975-85A A-2 Baykonur-Tyuratam | Total weight: Unavailable. Objective: "Investigation of upper atmosphere and outer space." Description: Unavailable. | 316 | 170 | 89.4 | 65.0 | Reentered 30 Sept. |
| 17 Sept. | Cosmos 761 U.S.S.R. 1975-86A C-1 Plesetsk and | Total weight: Unavailable. Objective: "Investigation of upper atmosphere and outer space." Description: Unavailable. | 1484 | 1401 | 114.6 | 74.0 | Probable 8 comsats on single carrier; still in orbit. |
| | Cosmos 762 U.S.S.R. 1975-86B C-1 Plesetsk and | Total weight: Unavailable. Objective: "Investigation of upper atmosphere and outer space." Description: Unavailable. | 1487 | 1439 | 115.1 | 74.0 | |
| | Cosmos 763 U.S.S.R. 1975-86C C-1 Plesetsk | Total weight: Unavailable. Objective: "Investigation of upper atmosphere and outer space." Description: Unavailable. | 1511 | 1475 | 115.8 | 74.0 | |
| | Cosmos 764 U.S.S.R. 1975-86D C-1 Plesetsk | Total weight: Unavailable. Objective: "Investigation of upper atmosphere and outer space." Description: Unavailable. | 1527 | 1480 | 116.0 | 74.0 | |

| | | | | | | |
|---|---|---|---|---|---|---|
| and | | | | | | |
| | Cosmos 765<br>U.S.S.R.<br>1975–86E<br>C–1<br>Plesetsk | Total weight: Unavailable.<br>Objective: "Investigation of upper atmosphere and outer space."<br>Description: Unavailable. | 1552 | 1480 | 116.3 | 74.0 | |
| and | | | | | | |
| | Cosmos 766<br>U.S.S.R.<br>1975–86F<br>C–1<br>Plesetsk | Total weight: Unavailable.<br>Objective: "Investigation of upper atmosphere and outer space."<br>Description: Unavailable. | 1486 | 1419 | 114.9 | 73.9 | |
| and | | | | | | |
| | Cosmos 767<br>U.S.S.R.<br>1975–86G<br>C–1<br>Plesetsk | Total weight: Unavailable.<br>Objective: "Investigation of upper atmosphere and outer space."<br>Description: Unavailable. | 1488 | 1457 | 115.3 | 73.9 | |
| and | | | | | | |
| | Cosmos 768<br>U.S.S.R.<br>1975–86H<br>C–1<br>Plesetsk | Total weight: Unavailable.<br>Objective: "Investigation of upper atmosphere and outer space."<br>Description: Unavailable. | 1492 | 1473 | 115.5 | 73.9 | |
| 18 Sept. | Meteor 1<br>U.S.S.R.<br>1975–87A<br>A–1<br>Plesetsk | Total weight: Unavailable.<br>Objective: Collect meteorological information for operational weather service.<br>Description: Unavailable. | 899 | 835 | 102.3 | 81.3 | Twenty-second of series; still in orbit. |
| 23 Sept. | Cosmos 769<br>U.S.S.R.<br>1975–88A<br>A–2<br>Plesetsk | Total weight: Unavailable.<br>Objective: "Investigation of upper atmosphere and outer space."<br>Description: Unavailable. | 305 | 201 | 89.6 | 72.8 | Reentered 5 Oct. |
| 24 Sept. | Cosmos 770<br>U.S.S.R.<br>1975–89A<br>C–1<br>Plesetsk | Total weight: Unavailable.<br>Objective: "Investigation of upper atmosphere and outer space."<br>Description: Unavailable. | 1195 | 1168 | 109.2 | 83.0 | Probable navigational and geodetic satellite; still in orbit. |

| Launch Date | Spacecraft, Country, Int'l Designation, Vehicle, Launch Site | Payload Data | Apogee (km) | Perigee (km) | Period (min) | Inclination (degrees) | Remarks |
|---|---|---|---|---|---|---|---|
| 25 Sept. | Cosmos 771 U.S.S.R. 1975–90A A–2 Plesetsk | Total weight: Unavailable. Objective: "Investigation of upper atmosphere and outer space." Description: Unavailable. | 226 | 214 | 88.9 | 81.3 | Reentered 8 Oct. |
| 26 Sept. | Intelsat IVA–F1 U.S. 1975–91A Atlas Centaur ETR | Total weight: 1500 kg. Objective: Provide combination of commercial communications services. Description: Rotating solar-cell-covered cylinder topped by despun platform carrying 20 transponders and other communications equipment. | 35 821 | 35 294 | 1424.4 | 0.1 | First of improved IVA comsats, twice the capacity of previous series, launched by NASA for ComSatCorp.; commercial service to begin in 1976. |
| 27 Sept. | Aura (D2–B1) France 1975–92A Diamant BP4 Kourou, French Guiana | Total weight: 106.6 kg. Objective: Study stellar and solar ultraviolet radiation in galaxy. Description: Cylinder, 0.8 m long, 0.7 m in dia., with 4 vanes. | 722 | 500 | 96.9 | 37.2 | Final flight of Diamant launcher; still in orbit; first French satellite program managed entirely by industry. |
| 29 Sept. | Cosmos 772 U.S.S.R. 1975–93A A–2 Baykonur-Tyuratam | Total weight: Unavailable. Objective: "Investigation of upper atmosphere and outer space." Description: Unavailable. | 328 | 196 | 89.8 | 51.8 | New type of Soyuz ferry to be used as unmanned supply ship for space stations; reentered 2 Oct. |
| 30 Sept. | Cosmos 773 U.S.S.R. 1975–94A C–1 Plesetsk | Total weight: Unavailable. Objective: "Investigation of upper atmosphere and outer space." Description: Unavailable. | 806 | 788 | 100.8 | 74.0 | Probable comsat; still in orbit. |
| 1 Oct. | Cosmos 774 U.S.S.R. 1975–95A A–2 Baykonur-Tyuratam | Total weight: Unavailable. Objective: "Investigation of upper atmosphere and outer space." Description: Unavailable. | 311 | 204 | 89.7 | 71.3 | Reentered 15 Oct. |

| Date | Spacecraft | Description | Perigee (km) | Apogee (km) | Period (min) | Remarks |
|---|---|---|---|---|---|---|
| 6 Oct. | Explorer 54 (AE-D) U.S. 1975-96A Thorad-Delta WTR | Total weight: 675 kg. Objective: Investigate process controlling structure and behavior of earth atmosphere and ionosphere. Description: Cylinder, 16 sides, 115 cm high, 135 cm in dia., solar-cell-covered; 12 scientific instruments. | 3816 | 154 | 126.9 | 90.1 | Second of 3 second-generation atmosphere explorers; elliptical orbit, dipping in and out of atmosphere; all 12 science instruments active; reentered 12 Mar. 76 (power-supply failure). |
| 8 Oct. | Cosmos 775 U.S.S.R. 1975-97A D-1-e Baykonur-Tyuratam | Total weight: Unavailable. Objective: "Investigation of upper atmosphere and outer space." Description: Unavailable. | Circular orbit at 35 900 km | | 632.2 | 0.1 | Probable early-warning satellite; still in orbit. |
| 9 Oct. | DOD spacecraft U.S. 1975-98A Titan IIIB-Agena WTR | Total weight: Unavailable. Objective: "Develop space flight techniques and technology." Description: Unavailable. | 357 | 119 | 96.4 | 89.3 | Second lower-resolution photo-reconnaissance spacecraft; reentered 30 Nov. |
| 12 Oct. | Triad 2 (TIP-2) U.S. 1975-99A Scout WTR | Total weight: 94 kg. Objective: Part of Navy's Transit Improvement Program (navigation). Description: Unavailable. | 703 | 359 | 95.3 | 90.7 | Still in orbit; solar paddles failed to deploy; Navy salvage attempts continuing. |
| 16 Oct. | Goes 1 U.S. 1975-100A Delta ETR | Total weight: 243 kg (empty). Objective: Geostationary operational environmental satellite to provide day-night data to operational metsat system. Description: Spin-stabilized solar-cell-covered cylinder 344 cm long, 190 cm in dia., carrying radiometer, telescope, and data-transmission and collection system. | 36 082 | 35 775 | 1443.4 | 1.0 | First of new series launched by NASA for NOAA; working with Sms 2, will provide western-hemisphere 24-hr coverage beginning 1976. |
| 17 Oct. | Cosmos 776 U.S.S.R. 1975-101A A-2 Plesetsk | Total weight: Unavailable. Objective: "Investigation of upper atmosphere and outer space." Description: Unavailable. | 284 | 198 | 89.3 | 62.8 | Reentered 29 Oct. |
| 29 Oct. | Cosmos 777 U.S.S.R. 1975-102A F-1-m Baykonur-Tyuratam | Total weight: Unavailable. Objective: "Investigation of upper atmosphere and outer space." Description: Unavailable. | 444 | 428 | 93.3 | 65.0 | Reentered 3 June 76; probable electronic ferret. |

# Appendix A

| Launch Date | Spacecraft, Country, Int'l Designation, Vehicle, Launch Site | Payload Data | Apogee (km) | Perigee (km) | Period (min) | Inclination (degrees) | Remarks |
|---|---|---|---|---|---|---|---|
| 4 Nov. | Cosmos 778<br>U.S.S.R.<br>1975-103A<br>C-1<br>Plesetsk | Total weight: Unavailable.<br>Objective: "Investigation of upper atmosphere and outer space."<br>Description: Unavailable. | 1004 | 974 | 104.9 | 83.0 | Probable navigation satellite; still in orbit. |
| 4 Nov. | Cosmos 779<br>U.S.S.R.<br>1975-104A<br>A-2<br>Plesetsk | Total weight: Unavailable.<br>Objective: "Investigation of upper atmosphere and outer space."<br>Description: Unavailable. | 322 | 178 | 89.5 | 62.8 | Reentered 18 Nov. |
| 14 Nov. | Molniya 3<br>U.S.S.R.<br>1975-105A<br>A-2-e<br>Plesetsk | Total weight: Unavailable.<br>Objective: Operate long-range communications; transmit Soviet TV programs to stations in Orbita network.<br>Description: Unavailable. | 40 788 | 526 | 737.3 | 62.9 | Third of series; still in orbit. |
| 17 Nov. | Soyuz 20<br>U.S.S.R.<br>1975-106A<br>A-2<br>Baykonur-Tyuratam | Total weight: Unavailable.<br>Objective: Dock with Salyut 4, make comprehensive check of on-board spacecraft systems under various flight conditions.<br>Description: Unavailable. | 342 | 315 | 91.1 | 51.6 | Unmanned biological satellite docked 19 Nov. with Salyut 4; reentered 16 Feb. 76; bioscience experiments similar to those on Cosmos 782. |
| 20 Nov. | Explorer 55<br>(AE-E)<br>U.S.<br>1975-107A<br>Thored-Delta<br>ETR | Total weight: 720 kg.<br>Objective: Investigate processes controlling structure and behavior of earth atmosphere and ionosphere.<br>Description: Cylinder, 16 sides, solar-cell-covered, 115 cm high, 135 cm in dia; 12 scientific instruments. | 3025 | 157 | 118.0 | 19.7 | Last of 3 second-generation atmosphere explorers; elliptical equatorial orbit, dipping in and out of atmosphere; all 12 science instruments active; still in orbit. |
| 21 Nov. | Cosmos 780<br>U.S.S.R.<br>1975-108A<br>A-2<br>Baykonur-Tyuratam | Total weight: Unavailable.<br>Objective: "Investigation of upper atmosphere and outer space."<br>Description: Unavailable. | 272 | 199 | 89.2 | 65.0 | Reentered 3 Dec. |
| 21 Nov. | Cosmos 781<br>U.S.S.R.<br>1975-109A<br>C-1<br>Plesetsk | Total weight: Unavailable.<br>Objective: "Investigation of upper atmosphere and outer space."<br>Description: Unavailable. | 548 | 505 | 95.2 | 74.0 | Probable electronic ferret; still in orbit. |

ASTRONAUTICS AND AERONAUTICS, 1975  Appendix A

| Date | Name | Details | | | | Remarks |
|---|---|---|---|---|---|---|
| 25 Nov. | Cosmos 782 U.S.S.R. 1975–110A A-2 Plesetsk | Total weight: Unavailable. Objective: Continuation of investigation of effects of space environment on living organisms. Description: Unavailable. | 383 | 217 | 90.5 | 62.8 | Biosputnik; carried experiments from U.S.S.R., U.S., and 5 other countries; reentered 15 Dec. |
| 26 Nov. | China 4 Peoples Republic of China 1975–111A China B Shuang-Cheng-tze | Total weight: Unavailable. Objective: Unavailable. Description: Unavailable. | 479 | 179 | 91.1 | 62.9 | Reentered 29 Dec.; China claimed capsule recovery 2 Dec. |
| 28 Nov. | Cosmos 783 U.S.S.R. 1975–112A C-1 Plesetsk | Total weight: Unavailable. Objective: "Investigation of upper atmosphere and outer space." Description: Unavailable. | 813 | 794 | 101.0 | 74.1 | Probable comsat; still in orbit. |
| 3 Dec. | Cosmos 784 U.S.S.R. 1975–113A A-2 Plesetsk | Total weight: Unavailable. Objective: "Investigation of upper atmosphere and outer space." Description: Unavailable. | 229 | 213 | 88.9 | 81.3 | Reentered 15 Dec. |
| 4 Dec. | DOD spacecraft U.S. 1975–114A, B Titan IIID WTR | Total weight: Unavailable. Objective: "Develop space flight techniques and technology." Description: Unavailable. | (A) 275 (B) 1513 | 156 226 | 88.7 102.3 | 96.2 96.3 | Reentered 1 Apr. 76. Reentered 1 May 78. |
| 11 Dec. | Intercosmos 14 U.S.S.R. 1975–115A C-1 Plesetsk | Total weight: Unavailable. Objective: Study low-frequency electromagnetic vibrations in earth's magnetosphere, structure of ionosphere, intensity of micrometeorite fluxes: international cooperation. Description: Unavailable. | 1682 | 333 | 105.3 | 74.0 | Countries participating with U.S.S.R.: Bulgaria, Hungary, Czechoslovakia, and German Democratic Republic; still in orbit. |
| 12 Dec. | Cosmos 785 U.S.S.R. 1975–116A F-1-m Baykonur-Tyuratam | Total weight: Unavailable. Objective: "Investigation of upper atmosphere and outer space." Description: Unavailable. | 1021 | 898 | 104.3 | 65.1 | Probable ocean-surveillance satellite; still in orbit. |

| Launch Date | Spacecraft, Country, Int'l Designation, Vehicle, Launch Site | Payload Data | Apogee (km) | Perigee (km) | Period (min) | Inclination (degrees) | Remarks |
|---|---|---|---|---|---|---|---|
| 13 Dec. | Satcom I (RCA I) U.S. 1975–117A Thorad-Delta ETR | Total weight: 463 kg (without apogee motor). Objective: Provide TV, voice, and high-speed data communications to continental U.S., Alaska, Hawaii. Description: Three-axis-stabilized box, 1.2 × 1.2 × 1.6m, with two solar panels and six feedhorns; 24-channel communications subsystem; geosynchronous orbit over equator. | 36 086 | 35 625 | 1439.7 | 0.3 | First of 3 domestic comsats launched by NASA for RCA; commercial operation to begin Mar. 76. |
| 14 Dec. | DOD satellite U.S. 1975–118A Titan III C ETR | Total weight: Unavailable. Objective: "Develop space flight techniques and technology." Description: Unavailable. | 35 800 | 35 700 | 1436.0 | 0.3 | Probable early-warning replacement—still in orbit. |
| 16 Dec. | China 5 Peoples Republic of China 1975–119A China B Shuang Cheng-ize | Total weight: Unavailable. Objective: Unavailable. Description: Unavailable. | 387 | 185 | 90.2 | 69.0 | Reentered 27 Jan. 76. |
| 16 Dec. | Cosmos 786 U.S.S.R. 1975–120A A-2 Baykonur-Tyuratam | Total weight: Unavailable. Objective: "Investigation of upper atmosphere and outer space." Description: Unavailable. | 324 | 172 | 89.5 | 65.0 | Reentered 29 Dec.; probable reconnaissance satellite. |
| 17 Dec. | Molniya 2 U.S.S.R. 1975–121A A-2-e Plesetsk | Total weight: Unavailable. Objective: Operate long-range communications system; transmit TV programs to stations in Orbita network. Description: Unavailable. | 40 855 | 434 | 736.8 | 62.8 | Fifteenth of series; still in orbit. |

| Date | Name/Designation | Description | Orbit params | | | Notes |
|---|---|---|---|---|---|---|
| 22 Dec. | Prognoz 4<br>U.S.S.R.<br>1975-122A<br>A-2-e<br>Baykonur-Tyuratam | Total weight: Unavailable. Objective: Investigate corpuscular and electromagnetic radiation from the sun and solar fluxes; study magnetic field in earth area to determine effects of solar activity on interplanetary medium and on earth's magnetosphere. Description: Unavailable. | 199 000 | 634 | 91.5 | 65.0 | Still in orbit; scientific satellite in highly elliptical orbit. |
| 22 Dec. | Raduga (Statsionar 1)<br>1975-123A<br>D-1-e<br>Baykonur-Tyuratam | Total weight: Unavailable. Objective: Provide uninterrupted day-night telephone and telegraph radio-communications in U.S.S.R. and simultaneous transmission of color and black-and-white Soviet central TV to stations in Orbita network. Description: Unavailable. | Circular orbit at 35 800 km | | 1434.0 | 0.3 | First fully operational Soviet geostationary comsat; still in orbit. |
| 25 Dec. | Meteor 1<br>U.S.S.R.<br>1975-124A<br>A-1<br>Plesetsk | Total weight: Unavailable. Objective: Acquire meteorological information for use in operational weather system. Description: Unavailable. | 900 | 840 | 102.4 | 81.3 | Twenty-third of series; still in orbit. |
| 27 Dec. | Molniya 3<br>U.S.S.R.<br>1975-125A<br>A-2-e<br>Plesetsk | Total weight: Unavailable. Objective: Operate long-range radio communications system; transmit Soviet TV programs to stations in Orbita network. Description: Unavailable. | 40 762 | 443 | 735.1 | 62.8 | Fourth of series; still in orbit. |

# Appendix B

## MAJOR NASA LAUNCHES, 1975

The following table of major NASA launches includes payloads carried by all rocket vehicles larger than sounding rockets launched in 1975 by NASA or under NASA direction.

During 1975, the U.S. had 34 launches of which 28 were successful. Of these, 7 were launches by DOD. The remaining 21 were NASA launches, 10 for its own program, including ASTP in July; the two Viking spacecraft launched toward Mars in Aug. and Sept.; 3 applications satellites, *Landsat 2*, *Geos 3*, and *Nimbus 6*; and 4 scientific satellites—*Oso 8*, Sas-C (*Explorer 53*, launched by an Italian crew from the San Marco platform into an equatorial orbit); AE–D (*Explorer 54*), and AE–E (*Explorer 55*). In 1974, NASA had launched only one all-U.S. scientific payload. Launches for others included 3 spacecraft for DOD; 3 comsats—*Intelsat IV* in May and *Intelsat IVA* in Sept., and the RCA *Satcom 1*; 2 for NOAA, *Sms 2* in Feb. and *Goes 1* in Oct.; and 3 international satellites—*Anik* for Canada in May, *Cos-B* for ESA in Aug., and *Symphonie 2* in Aug. for the West German-French consortium.

The table includes vehicle and payload performance under categories S for successful, P for partially successful, and U for unsuccessful. A fourth category (Unk) indicates payloads that did not operate because of vehicle failure. The categories, which are unofficial, do not take into account that U missions may produce valuable information, or that payloads with a long-life design may fail to meet design requirements and become officially unsuccessful at a later date. Further information on these launches appears in Appendix A and in the indexed entries in the text.

## MAJOR NASA LAUNCHES, 1975

| Date | Name (NASA Code) | General Mission | Launch Vehicle (Site) | Performance Vehicle | Performance Payload | Remarks |
|---|---|---|---|---|---|---|
| 22 Jan. | *Landsat 2* (ERTS-B) | Earth resources technology satellite | Thorad-Delta (WTR) | S | S | Circular sun-synchronous near-polar orbit; companion to *Erts 1*, launched 1972. |
| 6 Feb. | *Sms 2* | Operational prototype meteorological satellite | Thorad-Delta (ETR) | S | S | Operated by NOAA after 10 Mar.; used with *Sms 1* for 24-hr coverage of western hemisphere; to be used with *Goes 1* in future operational system. |
| 10 Mar. | DOD spacecraft (SDS–1) | Develop space flight techniques and technology | Titan IIIB-Agena D (WTR) | — | — | Classified payload |
| 9 Apr. | *Geos 3* | Geodynamics experimental ocean satellite | Thorad-Delta (WTR) | S | S | Circular orbit; demonstrated use of satellite altimeters for mapping ocean surfaces; conducted successful satellite-to-satellite tracking with *Ats 6*. |
| 18 Apr. | DOD spacecraft | Develop space flight techniques and technology | Titan IIIB-Agena D (WTR) | — | — | Decayed 29 May; classified payload. |
| 7 May | *Explorer 53* (SAS–C) | Monitor galactic x-ray sources | Scout (San Marco) | S | S | Third small astronomy satellite; extended spectral range and longer continuous observation. |
| 7 May | *Anik 3* (Telesat–C) | Transmit TV, voice, and other data throughout Canada | Thorad-Delta (ETR) | S | S | Last of 3 comsats launched by NASA for Canada under a 1971 agreement. |
| 22 May | *Intelsat IV F–1* | Operational commercial communications satellite | Atlas-Centaur (ETR) | S | S | Seventh and last successful Intelsat IV satellite; launched by NASA for Comsat General Corp; commercial operations began June 1975. |
| 12 June | *Nimbus 6* (Nimbus F) | Develop instrumentation for expanding capabilities of remote sensing of atmosphere | Thorad-Delta (WTR) | S | S | Most sophisticated metsat yet developed; near-polar circular orbit; 9 experiments to obtain data for numerical model. |
| 21 June | *Oso 8* | Scientific satellite, solar studies | Thorad-Delta (ETR) | S | S | Eighth OSO spacecraft, larger and more sophisticated than previous ones; circular orbit; all 8 experiments active. |
| 15 July | Apollo (ASTP) | Spacecraft rendezvous and docking, intervehicular transfer of crews, interaction of U.S. and U.S.S.R. control centers | Saturn 1B (ETR) | S | S | First international space flight; 4 crew transfers during joint flight of nearly 2 days; crew returned safely 24 July. |
| 9 Aug. | *Cos–B* | Study extraterrestrial gamma radiation | Thorad-Delta (WTR) | S | S | European Space Agency satellite, launched by NASA into highly elliptical near-polar orbit; all spacecraft systems and experiments active. |

| Date | Name | Purpose | Launch vehicle (range) | Results | Remarks |
|---|---|---|---|---|---|
| 20 Aug. | Viking 1 (Viking B) | Direct measurement of Mars from orbit and on surface | Titan IIIE-Centaur (ETR) | S | First on-site investigation of surface material of another planet; successful soft landing on Mars 20 July 76. |
| 27 Aug. | Symphonie 2 (Symphonie B) | Experimental comsat | Thorad-Delta (ETR) | S | Second of 2 comsats developed jointly by France and Germany; all systems active. |
| 9 Sept. | Viking 2 (Viking A) | Direct measurement of Mars from orbit and on surface | Titan IIIE-Centaur (ETR) | S | Second spacecraft successfully launched to Mars; successful soft landing 3 Sept. 76; successfully returned scientific data. |
| 26 Sept. | Intelsat IVA F–1 | Operational commercial communications satellite | Atlas-Centaur (ETR) | S | First of improved IVA series, twice the communications capacity of Intelsat IV series; launched by NASA for Comsat General Corp.; commercial operations to begin in 1976. |
| 6 Oct. | Explorer 54 (AE–D) | Scientific satellite, atmospheric research | Thorad-Delta (WTR) | S | Second of a series of 3 second-generation atmosphere explorers; elliptical polar orbit, dipping in and out of atmosphere; all 12 science instruments functioning normally; reentered 12 Mar. 76. |
| 12 Oct. | TIP 2 | Test developments in spacecraft for Transit Improvement Program | Scout (WTR) | U | Second launch in Navy's Transit Improvement Program; NASA launched into elliptical polar transfer orbit; solar paddles failed to deploy, Navy working to salvage mission. |
| 16 Oct. | Goes 1 (SMS-C) | Geostationary operational environmental satellite | Thorad-Delta (ETR) | S | First operational spacecraft in a series launched by NASA for NOAA; synchronous orbit, to operate with Sms 2 in providing 24-hr coverage of western hemisphere. |
| 20 Nov. | Explorer 55 (AE–E) | Scientific satellite, atmospheric research | Thorad-Delta (ETR) | S | Last of 3 second-generation atmosphere explorer; elliptical equatorial orbit, dipping in and out of atmosphere; all 12 scientific instruments functioning normally. |
| 13 Dec. | Satcom 1 (RCA–A) | Commercial domestic communications satellite | Thorad-Delta (ETR) | S | First of a series of 3 domestic comsats to be launched by NASA for RCA Corp.; stationary orbit at 119°W; commercial operations to begin March 1976. |

# Appendix C
# MANNED SPACE FLIGHTS, 1975

A high point in NASA's activity in 1975 was the Apollo-Soyuz Test Project, in which three astronauts in the Apollo (ASTP) spacecraft and two cosmonauts in the U.S.S.R. *Soyuz 19* flight-tested a mechanism for joining two orbiting manned spacecraft from different countries, in history's first international manned space flight. The ASTP achieved a number of firsts: First space flight involving crews and spacecraft of more than one nation; first linking in orbit of spacecraft of more than one nation; first detection of a pulsar outside earth's galaxy; first separation of live biological materials in space, by electrophoresis; first communications between a manned orbiting spacecraft and ground controllers through an orbiting unmanned satellite (*Ats 6*). The *Soyuz 19* liftoff for ASTP was the first Soviet launch shown on live television, and its parachute descent and landing were the first to be televised live. Of the 28 ASTP experiments, 21 were U.S., 5 were joint U.S.-U.S.S.R., and 2 were West German. All primary mission objectives were met. ASTP was the only U.S. manned flight of the year, and would be the last until the Space Shuttle flights scheduled for the 1980s.

The U.S.S.R. attempted 4 manned launches including *Soyuz 19* for ASTP; 3 of these were successful. The April anomaly was attributed to a booster failure, and the crew was retrieved safely; the incident called into question the safety of crews during ASTP, but the latter ended successfully.

By the end of 1975, the U.S. had made 31 manned space flights: 2 suborbital, 20 in earth orbit, 3 in lunar orbit, and 6 lunar landings, with a total of 43 different crewmen. The U.S.S.R. had made 27 manned flights, all in earth orbit, with 34 cosmonauts. The U.S. total of manned-spacecraft hours in flight was 7681 hr 10 min; the U.S.S.R. total, 5265 hr 3 min. Total cumulative man-hours in space for the U.S. were 22 503 hr 39 min; for the U.S.S.R., 10 734 hr 48 min.

## MANNED SPACE FLIGHTS, 1975

| Date Launched | Date Recovered | Designation, Crew | Weight (kg) | Duration, Revolutions | Remarks |
|---|---|---|---|---|---|
| 10 Jan. | 9 Feb. | Soyuz 17<br>Aleksey A. Gubarov<br>Georgy M. Grechko | 6570 (est.) | 709 hr 20 min<br>467.6 rev. | Docked with Salyut 4 on 12 Jan. |
| 5 Apr. | 5 Apr. | Anomaly<br>Vasily Lazarev<br>Oleg Makarov | Not available | 1600 km from launch site | Mission aborted, stage-separation malfunction of A-2 booster; crew landed in Siberia. |
| 24 May | 26 July | Soyuz 18<br>Pyotr Klimuk<br>Vitaly Sevastyanov | 6570 (est.) | 1511 hr 40 min<br>992.3 rev. | Docked with Salyut 4 on 25 May; communicated with Soyuz 19 during ASTP; broke Soviet duration record. |
| 15 July | 21 July | Soyuz 19 (ASTP)<br>Alekesy A. Leonov<br>Valery N. Kubasov | 5800 | 142 hr 31 min<br>96.6 rev. | Docked with Apollo (ASTP) 17 July; after 19 July undocking, performed biological experiment; launch and soft landing first televised by Soviets in real time. |
| 15 July | 23 July | Apollo (ASTP)<br>Thomas P. Stafford<br>Vance D. Brand<br>Donald K. Slayton | 14 856 (combined command service module and docking module) | 217 hr 28 min<br>148.9 rev. | Docked with Soyuz 19 on 17 July; after 19 July undocking, simulated solar eclipse for other spacecraft and performed science experiments, including location of extreme UV sources and first pulsar discovered outside galaxy, and 110 earth-observation tasks; landed upside down in last ocean landing planned for U.S. manned space flight. |

# Appendix D
# NASA SOUNDING ROCKET LAUNCHES, 1975

The table following lists the 83 sounding rockets of the Arcas class and above launched by NASA in 1975. The launches took place in Peru, Greenland, Canada, and the Kerguelen Islands of the Indian Ocean as well as in the United States. Payloads were carried for the Naval Research Laboratory, the Smithsonian Astrophysical Observatory, and three NASA centers; fifteen universities; and two U.S. corporations. Onboard experiments included 35 physics; 18 meteorology; 11 each astronomy and aeronomy; 7 performance tests; and 1 space-processing experiment.

Information for the table came from Goddard Space Flight Center's Quick Look Sounding Rocket Data sheets, issued after launches, with additional information from some of the experimenters concerned. The Kerguelen Islands launches in January and February were part of a joint Soviet-French exercise called ARAKS—artificial aurora between Kerguelen and Sogra (U.S.S.R.)—from which the usual telemetry reports were not available. Launch dates in the table are local time, with the date by Greenwich Mean Time given if different.

# NASA-LAUNCHED SOUNDING ROCKETS, 1975

| Launch Date (local time) | Rocket, NASA Designation, Launch Site | Apogee (km) | Remarks |
|---|---|---|---|
| 20 Jan. | Black Brant VC 21.024US White Sands, N.M. | 281.6 | Univ. of Calif. galactic astronomy experiment. |
| 26 Jan. | Super Arcas 15.129UE Kerguelen Is. | — | Univ. of Houston magnetospheric physics experiment in conjunction with ARAKS. |
| 26 Jan. | Super Arcas 15.130UE Kerguelen Is. | — | Univ. of Houston magnetospheric physics experiment in conjunction with ARAKS. |
| 31 Jan. (1 Feb. Z) | Aerobee 200A 26.027GG White Sands, N.M. | 222.0 | GSFC galactic astronomy experiment. |
| 3 Feb. | Aerobee 350 17.013CG White Sands, N.M. | 253.0 | Lockheed x-ray astronomy experiment. |
| 3 Feb. | Super Arcas 15.138UE Kerguelen Is. | — | Univ. of Houston magnetospheric physics experiment in conjunction with ARAKS. |
| 10 Feb. | Nike-Apache 14.536UA Ft. Churchill, Canada | 142.5 | Univ. of Pittsburgh aeronomy experiment. |
| 10 Feb. (11 Feb. Z) | Javelin 8.056UE Ft. Churchill, Canada | 791 | Univ. of Wisc. magnetospheric physics experiment (auroral). |
| 13 Feb. | Aerobee 200A 26.040UE White Sands, N.M. | 295 | Univ. of Mich. magnetospheric physics experiment. |
| 12 Feb. (13 Feb. Z) | Nike-Apache 14.535UA Ft. Churchill, Canada | 135 | Univ. of Pittsburgh aeronomy experiment. |
| 15 Feb. | Super Arcas 15.131UE Kerguelen Is. | — | Univ. of Houston magnetospheric physics experiment. |

| Date | Vehicle | Altitude (km) | Experiment |
|---|---|---|---|
| 15 Feb. | Super Arcas 15.132UE Kerguelen Is. | — | Univ. of Houston magnetospheric physics experiment. |
| 24 Feb. (25 Feb. Z) | Nike-Apache 14.537UA Ft. Churchill, Canada | 138 | Univ. of Pittsburgh aeronomy experiment. |
| 28 Feb. (1 Mar. Z) | Nike-Tomahawk 18.159UE Poker Flat Range, Al. | 187.5 | Rice Univ. magnetospheric physics experiment (auroral). Component failure in magnetometer; experiment 50% successful. |
| 7 Mar. (8 Mar. Z) | Aerobee 200A 26.032UG White Sands, N.M. | 225.9 F | Univ. of Wis. galactic astronomy experiment. Telemetry failed at liftoff, no experiment data obtained. |
| 10 Mar. (11 Mar. Z) | Aerobee 150A-MII 4.334UA Ft. Churchill, Canada | 171 | Johns Hopkins Univ. aeronomy experiment. |
| 11 Mar. (12 Mar. Z) | Nike-Apache 14.416GA Ft. Churchill, Canada | 137 | GSFC aeronomy experiment. |
| 13 Mar. | Super Arcas 15.137GM White Sands, N.M. | 77.85 | GSFC meteorology experiment. |
| 15 Mar. | Aerobee 200A 26.039UG White Sands, N.M. | 211.3 | Johns Hopkins Univ. galactic astronomy experiment. |
| 14 Mar. (15 Mar. Z) | Aerobee 350 17.015UG White Sands, N.M. | 224.4 | Smithsonian Astrophysical Observatory galactic astronomy experiment. |
| 17 Mar. (18 Mar. Z) | Nike-Tomahawk 18.165UE Poker Flat Range, Al. | 206.0 (predicted 200 km) | Univ. of N.H./Univ. of Minn. experiment in magnetospheric physics (auroral). No payload beacon. |
| 23 Mar. | Nike-Tomahawk 18.166 UE Poker Flat Range, Al. | 230 | Univ. of N.H./Univ. of Minn. experiment in magnetospheric physics (auroral). |
| 4 Apr. | Nike-Tomahawk 18.172GE Poker Flat Range, Al. | F (predicted 299 km) | GSFC magnetospheric physics experiment (no release of barium or trimethylaluminum). |

Appendix D  ASTRONAUTICS AND AERONAUTICS, 1975

| Launch Date (local time) | Rocket, NASA Designation, Launch Site | Apogee (km) | Remarks |
|---|---|---|---|
| 8 Apr. | Nike-Cajun 10.377UM Wallops Island, Va. | 89.0 | Univ. of Mich. meteorology experiment. |
| 9 Apr. (10 Apr. Z) | Aerobee 170A 13.088GG White Sands, N.M. | 176.2 | GSFC galactic astronomy experiment. |
| 13 Apr. | Nike-Tomahawk 18.173GE Poker Flat Range, Al. | 289.6 (predicted 299 km) | GSFC magnetospheric physics experiment; released 4 barium and 1 trimethylaluminum clouds. |
| 12 Apr. (13 Apr. Z) | Nike-Tomahawk 18.174GE Poker Flat Range, Al. | 289.6 (predicted 299 km) | GSFC magnetospheric physics experiment; released 4 barium and 1 trimethylaluminum clouds. |
| 15 Apr. (16 Apr. Z) | Nike-Tomahawk 18.175GE Poker Flat Range, Al. | — (predicted 299 km) | GSFC magnetospheric physics experiment; released 4 barium and 1 trimethylaluminum clouds. |
| 15 May | Astrobee F 25.003GT/NP White Sands, N.M. | 139.5 | GSFC test of performance characteristics of Astrobee F; JPL & MSFC test of s/c components under zero gravity. Rocket-motor performance successful, all payload functions successful. |
| 23 May (24 May Z) | Super Arcas 15.139UE Chilca Range, Peru | 92.0 | Penn. State Univ. magnetospheric physics experiment. |
| 23 May (24 May Z) | Super Arcas 15.133GM Chilca Range, Peru | 68.0 | GSFC meteorology experiment. |
| 24 May (25 May Z) | Super Arcas 15.140UE Chilca Range, Peru | 92.0 | Penn. State Univ. magnetospheric physics experiment. |
| 24 May | Super Arcas 15.134GM Chilca Range, Peru | 71.0 | GSFC meteorology experiment; chute failure. |

ASTRONAUTICS AND AERONAUTICS, 1975　　　　Appendix D

| Date | Vehicle | Altitude | Experiment |
|---|---|---|---|
| 24 May | Super Arcas 15.135GM Chilca Range, Peru | 69.0 | GSFC meteorology experiment. |
| 23 May (24 May Z) | Nike-Tomahawk 18.170GE Chilca Range, Peru | 338 | GSFC magnetospheric physics experiment; all systems functioned properly except lunar sensor. |
| 24 May | Nike-Tomahawk 18.171GE Chilca Range, Peru | 324 | GSFC magnetospheric physics experiment. |
| 27 May | Nike-Apache 14.530UA Chilca Range, Peru | 126.7 | Dudley Observatory (Albany, N.Y.) aeronomy experiment. |
| 27 May | Nike-Apache 14.531UA Chilca Range, Peru | 127.0 | Dudley Observatory (Albany, N.Y.) aeronomy experiment. |
| 28 May | Nike-Apache 14.141UE Chilca Range, Peru | 82.0 | Penn. State Univ. magnetospheric physics experiment. |
| 28 May | Super Arcas 15.142UE Chilca Range, Peru | 86.0 | Penn. State Univ. magnetospheric physics experiment. |
| 28 May | Nike-Apache 14.532UI Chilca Range, Peru | 190.6 | Univ. of Ill. magnetospheric physics experiment. |
| 29 May (30 May Z) | Nike-Apache 14.524UI Chilca Range, Peru | 187.8 | Univ. of Ill. magnetospheric physics experiment. |
| 31 May (1 June Z) | Nike-Apache 14.538UA Chilca Range, Peru | 130.5 | Univ. of Pittsburgh aeronomy experiment. |
| 2 June | Nike-Apache 14.525UI Chilca Range, Peru | 187.0 | Univ. of Ill. magnetospheric physics experiment. |
| 3 June | Nike-Tomahawk 18.149GE Chilca Range, Peru | F | GSFC magnetospheric physics experiment; vehicle failed at T+20. |
| 3 June | Nike-Apache 14.540CA Chilca Range, Peru | 204.0 | Geophysics Corp. of America aeronomy experiment. |

# Appendix D

| Launch Date (local time) | Rocket, NASA Designation, Launch Site | Apogee (km) | Remarks |
|---|---|---|---|
| 7 June | Nike-Tomahawk 18.150GE Chilca Range, Peru | 256.0 | GSFC magnetospheric physics experiment; tone ranging failed, peak altitude close to predicted 259 km. |
| 7 June | Nike-Apache 14.541CA Chilca Range, Peru | near nominal (predicted 194.0) | Geophysics Corp. of America aeronomy experiment. |
| 9 June | Aerobee 170 13.112UA White Sands, N.M. | 29.0 F | Univ. of Colo. aeronomy experiment; radar loss, experiment failed. |
| 27 June | Terrier Malemute 29.002GT Wallops Island, Va. | 283.0 F | Failure of structure during test at T+30 sec.; cause of failure under investigation. |
| 10 July | Aerobee 170 13.073GG White Sands, N.M. | 178.6 | GSFC galactic astronomy experiment. |
| 15 July | Arcas 15.048WT White Sands, N.M. | 68.0 | Wallops test of recovery system; magnetospheric physics experiment. |
| 15 July | Astrobee D 23.002UE White Sands, N.M. | 76.2 | Penn. State Univ. magnetospheric physics experiment. |
| 21 July | Black Brant VC 21.025DS White Sands, N.M. | 209.7 | NRL solar physics experiment. |
| 23 July (24 July Z) | Super Arcas 15.143CM Wallops Island, Va. | 84.7 | GSFC meteorology experiment. |
| 24 July | Nike-Cajun 10.413CM Wallops Island, Va. | 119.5 | GSFC meteorology experiment. |
| 28 July | Black Brant VC 21.029US White Sands, N.M. | 277.6 | Univ. of Colo. solar physics experiment. |

| Date | Vehicle | Altitude (km) | Experiment |
|---|---|---|---|
| 29 July | Super Arcas 15.144CM Wallops Island, Va. | 81.2 | GSFC meteorology experiment. |
| 29 July | Super Arcas 15.136CM Wallops Island, Va. | 73.1 | GSFC meteorology experiment. |
| 7 Aug. | Nike-Cajun 10.407CM Wallops Island, Va. | 118.9 | GSFC meteorology experiment. |
| 16 Aug. | Aerobee 170 13.050UH White Sands, N.M. | 172.1 | MIT high-energy astrophysics experiment. |
| 18 Aug. | Aries 12.026GT/TT White Sands, N.M. | 416.9 | GSFC vehicle systems test; recovery system failed. |
| 26 Aug. | Aerobee 170 13.074UA White Sands, N.M. | 180 | Univ. of Colo. UV in solar spectrum experiment. |
| 9 Sept. | Nike-Hawk 12.028WT Wallops Island, Va. | 163.0 | Wallops vehicle-system test; vehicle performance successful, piggyback experiment successful. |
| 24 Sept. | Nike-Tomahawk 18.163UE Wallops Island, Va. | 273 | Univ. of Texas ionospheric measurements experiment (magnetospheric physics). |
| 2 Oct. (3 Oct. Z) | Aerobee 200A 26.047DH White Sands, N.M. | 212 | NRL high-energy astrophysics (x-ray scans of galactic nuclei); roll control faulty. |
| 9 Oct. | Aerobee 170 13.117UL White Sands, N.M. | 203 | Univ. of Colo. lunar and planetary astronomy experiment. |
| 17 Oct. (18 Oct. Z) | Aerobee 170 13.108UH White Sands, N.M. | 182 | MIT high-energy astrophysics experiment (soft x-rays). |
| 28 Oct. (29 Oct. Z) | Aerobee 170A 13.048GG White Sands, N.M. | 185 | GSFC galactic astronomy experiment (UV spectrum). |
| 7 Nov. (8 Nov. Z) | Aerobee 170 13.049UH White Sands, N.M. | 178.7 | Univ. of Wis. high-energy astrophysics experiment (x-ray mapping). |

## Appendix D

| Launch Date (local time) | Rocket, NASA Designation, Launch Site | Apogee (km) | Remarks |
|---|---|---|---|
| 18 Nov. | Nike-Cajun 10.409GM Wallops Island, Va. | F | GSFC meteorology experiment; 2nd stage failed to ignite, maximum altitude 21 km. |
| 18 Nov. | Nike-Javelin 32.001WT Wallops Island, Va. | 164.0 | Wallops vehicle-system test; low apogee; vehicle, recovery system, Dudley experiment all functioned normally. |
| 19 Nov. | Nike-Cajun 10.415GM Wallops Island, Va. | 120.0 | GSFC meteorology experiment. |
| 20 Nov. (21 Nov. Z) | Nike-Cajun 10.423GM Wallops Island, Va. | 121.0 | GSFC meteorology experiment. |
| 3 Dec. | Nike-Cajun 10.424GM Wallops Island, Va. | 121.9 | GSFC meteorology experiment. |
| 4 Dec. | Nike-Cajun 10.410GM Wallops Island, Va. | 119.4 | GSFC meteorology experiment. |
| 5 Dec. (6 Dec. Z) | Aerobee 350 17.016UH White Sands, N.M. | 221.2 | SAO high-energy astrophysics experiment (x-ray telescope). |
| 6 Dec. | Aerobee 170 13.118DG White Sands, N.M. | 186.2 | NRL galactic astronomy experiment (UV). |
| 11 Dec. | Black Brant VC 21.032NP White Sands, N.M. | 206.3 | MSFC space-processing experiment. |
| 15 Dec. | Super Arcas 15.110GM Søndre Strømfjord, Greenland | 70.0 | GSFC meteorology experiment. |
| 15 Dec. | Super Arcas 15.145GM Søndre Strømfjord, Greenland | 81.0 | GSFC meteorology experiment. |

| | | | |
|---|---|---|---|
| 16 Dec. | Nike-Tomahawk<br>12,031 GT<br>White Sands, N.M. | 241.6 | GSFC vehicle-launcher test. |
| 21 Dec. | Super Arcas<br>15.146 GM<br>Søndre Strømfjord,<br>Greenland | nominal<br>(70 to<br>80 km) | GSFC meteorology experiment. |

# Appendix E
## X-24B LIFTING-BODY FLIGHTS, 1975

Lifting bodies are wingless vehicles that receive aerodynamic lift from their shape alone. NASA undertook in the early 1950s a program to develop concepts for a maneuverable manned vehicle that could reenter the atmosphere from space and to provide a technological basis for hypersonic cruise aircraft. Since 1963, NASA's Flight Research Center (called Dryden Flight Research Center after January 1976) has flight-tested the M2 configuration designed by Ames Research Center, the HL-10 designed by the Langley Research Center, and the X-24A and X-24B in a joint program with the Air Force.

The X-24B program began with a captive flight in 1973 and ended with the 36th and final flight on 26 Nov. 1975. The X-24B was launched from a B-52 aircraft, usually at an altitude of 13 700 meters; in captive flight, the vehicle remained attached to the B-52. In unpowered free flight, the vehicle glided through maneuvers to landings on a dry lake bed; in powered flight, the rocket engine was ignited to take the vehicle to higher altitudes and achieve higher speeds.

The following table lists all X-24B missions that left the ground in 1975, even those eventually aborted because of weather or mechanical problems. The flight number includes vehicle designation (B for X-24B vehicles)—free-flight number—B-52 carry number. "B-19-30" was the 19th free flight of the X-24B and the 30th time it was carried by the B-52. C or A appearing instead of a free-flight number indicates a captive or an aborted flight.

Information for the table appears in flight reports issued by the Office of Aeronautics and Space Technology at NASA Headquarters and in the X-24B flight summaries issued by the Flight Research Center.

### X-24B LIFTING-BODY FLIGHTS, 1975

| Flight Date and Number | Pilot, Maximum Altitude, Maximum Speed | Remarks |
|---|---|---|
| 14 Jan. 75 (B-19-30) | John A. Manke<br>22 098 m<br>Mach 1.76 | Objectives—to evaluate longitudinal stability and control at mach 1.8; perform with upper flap at -30° and approach with upper flap at -24°; and perform boundary-layer noise and vibration experiment—were attained. |
| 31 Jan. 75 (B-A-31) | Maj. Michael V. Love | Flight aborted because of weather. |
| 20 Mar. 75 (B-20-32) | Maj. Michael V. Love<br>22 860 m<br>Mach 1.40 | Objectives—to evaluate stability and control at speeds up to mach 1.30; perform with upper flap at -35°; approach with upper flap at -24°; survey fin, rudder, and flap pressure; and make boundary-layer noise and vibration experiments—were attained. |

Appendix E　　　　　　　　ASTRONAUTICS AND AERONAUTICS, 1975

| Flight Date and Number | Pilot, Maximum Altitude, Maximum Speed | Remarks |
|---|---|---|
| 18 April 75 (B-21-33) | John A. Manke<br>16 764 m<br>Mach 1.17 | Primary objectives were to evaluate stability and control at speeds up to mach 1.60; operational checkout of Ay feedback system; to approach with upper flap at −24°; to survey fin, rudder, and flap pressure; to make TPS qualification test; and to check handling qualities with all dampers at zero gain. Three chambers of XLR-11 engine ignited, not the four planned for this flight. All objectives were achieved except for the stability and control evaluation, which was limited to data on speeds up to mach 1.17. |
| 6 May 75 (B-22-34) | Lt. Col. Michael V. Love<br>22 250 m<br>Mach 1.5 | Shutdown in #3 engine B-52 30 min. after takeoff attributed to fire; launch altitude 12 800 m instead of 13 700 m. All objectives—to evaluate stability and control at speeds up to mach 1.60, and with aileron bias at 3°; approach with upper flap at −24°; survey fin, rudder, and flap pressure; make TPS qualification test; and check performance and trim at 18° and with upper flap at 30°—were attained except for the stability and control evaluation, which was limited to data up to mach 1.15. |
| 22 May 75 (B-23-35) | John A. Manke<br>22 250 m<br>Mach 1.6 | Primary objectives—to evaluate stability and control at speeds up to mach 1.60, and with aileron bias at 3°; approach with upper flap at −24°; survey fin, rudder, and flap pressure; make TPS qualification test (SLA 220); check performance and trim at 20°, and with 30° upper flap; and check flying qualities with Ay feedback system—were attained, except for hinge-moment data on aileron bias performance. Hinge-moment gauges did not operate as desired. |
| 6 June 75 (B-24-36) | Lt. Col. Michael V. Love<br>22 250 m<br>Mach 1.68 | All objectives—to evaluate stability and control at speeds up to mach 1.70; approach with upper flap at −24°; survey fin, rudder, and flap pressure; make TPS qualification test; and check stability and control at 20°∞ with upper flap at 30°—were attained. |
| 24 June 75 (B-A-37) | John A. Manke | Flight aborted because of winds. |
| 25 June 75 (B-25-38) | John A. Manke<br>18 590 m<br>Mach 1.43 | All objectives—to evaluate power effects on directional stability; approach and land with upper flap at −28°; survey fin, rudder, and flap pressure; make TPS qualification test; and evaluate stability and control with aileron bias at 30°—were attained. |
| 15 July 79 (B-26-39) | Lt. Col. Michael V. Love<br>21 175 m<br>Mach 1.585 | All objectives—to evaluate stability and control at speeds up to mach 1.68; approach and land with upper flap at −28°; survey fin, rudder, and flap pressure; make TPS qualification test; and evaluate stability and control with aileron bias at 11°—were attained. First flight after realignment of engine in yaw axis. |
| 5 Aug. 75 (B-27-40) | John A. Manke<br>17 678 m<br>Mach 1.23 | First landing of X-24B on prepared concrete runway; touchdown was within 1.5 meters of marker. Objectives—landing on concrete runway, survey of body pressure, study of left-hand fin tuft, study of TPS qualification, and evaluation of stability and control with aileron bias at 3°—were attained although only three of engine's four chambers ignited. |
| 20 Aug. 75 (B-28-41) | Lt. Col. Michael V. Love<br>21 945 m<br>Mach 1.53 | Second successful landing on runway. Objectives—landing on concrete runway, body-pressure survey, left-hand fin study, TPS qualification study, and evaluation of stability and control with rudder bias at 5° toe-out—were attained. |
| 9 Sept. 75 (B-29-42) | William H. Dana<br>19 200 m<br>Mach 1.48 | Objectives—pilot checkout, evaluation of stability and control at speeds up to mach 1.45, body-pressure survey, TPS qualification test (high-density RSI), and study of left-hand fin tuft—were attained. First X-24B flight for this pilot, who had previously flown M2-F3 and HL-10. |
| 23 Sept. 75 (B-30-43) | William H. Dana<br>17 678 m<br>Mach 1.2 | Last of scheduled powered flights; six remaining flights before end of CY 75 to be glide flights. Objectives—evaluation of stability and control with rudder bias at 10°, evaluation of handling qualities with all damper gains at zero, survey of body pressure, HRSI qualification test, and study of left-hand fin tuft—were attained. Subsonic data attained for all objectives. |
| 9 Oct. 75 (B-31-44) | Einar Enevoldson<br>13 716 m<br>Mach 0.7 | Unpowered flight. Objectives—pilot checkout, evaluation of handling qualities, and TPS qualification test (HRSI)—were attained. |

| Flight Date and Number | Pilot, Maximum Altitude, Maximum Speed | Remarks |
|---|---|---|
| 21 Oct. 75 (B-32-45) | Capt. Francis R. Scobee<br>13 716 m<br>Mach 0.7 | Unpowered flight. Objectives—pilot checkout, evaluation of handling qualities, and TPS qualification test (HRSI)—were attained. |
| 3 Nov. 75 (B-33-46) | Thomas C. McMurtry<br>13 716 m<br>Mach 0.7 | Unpowered flight. Objectives—pilot checkout, evaluation of handling qualities, and TPS qualification test (HRSI)—were attained. |
| 12 Nov. 75 (B-34-47) | Einar Enevoldson<br>13 716 m<br>Mach 0.7 | Unpowered flight. Objectives—evaluation of handling qualities with zero damper, TPS qualification test (HRSI), approach with upper flap at 24°, and landing with upper flap at 22°—were attained. |
| 19 Nov. 75 (B-35-48) | Capt. Francis R. Scobee<br>13 716 m<br>Mach 0.7 | Unpowered flight. Objectives—evaluation of handling qualities with zero damper gains, TPS qualification test (HRSI), approach with upper flap at 24°, landing with upper flap at 22°, and survey of left-hand fin pressure—were attained. |
| 26 Nov. 75 (B-36-49) | Thomas C. McMurtry<br>13 716 m<br>Mach 0.7 | Unpowered flight. Final flight of X-24B in NASA/Air Force lifting-body flight research program. Objectives—evaluation of handling qualities with zero damping gains, TPS qualification test (HRSI), approach with upper flap at 24°, landing with upper flap at 22°, and survey of left-hand fin pressure—were attained. |

# Appendix F
## ABBREVIATIONS OF REFERENCES

Listed here are the abbreviations used for citing sources in the text. Not all the sources are listed, only those that are abbreviated.

| | |
|---|---|
| *AAAS Bull* | American Association for the Advancement of Science's *AAAS Bulletin* |
| *A&A* | American Institute of Aeronautics and Astronautics' magazine, *Astronautics & Aeronautics* |
| *A&A 1974* | NASA's *Astronautics and Aeronautics, 1974: A Chronology* (this publication) |
| ABC | American Broadcasting Company |
| AEC Release | Atomic Energy Commission news release |
| *Aero Daily* | *Aerospace Daily* newsletter |
| *Aero Med* | *Aerospace Medicine* magazine |
| *AF Mag* | Air Force Association's *Air Force Magazine* |
| *AFHF Newsletter* | *Air Force Historical Foundation Newsletter* |
| *AFJ* | *Armed Forces Journal* magazine |
| AFSC *Newsreview* | Air Force Systems Command's *Newsreview* |
| AFSC Release | Air Force Systems Command news release |
| AIA Release | Aerospace Industries Association of America news release |
| AIAA *Facts* | American Institute of Aeronautics and Astronautics' *Facts* |
| AIAA Release | American Institute of Aeronautics and Astronautics news release |
| AIP *Newsletter* | *American Institute of Physics Newsletter* |
| AP | Associated Press news service |
| ARC *Astrogram* | NASA Ames Research Center's *Astrogram* |
| *Astro Journ* | American Astronomical Society's *Astrophysical Journal* |
| *Atlanta JC* | *Atlanta Journal Constitution* newspaper |
| *Av Wk* | *Aviation Week & Space Technology* magazine |
| *B News* | *Birmingham News* newspaper |
| *B Sun* | Baltimore *Sun* newspaper |
| *Bull Atom Sci* | Education Foundation for Nuclear Science's *Bulletin of the Atomic Scientists* |
| *Bus Wk* | *Business Week* magazine |
| *C Daily News* | *Chicago Daily News* newspaper |
| *C Trib* | *Chicago Tribune* newspaper |
| Can Press | Canadian Press news service |
| CBS | Columbia Broadcasting System |
| *C&E News* | *Chemical & Engineering News* magazine |
| *Cl PD* | Cleveland *Plain Dealer* newspaper |
| *Cl Press* | *Cleveland Press* newspaper |
| *Columbia J Rev* | *Columbia Journalism Review* magazine |
| ComSatCorp Release | Communications Satellite Corporation news release |
| *CQ* | *Congressional Quarterly* |
| *CR* | *Congressional Record* |
| *CSM* | *Christian Science Monitor* newspaper |
| CTNS | Chicago Tribune News Service |
| *D News* | *Detroit News* newspaper |
| *D Post* | *Denver Post* newspaper |
| DASA Release | Defense Atomic Support Agency news release |

| | |
|---|---|
| DFRC | See FRC. |
| DJ | Dow Jones news service |
| DOC PIO | Department of Commerce Public Information Office |
| DOD Release | Department of Defense news release |
| DOT Release | Department of Transportation news release |
| EOP Release | Executive Office of the President news release |
| FAA Release | Federal Aviation Administration news release |
| FBIS–Sov | Foreign Broadcast Information Service, Soviet number |
| FonF | *Facts on File* |
| FRC Release | Flight Research Center news release (after 8 Jan. 1976, became Dryden Flight Research Center news release) |
| FRC *X–Press* | NASA Flight Research Center's *X-Press* |
| *GE Forum* | *General Electric Forum* magazine |
| *Goddard News* | NASA Goddard Space Flight Center's *Goddard News* |
| GSFC Release | NASA Goddard Space Flight Center news release |
| GSFC *SSR* | NASA Goddard Space Flight Center's *Satellite Situation Report* |
| GT&E Release | General Telephone & Electronics news release |
| *H Chron* | *Houston Chronicle* newspaper |
| *H Post* | *Houston Post* newspaper |
| *JA* | *Journal of Aircraft* magazine |
| JPL *Lab-Oratory* | Jet Propulsion Laboratory's *Lab-Oratory* |
| JPL Release | Jet Propulsion Laboratory news release |
| JPRS | Department of Commerce Joint Publications Research Service |
| JSC Release | NASA Lyndon B. Johnson Space Center (Manned Spacecraft Center until 17 Feb. 1973) news release |
| JSC *Roundup* | NASA Lyndon B. Johnson Space Center's *Space News Roundup* |
| *JSR* | American Institute of Aeronautics and Astronautics' *Journal of Spacecraft and Rockets* magazine |
| *KC Star* | *Kansas City Star* newspaper |
| *KC Times* | *Kansas City Times* newspaper |
| KSC Release | NASA John F. Kennedy Space Center news release |
| LA *Her-Exam* | Los Angeles *Herald-Examiner* newspaper |
| *LA Times* | *Los Angeles Times* newspaper |
| *Langley Researcher* | NASA Langley Research Center's *Langley Researcher* |
| LARC Release | NASA Langley Research Center news release |
| LATNS | Los Angeles Times News Service |
| LERC Release | NASA Lewis Research Center news release |
| *Lewis News* | NASA Lewis Research Center's *Lewis News* |
| *M Her* | *Miami Herald* newspaper |
| *M News* | *Miami News* newspaper |
| *M Trib* | *Minneapolis Tribune* newspaper |
| *Marshall Star* | NASA George C. Marshall Space Flight Center's *Marshall Star* |
| *MJ* | *Milwaukee Journal* newspaper |
| MSFC Release | NASA George C. Marshall Space Flight Center news release |
| *N Hav Reg* | *New Haven Register* newspaper |
| *N News* | *Newark News* newspaper |
| *N Va Sun* | *Northern Virginia Sun* newspaper |
| NAA *News* | National Aeronautic Association *News* |
| NAA Record Book | National Aeronautic Association's *World and U.S.A. National World Aviation–Space Records* |
| NAC Release | National Aviation Club news release |
| NAE Release | National Academy of Engineering news release |
| NANA | North American Newspaper Alliance |
| NAS Release | National Academy of Sciences news release |
| NAS-NRC Release | National Academy of Sciences–National Research Council news release |
| NAS–NRC–NAE *News Rpt* | National Academy of Sciences–National Research Council–National Academy of Engineering *News Report* |
| NASA anno | NASA announcement |

| | |
|---|---|
| NASA Gen Mgmt Rev Rpt | NASA Headquarters "General Management Review Report" |
| NASA HHR-39 | NASA Historical Report No. 39 |
| NASA Hist Off | NASA History Office |
| NASA Hq *WB* | NASA Headquarters *Weekly Bulletin* |
| NASA Int Aff | NASA Office of International Affairs |
| NASA *LAR*, XIII/8 | NASA *Legislative Activities Report*, Vol. XIII, No. 8 |
| NASA Leg Off | NASA Office of Legislative Affairs |
| NASA MOR | NASA Headquarters Mission Operations Report, preliminary prelaunch and postlaunch report series; information may be revised and refined before publication |
| NASA prog off | NASA program office (for the program reported) |
| NASA proj off | NASA project office (for the project reported) |
| NASA Release | NASA Headquarters news release |
| NASA Rpt SRL | NASA report of sounding rocket launching |
| NASA SP-4019 | NASA Special Publication No. 4019 |
| *Natl Obs* | *National Observer* magazine |
| *Nature* | *Nature Physical Science* magazine |
| NBC | National Broadcasting Company |
| NGS Release | National Geographic Society news release |
| NMI | NASA Management Instruction |
| NN | NASA Notice |
| NOAA Release | National Oceanic and Atmospheric Administration news release |
| NRL Release | Naval Research Laboratory news release |
| NSC Release | National Space Club news release |
| NSC *Letter* | National Space Club *Letter* |
| NSC *News* | National Space Club *News* |
| NSF *Highlights* | National Science Foundation's *Science Resources Studies Highlights* |
| NSF Release | National Science Foundation news release |
| NSTL Release | NASA National Space Technology Laboratories news release |
| *NY News* | *New York Daily News* newspaper |
| *NYT*, 5:4 | *New York Times* newspaper, section 5, page 4 |
| NYTNS | New York Times News Service |
| *O Sen Star* | *Orlando Sentinel Star* newspaper |
| *Oakland Trib* | *Oakland Tribune* newspaper |
| *Omaha W-H* | *Omaha World-Herald* newspaper |
| ONR *Rev* | Navy's Office of Naval Research *Reviews* |
| P *Bull* | Philadelphia *Evening* and *Sunday Bulletin* newspaper |
| P *Inq* | Philadelphia *Inquirer* newspaper |
| PAO | Public Affairs Office |
| PD | National Archives and Records Service's *Weekly Compilation of Presidential Documents* |
| PIO | Public Information Office |
| PMR *Missile* | USN Pacific Missile Range's *Missile* |
| PMR Release | USN Pacific Missile Range news release |
| Pres Rpt 74 | *Aeronautics and Space Report of the President: 1974 Activities* |
| SAO Release | Smithsonian Astrophysical Observatory news release |
| SBD | *Defense/Space Business Daily* newspaper |
| *Sci Amer* | *Scientific American* magazine |
| Sci & Govt Rpt | *Science & Government Report*, independent bulletin of science policy |
| SciServ | Science Service News service |
| SD | *Space Digest* magazine |
| *SD Union* | *San Diego Union* newspaper |
| *SET Manpower Comments* | Scientific Manpower Commission's *Scientific, Engineering, Technical Manpower Comments* |
| *SF* | British Interplanetary Society's *Spaceflight* magazine |
| *SF Chron* | *San Francisco Chronicle* newspaper |
| *SF Exam* | *San Francisco Examiner* newspaper |
| *Sov Aero* | *Soviet Aerospace* newsletter |
| *Sov Rpt* | Center for Foreign Technology's *Soviet Report* (translations) |

| | |
|---|---|
| SP | *Space Propulsion* newsletter |
| Spaceport News | NASA John F. Kennedy Space Center's *Spaceport News* |
| Spacewarn | IUWDS World Data Center A for Rockets and Satellites' *Spacewarn Bulletin* |
| SR list | NASA compendium of sounding rocket launches |
| SSN | *Soviet Sciences in the News*, publication of Electro-Optical Systems, Inc. |
| St Louis G–D | *St Louis Globe-Democrat* newspaper |
| St Louis P–D | *St Louis Post-Dispatch* newspaper |
| T–Picayune | New Orleans *Times-Picayune* newspaper |
| Tech Rev | Massachusetts Institute of Technology's *Technology Review* |
| Testimony | Congressional testimony, prepared statement |
| Text | Prepared report or speech text |
| Transcript | Official transcript of news conference or congressional hearing |
| UN Reg | United Nations Public Registry of Space Flight |
| UPI | United Press International news service |
| USGS Release | U.S. Geological Survey news release |
| USPS Release | U.S. Postal Service news release |
| W Post | *Washington Post* newspaper |
| W Star-News | *Washington Star-News* newspaper |
| WFC Release | NASA Wallops Flight Center news release (after 26 April 1974; formerly Wallops Station News Release) |
| WH Release | White House News Release |
| WJT | *World Journal Tribune* newspaper |
| WS Release | NASA Wallops Station news release (see also WFC) |
| WSJ | *Wall Street Journal* newspaper |

# INDEX AND LIST OF ABBREVIATIONS AND ACRONYMS

## A

A-300B (European airbus), 95
AAAS. See American Association for the Advancement of Science.
AAS. See American Astronautical Society; American Astronomical Society.
Abbott Laboratories, 4
ABM (antiballistic missile), 60
ABMA (Army Ballistic Missile Agency), 125
Accident on *Soyuz 11*, 92
ACF. See air combat fighter.
Acoustic grenade payloads, 237
Acoustic tests, 147
ACPL (Atmosphere Cloud Physics Laboratory), 182
ACS. (Attitude control systems)
Adenosine triphosphate (ATP), 176
Advanced medium short-takeoff-and-landing aircraft (AMST), 178
Advanced missions appropriations, 45
Advanced Technology Light Twin (ATLIT), 53
AE (Atmospheric Explorer satellites), 13
AEC. See Atomic Energy Commission.
AE-D (Atmospheric Explorer), 1
AEDC. See Arnold Engineering Development Center.
AEM (Atmospheric Explorer Mission), 229
AEM-A (Applications Explorer Mission), 185
Aero Company, 66, 124
*Aero Daily*, 217, 231
Aero Propulsion Laboratory (USAF), 67
Aerobee (sounding rocket)
  Aerobee 150A-MII, 285
  Aerobee 170, 288, 290
  Aerobee 170A, 286, 289
  Aerobee 180, 289
  Aerobee 200A, 284, 285, 289
  Aerobee 350, 290
Aerojet Electrosystems Co., 26
Aerojet General Corporation, 136
Aerojet mechanical-thermal simulator, 26
Aeronautical and Space Sciences, Senate Committee on, 25, 80, 116, 183, 185, 218, 234
Aeronautical research, 19, 101
*Aeronautics and Space Report of the President: 1974 Activities*, 106
*Aeronautics and Space Report of the President: 1975 Activities*, 251
Aeronautics and Space Technology, NASA Office of (OAST), 22, 293

Aeronutronic Ford Corporation, 94, 211, 219
Aerosat (Canadian-ESRO-U.S. comsat program), 48, 116
Aerosol, 159, 229, 236
Aerosol Education Bureau, 247
Aerospace Corporation Vehicle Systems Division, 201, 237
Aerospace Day, 35
Aerospace Defense Command (USAF), 215
Aerospace Industries Association (AIA), 44, 220
Aerospace industry, 36
*Aerospace* magazine, 250
Aetna Life and Casualty Company, 197, 245
AFCRL. See Cambridge Research Laboratories (USAF).
Africa, 38, 140
AFSATCOM (Air Force Satellite Communications System), 43
AFSC. See Air Force Systems Command.
Agriculture, U.S. Dept. of, 1, 9, 29, 59, 232
Agriculture, solar power for, 250
AIA. See Aerospace Industries Association.
AIAA. See American Institute of Aeronautics and Astronautics.
AIDJEX (Arctic ice dynamics joint experiments), 71
Air and Space Museum, Smithsonian Institution, 128, 159, 179, 243
Air combat fighter F-16, 4, 99, 107, 138
Air Force, U.S. See U.S. Air Force (USAF).
Air Force Systems Command (AFSC), 24, 66, 67, 81, 120, 124, 154, 189, 192
*Air Forces Journal International*, 70
Air freight, 24
Air Products and Chemicals, Inc., 123
Air rescue, 54-55, 74
Air Sampling Program, Global (GASP), 41
Air Show, Paris, 95, 106
Air traffic safety, 23
Air Transport Association of America (ATA), 185
Air transportation research budget, 19
Airborne Infrared Observatory, 29, 88
Airborne Science Spacelab Experiments System Simulation (ASSESS), 100
Airborne Warning and Control System (AWACS), USAF, 59
Airbus, 107
Aircraft, 4, 6, 8, 13, 23, 57, 115, 211
Airflow, supersonic, 149

Airfoil, General-Aviation-Whitcomb, 53
Alabama, 232
ALADDIN (Atmospheric Layering and Density Distribution of Ions and Neutrals program), 6
Alaska, 9, 157, 231–232, 246
Alaskan Federation of Natives, 36
Albedo, 142
Albert, Rep. Carl B., 106
Albuquerque, N. Mex., 127
Alcator (reactor at MIT), 220
Alexandrov, Anatoly, 230
Algeria, 3
All Nippon Airways, 218
Allen, Dr. Joseph P., 139
Allison, L/C Edgar L., 74
Alloy, heat-resistant, developed in China, 236
All-Women Transcontinental Air Race, 249
Alma-Ata, U.S.S.R., 44
Alpha Jet (aircraft), 95
Aluminum Company of America, 129
Aluminum vapor, 6
Ambler, Ernest, 249
American Airlines, Inc., 2
American Association for the Advancement of Science (AAAS), 226
American Astronomical Society (AAS), 168
American Defense Preparedness Assn., 205–206
American Geophysical Union, 191
American Institute of Aeronautics and Astronautics (AIAA), 35, 160, 190, 191
American Mining Congress, 222
American Satellite Corporation, 51, 229
American Society of Mechanical Engineers (ASME), 235
American Telephone & Telegraph Company, 10
Ames *Astrogram*, 215
Ames Research Center (ARC), 22, 29, 36, 69, 80, 88, 100, 108, 114, 116, 128, 138, 144, 159, 161, 167, 174, 198, 214, 228, 231, 245, 293
AMOS (antireflection-coated metal oxide semiconductor), 80
AMPS (Atmospheric, Magnetospheric and Plasmas-in-Space) Spacelab payload, 13, 160, 221
AMST (advanced medium STOL aircraft), 178
Ancient Mariner, 206
Anders, William A., 6
Anderson, Jack, 77
Anderson, Robert, 173
Andreyev, Boris D., 28
*Anik* (Canadian comsat), 79, 84, 259, 277, 278
Animals in space flight, 121
*Annandale* (research vessel), 187
*Ans 1*. See Netherlands astronomical satellite.
Antarqui (project), 90
Antennas, 3, 231, 232
Antiballistic missile (ABM), 60
Antireflection-coated metal-oxide semiconductor (AMOS), 80
Apathy toward space, U.S., 146
Apogee motor, solid-fuel, 195
Apollo-Salyut proposal, 131
Apollo-Saturn hardware storage, 61
Apollo-Soyuz Test Project (ASTP; see also Apollo, Soyuz), 1, 11, 21, 31, 51, 59, 76, 130, 131, 167, 189, 195, 225, 281, 282
*Apollo-Soyuz Test Project: Information for Press*, 136
Apollo spacecraft launch, 1, 131, 278
art exhibit, 137
astronauts visit U.S.S.R., 168, 191
*Ats 6* coverage, 65, 87, 126, 157
backup mission, 188
briefings, 37, 137, 145, 147
CIA warning, 127
comment, 140, 142, 145, 152, 153
communications, 7, 12, 104, 132, 157
control centers (Houston and Moscow), 82–83, 122
cosmonauts visit U.S., 168, 205
cost, 65
crew members, 5, 6, 28, 178
docking, 5, 12, 104, 139–140
experiments, 4, 37, 38, 133, 134–135, 169, 212
joint activities in orbit, 133
joint training, 7, 12, 28, 83
languages used, 28, 132
launch sites, 214, 265
medical conditions, 120, 134, 147, 149, 171, 214
meetings, 7, 65
military potential, 164
objectives, 135
photography, 134–135, 243
preparations, 4–5
Proxmire, Sen., 61, 102, 126
readiness review, 12, 110
recovery, 104, 135, 145
safety risks, 94, 102, 110, 144
Saturn IB (launch vehicle), 52, 92, 136, 160
simulations, 5, 51, 82, 83, 121
Soviet delegation, 222
space-food exhibition, 121
*Soyuz 19* launch, 265
technician housing, 81
telemetry coverage, 38
tracking, 12
training, 12, 28, 38
TV broadcasts, 132, 133, 136
working groups, 7, 77
Apollo spacecraft
*Apollo 9*, 67
*Apollo 10*, 178
*Apollo 12*, 94
*Apollo 14*, 11
*Apollo 15*, 67, 182
*Apollo 16*, 11, 188
*Apollo 17*, 11, 63
ASTP check, 123
atmospheres, 133, 214
corrosion, 110
design, 11, 12, 64, 104

Apollo spacecraft—continued
  hatches, 133
  launch site, 214
  life-support systems, 12
  malfunction on ASTP, 135
  oxygen supply, 135, 147
  propellants, 121
Apollo lunar missions, 11, 63, 143, 214
Apollo Telescope Mount (ATM), 14, 191
Appalachia, 157
Applications Explorer Mission (AEM-A), 185
Applications Systems Verification Test (ASVT), 5, 9
Applications Technology Satellite (ATS) (See also *Ats 6*), 26, 54, 87, 109
Appropriations, House Committee on, 120
Appropriations, Senate Committee on, 143-144
Aqaba, Gulf of, 243
Aquatic plants, vascular, 39
Arab League, 140
Arabian peninsula, ASTP images, 243
ARAKS (artificial aurora between Kerguelen and Sogra), 39, 283
Arcas rocket (see also Super Arcas), 288
ARC. See Ames Research Center.
Arctic, 71, 207, 218, 237
Arctic Ice Dynamics Joint Experiment (AIDJEX), 71
Arecibo, Puerto Rico, 231, 232
Argentina, 122, 203
Ariane (launch vehicle), 48
*Ariel 5* (U.S.-U.K. cooperative satellite), 158, 196, 208
Aries (rocket), 289
Arizona, Univ. of, 29, 35, 36, 58, 88, 94
Arkalyk, U.S.S.R., 92
Arkhangelsk Oblast, U.S.S.R., 39
Army Corps of Engineers, U.S., 9, 213
Arnold Engineering Development Center (AEDC-AFSC), 120, 154, 189
Artificial aurora between Kerguelen and Sogra (ARAKS), 39, 283
*Aryabhata* (India science satellite), 67, 251, 259
Asia, 103, 243
ASME. See American Society of Mechanical Engineers.
ASSESS (Airborne Science Spacelab Experiments System Simulation), 100
ASTP. See Apollo-Soyuz Test Project.
Astrain, Santiago, 159
Astrobee F, 286
Astronaut Office, JSC Science and Applications Directorate, 164
Astronautics programs, military, 19
Astronauts, 11, 33-34, 58, 88, 147-148, 157, 167, 171
Astronomers, microwave signals picked up by, 178
Astronomical Association of Northern California, 163
Astronomy, 141, 164, 225
*Astrophysical Journal*, 30, 181, 248
*Astrophysical Journal Letters*, 217

ASVT (Applications Systems Verification Test), 5, 9
ATA. See Air Transport Association.
Atlantic Ocean communications services, 180
Atlas-Centaur launch vehicles, 2, 33, 89, 194, 195, 223
ATLIT (advanced technology light twin aircraft), 53
ATM. See Apollo Telescope Mount.
Atmosphere modification, 183
Atmosphere Cloud Physics Laboratory (ACPL), 182
Atmospheric Explorer (AE) satellites, 1, 13, 227, 229
Atmospheric layering and density distribution of ions and neutrals (ALADDIN) program, 6, 15
Atmospheric, Magnetospheric and Plasmas-in-Space (AMPS), Spacelab payload, 13, 221
Atmospheric Sciences, Interdepartmental Committee for, 29
Atmospheric variability experiments (AVE), 71
Atomic Energy Commission (AEC), 6
Atomic Research Institute, Moscow, 230
ATP. See adenosine triphosphate.
*Ats 6* (Applications Technology Satellite), 38, 54-55, 62, 64-65, 87, 126-127, 130, 157, 233
AT&T. See American Telephone and Telegraph Company.
AU: astronomical unit.
Auburn, Ala., 243
Auburn, Mass., historic site acquisition, 12
*Aura* (French satellite), 251, 270
Aurora, artificial, 283
Australia, 207, 232
Australia-London news satellite link, 122
Austria, 27, 216
"Automobile Power Systems Evaluation Study," 181
Avco Corporation, 108
AVE. See atmospheric variability experiment.
Aviation and Transportation R&D, House Subcommittee on, 220
Aviation Safety Reporting Program, 204
*Aviation Week and Space Technology*, 23, 35, 48, 60, 67, 76, 82, 95, 99, 106, 107, 140, 157, 244
Avionics control system, automated, 129
Avionics Laboratory, USAF, 81
AWACS. See Airborne Warning and Control System.
Awards
  AIAA annual, 36
  ASME, 235
  ASTP, 189
  ATS, 64
  Collier, Robert J., Trophy, 74
  de Florez Award, 36
  Harmon Aviation, 74
  Landsat, 212
  Lindsay, John C., Award, 124

Awards—continued
  Losey, Robert M., award, 36
  Mariner-Venus team award, 77
  NASTRAN, 7
  National Space Club, 64
  Pecora, Wm. T., Award, 212
  Reed, Sylvanus A., Award, 36
  Space Science Award, 37
  Spacecraft Design Award (AIAA), 37
  Waterman, Alan T., Award, 84
  White, Thomas D., trophy, 187
  Wright Brothers, 190
  Wyld, James A., 190
Azcarraga, Luis de, 120

B

B-1 (strategic bomber), 10, 19, 64, 106, 180, 182
Babcock, Alice K., 64
Bacteria, 177, 223
Bahamas, Central, 178-179
Balch, Jackson, 143
Ball-bearings, 193
Ballistic Missile Agency (U.S. Army), 125
Ballistic motors, reusable, 154
Balloons, 13, 74, 89-90, 109, 207, 242
Baltimore *Sun*, 41, 54, 58, 112, 137, 149, 166, 210, 211, 236
Bangui Anomaly, 233
Barium cloud, 212
Barnes, James C., 168
Barstow, Calif., 203
Basilevsky, Aleksandr, 210
Bathymetry, 178-179
Batteries, solar, 103
Bay St. Louis, Miss., 39
Baykonur cosmodrome (U.S.S.R.), 3, 28, 59, 66, 83, 91, 105, 131-132, 134, 149, 225
Beame, Mayor Abraham D., 11
Bean, Alan L., 6, 64, 65, 123
Beartooth highway, Custer National Forest, Montana, 211
*Beayondan* (Johns Hopkins University research vessel), 179
Beechcraft 99 (commuter aircraft), 129
Beech-Hawker 125 (aircraft), 65
Beer, Reinhard, 205
Belgium, 31, 60, 99, 107
Bell Telephone Research Division, 189
Bendix Field Engineering Corporation, 15, 102, 201, 213, 244
Bends, 167, 214
Bennett, Gary L., 123
Bering Sea Marine Mammal Experiment (BESMEX), 71
Betelgeuse (star), 138
bev: billion electron volts.
Bicentennial, U.S., 172
"Big bang" theory, 178
Big Bird (reconnaissance satellites), 67, 86
Billingham, John, 163-164
Biological experiments in space, 3, 197-198
Biology, molecular, 190

*Biosatellite 3*, 229
BIS. See British Interplanetary Society.
Black Brant VC (sounding rocket), 11, 210, 217, 240, 284, 288, 290
Black holes, 48, 158, 196, 208
Blackout of communications, 73-74
Blagonravov, Anatoly A., 25
Blinov, N., 238
Blood pressure, cosmonauts', 92
Blount Brothers Construction Company, 161
"Blue Planet," at Paris Air Show, 95
Blum, Haywood, 60
BN2 Islander (British aircraft), 95
Bobko, Karol J., 65
Body size, pilots', 88
Boeing Company, 2, 44, 54, 78, 95, 124, 131, 154, 161, 185, 226, 241
Boeing 707, noise from, 32
Boeing 747, Global Air Sampling Program, 41
Boeing 747 Space Shuttle carrier, 34, 214, 240
Boeing 747SP aircraft, 224
Boeing Space Center, Kent, Wash., 206, 230
Boikov, Nikolai, 247
Bolivia, 203, 212
Bolling, Richard, 216
Bomber force, U.S., 216
Booster separation motors (BSM), 162
Booster, solid-propellant, 189
Boretsky, Michael, 68
Boulder, Colo., NOAA laboratories, 206
Bow shock wave at Venus, 52
Bowden, Donald R., 116
Bowyer, Stuart, 169
Brand, Vance D., 1, 5, 65, 83, 120, 121, 123, 132, 133, 134, 135, 147, 149, 168, 189, 191, 192, 265, 282
Brandt, John C., 186
Brayton gas-turbine engine, 181-182
Brazil, 3, 4, 94, 203
Breathing systems, 11, 12
Brezhnev, Leonid I., 192
British Association for the Advancement of Science, 178
British exhibits at Paris Air Show, 95
British Interplanetary Society (BIS), 169
British cooperation in AWACS, 59
British sounding rocket, 232
Brockett, H. R., 189
Brockmann, Carlos, 212
Bronk, Detlev W., 226
Brooks Air Force Base, Texas, 188
Brown, Harold, 115
Brussels, Univ. of, Belgium, 212
BSM (booster separation motors), 162
Bubble-domain technology, 90, 243
Bubbling behavior in melts in space, 210
Budget, NASA, 20, 21, 22, 45, 75, 76, 101-102, 107, 114, 143-145, 218
Bugging, Moscow ASTP control center, 157
Bulgaria, 153, 241
Bureau of Mines, U.S. Dept. of the Interior, 203

Burning, external exhaust, 154
Bushuyev, Konstantin D., 7, 12, 59, 77, 225
Business Aircraft Association, National, 65

## C

C−141 (Airborne Infrared Observatory), 29, 30, 35
C−5A (military cargo aircraft), 68, 246
Cable, flat conductor, 85, 86
Calcium loss by astronauts, 33−34
California, 167, 180
California Institute of Technology, 17, 115
California State University, observatory, 237
California, Univ. of, 111, 126, 166, 169, 177, 284
Callisto (Jupiter's moon), 43
*Calypso* (Cousteau Society research vessel), 179
Cambridge Research Laboratory (USAF), 6
Cameras, high-resolution, 86
Canada, 2, 4, 48, 54−55, 71, 78−79, 116, 203, 277, 283
Canadian National Exhibition, Toronto, 173
Canadian Overseas Telecommunications Corporation, 58
Canadian Science Writers Association, 34
Canberra, Australia, tracking station, 58
Cancer, 27, 28, 183, 223
Cannon, James, 90
Canopus lock for *Mariner 10*, 47
Cape Canaveral, 57, 241
Carbon dioxide on Mars, 171
Carbon monoxide detected in atmosphere, 205
Carbon tetrachloride, 247
Cardiac care systems, portable, 51
Cargo aircraft (YC-15), 178
Carmen (hurricane), 206
Carr, Gerald P., 64, 173, 191
Carrot-cell-culture experiment, 197
*Castor* and *Pollux* (French satellites), 86, 251, 260
CCD. See charge-coupled device.
Celestial bodies and space law, 165
Celestial Observation Satellite (*Cos−B*), 2, 162, 266, 277, 278
CEN (Centre d'Etudes Nucleaires), 212
Census, U.S. Bureau of, 1
Central African Republic, 233
Central Intelligence Agency (CIA), 102, 127, 130
Central Telephone and Utilities Corporation, 130
Centre d'Etudes Nucleaires (CEN), 212
Centre National d'Etudes Spatiales (CNES), 48, 177
Cernan, Eugene A., 65, 178
Cesium propellant, 93
CETI (communication with extraterrestrial intelligence), 232
Chamberlain Manufacturing Corporation, 12
Chapman, Dean R., 69
Charge-coupled device (CCD), 184

Charter Industries, Raleigh, N.C., 242
Charyk, Joseph V., 30, 116
Chase Econometric Associates, 157
Chemistry, 13, 68, 190, 226
Chicago, Ill., 112, 231, 235−236
Chilca Launch Range, Peru, 90
Chile, 58, 202, 203
China, People's Republic of, 129, 149, 231
  ancient supernova records, 186
  astronomers' discoveries, 78
  comment on ASTP, 140, 153
  heat-resistant alloy, 236
  recovery of earth satellite, 236
  satellite launches, 149, 176, 236, 243, 251
*China 1, 2, 3, 4* and *5* (satellites), 176, 266, 273, 274
Chlorine, 236
Chlorofluoromethanes, 13
*Christian Science Monitor*, 24, 67, 103, 105, 149, 161, 162, 169, 174, 236, 244
Christy-Sackmann, Juliana, 17
CIA. See Central Intelligence Agency.
CIAP. See Climatic Impact Assessment Program.
"City in space," 174
Civil aviation, 107
Civilian R&D programs, Federal, 23
Civilization(s), 73, 225, 231
Clark, John F., 29, 74
Climate, 50, 150
Climatic Impact Assessment Program (CIAP, DOT), 8
Cloud layer, Venus, 101, 210
Cloud research, 117
CNES (Centre National d'Etudes Spatiales), 48, 177
Coal, 8, 49, 203, 222
Coast Guard, U.S., 55, 201
Coastal disaster warnings, 2
Coating, anticorrosion, 148
Cockpit design for Space Shuttle, 88
Cockpit, ground-based instrumented, 167
Cole, Edward N., 24
Collier, Robert J., trophy, 74
Collins, Michael, 159
Collision, continental, 163
Colonization of space, Princeton Univ. conference, 250
Colorado, Univ. of, 118, 142, 288, 289
Columbia Univ., 248
Coma Berenices, dwarf star in, 169
Comets, 168
Command and data-acquisition centers, NOAA, 231
Commerce, U.S. Department of (DOC), 15, 26, 29, 55, 63, 70, 94, 211
Commercial activity and space, 235
Communications blackout on Intelsat, 73
Communications satellites (comsats: see individual satellites), 1, 2, 3, 4, 19, 43, 48, 87, 180
Communications Satellite Corporation, 1, 10, 30, 33, 74, 82, 89, 116, 148, 151, 180, 194, 230, 238, 245
Communist Party ideologist Mikhail Suslov, 230
Composite materials, 24, 179

Compression technology, 50
Computers, 13, 17, 53, 77, 117, 167, 241
Comsat General Corporation (subsidiary), 10, 230, 238. See also Communications Satellite Corporation.
ComSatCorp. See Communications Satellite Corporation.
Concept verification test (CVT), 164
"Conceptual Flight Possibilities for Saturn-Apollo Hardware," 188
Concorde (U.K.–France supersonic aircraft), 95, 107, 159, 218, 247
*Condor I* (balloon replica), 242
Conductor cable, flat, 85, 86
Congress, House of Representatives
  Appropriations, Committee on, 120
  Aviation and Transportation R&D, Subcommittee on, 220
  Science and Astronautics, Committee on. See Science and Technology, Committee on.
  Science and Technology, Committee on, 8, 45, 61, 108, 158, 250
  Space Science and Applications, Subcommittee on, 223, 224
Congress, Senate
  Aeronautical and Space Sciences, Committee on, 25, 80, 183, 185, 218, 234
  Appropriations, Committee on, 140–143
  Labor and Public Welfare, Committee on, 116
Congressional authorization for NASA transitional period, 102
*Congressional Record*, 130
Connecticut, 180
Connestype Fourier spectrometer, 205
Conrad, Charles P., 64, 74
Consiglio Nazionale delle Ricerche, Italy, 44
Consortium of European aircraft manufacturers, 99
Constellation (aircraft), 190
Construction and modification of facilities, budget, 101
Consumer Product Safety Commission, 247
Continental Airlines, Inc., 36
Contract, NASA, 44, 79, 112, 113, 137, 174, 201, 208, 226, 227, 240
Control Data Corporation, 78
Coos Bay, Oreg., 74
*Copernicus* (Orbiting Astronomical Observatory), 73, 113, 208
Copper-lead alloys, 205
Core reactors, 30
Cornell Univ., 63, 231
Cortisone, 147
Cortright, Edgar M., 129, 143
*Cos–B*: Celestial Observation Satellite.
Cosmic rays
  astronaut exposure, 11, 37, 94
  astronomy experiments, 164
  background component, 118
  degenerating solar cells, 103
  measurement, 11, 46
  sources, 141, 212
Cosmochemistry, 226
Cosmonauts, 91, 92, 132, 134, 205

Cosmos (U.S.S.R. satellites), 167
*Cosmos 122*, 66
*Cosmos 637*, 204
*Cosmos 702*, 253
*Cosmos 703*, 253
*Cosmos 704*, 253
*Cosmos 705*, 253
*Cosmos 706*, 254
*Cosmos 707*, 254
*Cosmos 708*, 254
*Cosmos 709*, 255
*Cosmos 710*, 255
*Cosmos 711*, 255
*Cosmos 712*, 255
*Cosmos 713*, 255
*Cosmos 714*, 255
*Cosmos 715*, 256
*Cosmos 716*, 256
*Cosmos 717*, 256
*Cosmos 718*, 256
*Cosmos 719*, 256
*Cosmos 720*, 257
*Cosmos 721*, 257
*Cosmos 722*, 257
*Cosmos 723*, 257
*Cosmos 724*, 257
*Cosmos 725*, 258
*Cosmos 726*, 258
*Cosmos 727*, 258
*Cosmos 728*, 258
*Cosmos 729*, 259
*Cosmos 730*, 259
*Cosmos 731*, 260
*Cosmos 732*, 261
*Cosmos 733*, 261
*Cosmos 734*, 261
*Cosmos 735*, 261
*Cosmos 736*, 261
*Cosmos 737*, 261
*Cosmos 738*, 262
*Cosmos 739*, 262
*Cosmos 740*, 262
*Cosmos 741*, 262
*Cosmos 742*, 262
*Cosmos 743*, 263
*Cosmos 744*, 263
*Cosmos 745*, 264
*Cosmos 746*, 264
*Cosmos 747*, 264
*Cosmos 748*, 264
*Cosmos 749*, 264
*Cosmos 750*, 265
*Cosmos 751*, 265
*Cosmos 752*, 265
*Cosmos 753*, 266
*Cosmos 754*, 266
*Cosmos 755*, 266
*Cosmos 756*, 267
*Cosmos 757*, 267
*Cosmos 758*, 267
*Cosmos 759*, 268
*Cosmos 760*, 268
*Cosmos 761*, 268
*Cosmos 762*, 268
*Cosmos 763*, 268
*Cosmos 764*, 268
*Cosmos 765*, 269

*Cosmos 766,* 269
*Cosmos 767,* 269
*Cosmos 768,* 269
*Cosmos 769,* 269
*Cosmos 770,* 269
*Cosmos 771,* 270
*Cosmos 772,* 270
*Cosmos 773,* 270
*Cosmos 774,* 270
*Cosmos 775,* 204, 251, 271
*Cosmos 776,* 271
*Cosmos 777,* 271
*Cosmos 778,* 272
*Cosmos 779,* 272
*Cosmos 780,* 272
*Cosmos 781,* 272
*Cosmos 782,* 229, 230, 251, 273
*Cosmos 783,* 273
*Cosmos 784,* 273
*Cosmos 785,* 273
*Cosmos 786,* 274
Cost containment, 36
Cost overruns, 246
Cotton, Paul E., 202
Cousteau, Jacques, 179
Cousteau, Philippe, 179
Cousteau Society, 178–179
CPK (creatine phosphokinase), 177
Crab Nebula, 186, 248
Creation of the earth, 178
Crimea Astrophysical Observatory, 225
Crippen, Robert L., 65
Cronkite, Walter, 146
Crop acreage, Landsat measurements, 9, 224
Crow, Duward L., 192
Crows Landing, Calif., 116
Crystal growth in zero-g, 213
CS (Japanese comsat), 139
Culbertson, Philip E., 48
Custer National Forest, Mont., 211
Cutler-Hammer, Inc., 69
CVT (concept verification test), 164
Cyprus, 233
Czechoslovakia, 229, 241

**D**

DACS (data-acquisition and control system), 231
DAD (Dual Air Density) spacecraft, 238
Dallas, Tex., 112
Dana, William H., 194, 294
Dash 7 (Canadian aircraft), 54
Dassault-Breguet Mirage F1E (French fighter aircraft), 99
Data-acquisition and control system (DACS), 231
Data analysis, Landsat, 10
Data Center, Federal, 9
Data processing on Shuttle Orbiter, 126
Data recorder, solid-state, 90, 243
Day, LeRoy E., 34
Days, Martian (sols), 111

DC–8 (jet transport), 32
Dead Sea, 233, 243
Debus, Kurt H., 12, 69
Decompression, 167, 214
Deep Space Communications Center, Moscow, 210
Defense, U.S. Dept. of, 2, 19, 23, 26, 29, 63, 96, 137, 209
Defense military effort and space, 180
Defense procurement, 20
Defense Research, Development, Test, and Evaluation, 19
*Defense Space Business Daily,* 176
deFlorez Award, 36
deGaulle, Charles, 66
Delhi Domestic News Service, 67
Deloffre, Bernard, 8, 66
Delta launch, 29, 51, 139, 217
Demerliac, Y., 93
Dendrite, 210
Denitrogenation, 214
Denmark, 99, 107
Denver, Univ. of, 91, 240
Desalination, 233
Desert geography, 135
Detente, 31, 175
Deutsche Presse Agentur, 43
Diagnostic system, 102
Diamant B/P.4 (French launcher), 27, 86
DIGITAC (digital flight control system), 27
Dirac, Paul, 166
Dirigibles, 99
Discoverer (satellites), 86
Disher, John H., 154
Disney World, 28
"Distorted Priorities in the R&D Budget," 41
Divers in decompression, 167
Dobrynin, Anatoly P., 132
DOC. See Commerce, U.S. Dept. of.
DOD. See Defense, U.S. Dept. of.
DOD spacecraft, 256, 259, 260, 263, 271, 273, 274, 278
DOI. See Interior, U.S. Dept. of.
Domes, geodesic, 242
d'Ornano, Michel, 95
Dornier company (West Germany), 233
DOT. See Transportation, U.S. Dept. of.
Dow Jones and Company, 228
Downey, Calif., 161
Drag reduction on trucks, 39
Drake, Frank B., 231
Dryden, Hugh L., 25
Dryden Flight Research Center, 293. See Flight Research Center.
Dryer, Murray, 37
Dual Air Density (DAD) spacecraft, 238
Duckett, Carl, 127
Dudley Observatory, Albany, N.Y., 91, 287
Duke, Charles M., Jr., 188
Duke University Medical Center, 34
Dwarf star in Coma Berenices, 169
Dynamic Test Facility (MSFC), 15
Dystrophy, muscular, 177
Dzhanebekov, Vladimir A., 28

## E

*Early Bird* (Intelsat I F-1), 89, 121, 159, 195
Earth
  galaxy, size of, 217
  gravity field, 63
  -moon relationship, 111
  -oriented space missions, 150
  resources, 1, 3, 23, 44, 90, 138
  stations, 51, 112
  -sun relationships, 118
  tremors simulation, 50
Earth Orbiting Teleoperator System (EOTS), 14
Earth radiation budget (ERB), 109
Earth Resources Experiment Package (EREP), 10, 168
Earth Resources Orbiting Satellite (EROS) Data Center, 212
Earth Resources Technology Satellite (ERTS), 1, 5, 74, 168. See Landsat.
ERTS-B *(Landsat 2)*, 1, 5, 8, 253
*Erts 1 (Landsat 1)*, 5, 163
Earth sciences budget, 18
East Africa, 126
Eastern Test Range (ETR), 25, 33, 60, 79, 87, 89, 117, 169, 177, 195, 227
*Echo 1*, commemoration, 165
Echo-free chambers, 146
Eclipse, solar, 37, 133
Ecliptic plane, 85
Econ, Inc., 30
Economic benefits of space program, 157
Educational TV by satellite, 87, 157
Edwards AFB, Calif., 178, 214, 215
"Effects of Stratospheric Pollution by Aircraft" (DOT study), 7, 18, 27
EIC Inc., 227
Eisenhower, President Dwight D., 125, 165, 226
el-Baz, Farouk, 243
ELDO (European Launcher Development Organization), 93
Electric power generation, 54, 89, 93, 126, 213, 219-220
Electric propulsion, 93, 190
Electrical fields, atmospheric, 94
Electrical system failure, 23
Electrically scanning microwave radiometer (ESMR), 109
Electricity furnished from space colony, 250
Electromagnetic wave energy-conversion (EWEC) concept, 71
Electron loss, ionospheric, 13, 97
Electronic Research Division (Rockwell), 90
Electrophoresis experiments, 4, 212
Emission from aircraft, 8
Employees, fewer Federal, 4
Employment, NASA, 122, 226
Energy (See also Fuels; Solar energy)
  budget, 17, 18, 22
  chemical, 68
  conservation, 68, 106
  electric autos, 182
  flywheel, 125, 234
  fuel efficiency, 186
  internal-combustion engine, 181, 182
  laser energy, 44, 63, 119, 229, 244
  national policy, 49
  nuclear, 219, 220
  power satellites, 30-31, 131, 161
  research and development, 18, 221
  satellites generating power, 30-31, 131, 161
  space communities, 250
  storage, 176
  technology, 45, 101
  thermionic conversion, 131
  wind energy, 127, 213
Energy Administration, Federal (FEA), 185
Energy Information Center, 127
Energy Institute at New Mexico State Univ., 219
Energy Reorganization Act of 1974, 6
Energy Research and Development Administration (ERDA), 18, 19, 21, 23, 24, 29, 38, 89, 111, 116, 119, 125, 213, 221, 246
  establishment, 6
  projects, 49, 208, 211, 222
  relocation, 43
Enevoldson, Einar, 167, 294, 295
Engine materials, composite, 179
Engine noise reduction, 2
Engineering, National Academy of, 68, 221, 246
Environment, 5, 9, 45, 71, 103, 142, 179, 213, 214
Environmental Defense Fund, 247
Environmental Protection Agency (EPA), 7, 9, 183
Environmental Quality, Council on, 29
Environmental Research and Technology, Inc., 168
Environmental Research Institute, Michigan, 179
Environmental Research Laboratories, 94
EOTS. See Earth Orbiting Teleoperator System.
EPA. See Environmental Protection Agency.
Epsilon Eridani (star), 73
Epsilon Indi (star), 73
Equator, 90, 233
Equatorial space station, 203
Equatorial undercurrent, 117
ERB (earth radiation budget) experiment, 109
ERDA. See Energy Research and Development Administration.
EREP (earth resources experiment package), 10, 168
Eridan (French sounding rockets), 39
ERNO Raumfahrttechnik GmbH, 35
EROS (Earth Resources Orbiting Satellite) Data Center, 212
ERTS. See Earth Resources Technology Satellite.
Erythrocytes decrease in cosmonauts, 92
ESA. See European Space Agency.
ESMR (electrically scanning microwave radiometer), 109
ESOC (ESA's space operations center), 162

ESRO. See European Space Research Organization.
*Essa 8*, 179
ESTEC (European Space Research and Technology Center), 8, 93, 163
ET. See External tank.
ETR. See Eastern Test Range.
Eurasia, 163
*Europa 2* (launch vehicle), 112
European Launcher Development Organization (ELDO), 93
European Space Agency (ESA), 2, 43, 44, 66, 95, 120, 121, 162, 212, 224, 277
"European Space Days," 93
European space programs, 48, 93, 116
European Space Research and Technology Center (ESTEC), 8, 93, 163
European Space Research Organization (ESRO), 8, 27, 44, 48–49, 60, 67, 85, 93, 95, 162. (See also ESA.)
European–U.S. cooperation, in space, 35, 85
Europeans, F–16 combat aircraft for, 107
Eurospace (Groupement Industriel Européen d'Etudes Spatiales), 93
Eutrophication, 183
EUV (extreme ultraviolet radiation), 169
ev: electron volt.
Evans, Ronald E., 6, 65, 123
Evans, William J., 181, 205
EWEC (electromagnetic wave energy-conversion) concept, 71
Exhaust pollution, 19, 154
Exosat (European x-ray astronomy satellite), 48
Expendable Launch Vehicle Programs, 198
Exploded stars, 248
Explorer satellites, 1
 *Explorer 1*, 47, 115
 *Explorer 29 (Geos 1)*, 63
 *Explorer 36 (Geos 2)*, 63
 *Explorer 42*, 79
 *Explorer 48* (Small Astronomy Satellite B), 79, 141
 *Explorer 49*, 100
 *Explorer 51 (Atmospheric Explorer–C)*, 90
 *Explorer 53* (SAS–C Small Astronomy Satellite), 1, 78–79, 84, 158, 259, 277, 278
 *Explorer 54*, 1, 271, 277, 279
 *Explorer 55*, 1, 227, 228, 272, 277, 279
 *Explorer 56*, 1
 *Explorer 57*, 1
Explorer Interplanetary Monitoring Platform (IMP), 124
*Explorers Journal*, 225
Explorers Society, International, 242
External tank (ET), 14, 108
Extraterrestrial life, intelligent, 164, 231, 232
Extreme ultraviolet radiation (EUV), 169
Exxon Corporation, 67, 82, 151
Eye-controlled switches, 51

F

F–1 (French fighter aircraft), 99
F–4 (Phantom jets), 28
F–14 (Tomcat fighter), 107
F–15 (Eagle fighters), 28, 32, 41, 107
F–16, 4, 99, 107, 138
F–18, 137, 138
F100 (engine), 4
F–104G (mach 2 fighter), 107, 126
F–111, 24, 187
F–111D (turbofan jet), 179, 215
FAA. See Federal Aviation Administration.
Facilities construction, 101
Faedo, Alessandro, 44
Faget, Maxime A., 235
Fairbanks, Alaska, 9
Fairchild Industries, 116
Fairchild Republic Company, 96
Fan blades, 24, 179
Far East Orbita stations, 103
FCC. See Federal Communications Commission.
FEA. See Federal Energy Administration.
Federal Aviation Administration (FAA), 23, 32, 127, 204
Federal Committee on Science and Technology, 151
Federal Communications Commission (FCC), 10, 112, 116, 148, 194, 230
Federal Council for Science and Technology, 29, 106
Federal Data Center, 9
Federal Energy Administration (FEA), 18, 218
Federal-industry liaison, 152
Federal Interagency Task Force on Inadvertent Modification of the Stratosphere (IMOS), 29
Federal Power Commission (FPC), 18
Federal Science and Technology Survey Committee, 151
Federal Solar Energy Research Institute, 221
Federal-state liaison, 152, 244
Fernandez, Gonzalo, 189
Ferri, Antonio, 36
Ferson telescope, 165
Fertilizers, chemical, 81, 223, 247
Fiber optics, 81
Fighter aircraft, 99
Filipchenko, Anatoly V., 28
Fink, Daniel J., 74
Finke, Wolfgang, 120
Finland, 153
Firefighters breathing system, 11–12
Firefighting module, 201
Firefly chemicals, 176–177
Fires, stellar, 17
Firnberg, Hertha, 27
Fish embryogenesis, 197
Fisk, Leonard A., 85
Flaps, 178
Flashes, stellar, 17
Flat conductor cable, 238
Fletcher, James C., 11, 13, 20, 21, 36, 44, 49, 50, 61, 123, 131, 141, 145, 150, 188, 198, 208, 212, 213, 221
Flight-control system, 129

Flight Research Center, 12, 22, 31, 39, 41, 65, 66, 73, 80, 87–88, 92, 115, 129, 167, 172, 187, 194, 240, 293
Florence, Univ. of Italy, 212
Florida, 23, 166, 179, 183, 243, 244, 249
FLTSATCOM (Air Force–Navy satcom system), 43
Fluids research, 50
Fluorocarbons, 183, 234, 236, 240
*Flying* magazine, 97
Flywheel, 125, 234
Food, 103, 121, 142
Forbes, Malcolm S., 74
Ford, President Gerald, 4, 6, 17, 21, 33, 49, 57, 64, 74, 77, 84, 90, 103, 104, 106, 113, 125, 132, 135, 136, 172, 179, 189, 190, 205, 249
Ford, Jack, 179
Ford Motor Company, 181
Forest management, 142
Forest Service, U.S., 211
Formica, Gianni, 66
Fort Stockton, Tex., 89
Fossil fuel search by satellite, 50
Fourier interferometer, 159
FPC (Federal Power Commission), 18
France (see also *Aura*, Mirage F1E, *Sret 2*, *Starlette*, *Symphonie 2*)
 Centre d'Etudes Nucleaires, 212
 Centre National d'Etudes Spatiales, 48, 177
 Centre National de la Recherche Scientifique, 118
 comsats, 4
 ESRO contract, 60
 experiments, 229
 Franco–Soviet cooperation, 103
 launchings, 2, 27, 86, 103, 177, 251
 nuclear reactors, 220
 space center in French Guiana, 111–112
Frank Briscoe Co., Inc., 87
*Frankfurter Allgemeine*, 152
FRC. See Flight Research Center.
Freeze-prediction model, contract for, 243
Freeze-temperature prediction research, 24
Freight carrier, giant, 24
French Guiana, 111–112
Freon, 13, 70, 247
"Frontier Technology and Shuttle Country," 35
Frutkin, Arnold W., iii
Fuels, 19, 21, 52, 54, 67–68, 92, 218, 220
Fuller, George A., Company, 240
Funding, 17, 18, 19, 148, 173
Fungi, zone-forming, 134
Fuqua, Don, 188, 218
Furguson, Ernest B., 84
Furnace, space experiments, 213
"Future Space Programs," 219

G

Gagarin, Yuri, 136, 191
Gagarin Cosmonaut Center, U.S.S.R., 65
Galapagos Islands, 87
Galactic center, 138
Galilee, Sea of, 243
*Galileo II* (instrumented Convair 990), 71, 100
Gamma rays, 11, 89, 141, 162, 181, 187
Ganymede (Jupiter's moon), 114–115
Garments, biological isolation, 51
GARP (Global Atmospheric Research Program), 109, 117
Garriott, Owen K., 64
Gas industry, 50
Gas, natural, 11, 99
Gas pipeline, Soviet, 244
Gas storage safety, 11
Gas, stratospheric, 159
GASP (Global Air Sampling Program), 41–42
GATE (GARP Atlantic Tropical Experiment), 117
Gates Learjet, 65
Gatland, Kenneth W., 67
GAW. See General Aviation-Whitcomb.
GCA Corporation, 91
Gemini program, 67, 128, 178
General aviation engines, noise investigations, 127
General Aviation Technology Advisory Panel (NASA), 98
General Aviation-Whitcomb (GAW) airfoil, 53
General Dynamics Corporation, 2, 4, 99, 107
General Electric Company, 89, 93, 109, 193, 202, 204, 211, 214, 233
General Services Administration (GSA), 114, 145
Generation of electricity, 54, 89
General Motors Corporation, 24, 36
Geochemistry and Analytical Chemistry, Moscow Institute of, 226
Geodesic domes, 242
GEODSS (ground electro-optical deep-space surveillance), 215
Geodynamic Experimental Ocean Satellite (GEOS), 1, 60, 201
Geography, ASTP observations, 135
Geological Survey, U.S. (USGS), 19, 71, 233
Geology, comparative, 175–176
Geophysical Union, 191
Geophysics Corporation, 287, 288
George A. Fuller Company, 240
Georgia Institute of Technology, 232
GEOS (Geodynamic Experimental Ocean Satellite), 1, 60, 201
 *Geos 1 (Explorer 29)*, 63
 *Geos 2 (Explorer 36)*, 63
 *Geos 3*, 62–63, 258, 277
Geostationary Operational Environmental Satellite *(GOES)*, 1, 26, 206, 271, 277–279
Geosynchronous orbit, *Ats 6*, 87
Getting, Ivan, 237
Ghana, 207
ghz: gigahertz (1 billion cycles per second)
Gibson, Edward G., 64, 173, 191
Gibson, Roy, 27, 43, 66, 93, 123
Gilruth, Robert R., 235
Giuli, Thomas, 37

Glenn, John, 136
Glennan, T. Keith, 125
Global Air Sampling Program (GASP), 41
Global Atmospheric Research Program (GARP), 109, 117, 207
GMS (Japanese meteorological satellite), 139
GNP (gross national product) U.S., 158
*Goddard News*, 124, 199
Goddard, Robert H., 12, 64
Goddard Space Flight Center, 2, 9, 22, 26, 36, 41, 46, 49, 53, 54, 55, 63, 71, 74, 79, 85, 91, 110, 119, 124, 126, 141, 148, 158, 163, 171, 177, 179, 186, 187, 201, 207, 208, 212, 213, 217, 244, 249, 251, 283
*GOES*. See Geostationary Operational Environmental Satellite.
Golan Heights, 243
Gold, Dr. Thomas, 63
Golden Gate Bridge, paint test, 148
Goldstein, Richard M., 114–115
Goldstone, Calif. (NASA facility), 9, 166, 203, 231
Goldwater, Senator Barry, 115
Goodyear Aerospace Corporation, 44, 59
Goodyear Rubber Corporation, 57
Gordon, Robert V., 147
Gorno-Altaisk, U.S.S.R., 59
Graphite, qualities of, 230
Gravitational field, earth, 63
Gravity, 11, 205, 230
Grechko, Georgy M., 3, 91, 253, 282, 283
Green Bank, W. Va., observatory, 231
Greenbelt Homes, Inc., 29
Greenbelt, Md., TDA station, 9
Greenberg, Daniel S., 103
Greenhouse effect, 183
Greenland, 207, 218, 246, 283
Greer, Robert E., 182
Griffin, Gerald D., 139
Grobecker, Alan J., 8
Gross national product (GNP), U.S., 158
Ground electro-optical deep-space surveillance (GEODSS), 215
Groupement Industriel Européen d'Etudes Spatiales (Eurospace), 93
Grumman Aerospace Corporation, 70, 94, 136
Grumman American Aviation Corp., 198
Grumman Gulfstream (aircraft), 65, 198
GSA (General Services Administration), 114, 145
GSFC. See Goddard Space Flight Center.
Gubarev, Aleksey A., 3, 91, 253, 282
Gum nebula, 225
Gwynn Island, Va., 74
*Gymnodinium* phytoplankton red-tide, 179, 249
Gyro anomaly investigation, 15
Gyroscopic systems, 214, 216

## H

Hackerman, Norman, 84
Halcon, Inc., 68
Haley Astronautics Award, 191
Hammarstrom, Ove, 66, 121
Hammud al-Khalid, Sulayman, 130
Handler, Philip, 34
Hansen, Grant L., 35, 36
Harmon Aviation Awards, 74
Harr, Karl G., 220
HASPA. See High-altitude superpressured powered aerostat.
Hawaii, 135, 137, 138, 195, 206, 246, 248
Hawkes, John, 8
HC–130H (rescue aircraft), 74
HCMM. See Heat Capacity Mapping Mission.
Health, Education, and Welfare, U.S. Dept. of (HEW), 29
HEAO. See High Energy Astronomy Observatory.
Heart attack, 177
Hearth, Donald P., 129
Heat Capacity Mapping Mission (HCMM), 185
Heat, 130, 175
Heavy-particle cosmic ray exposure received by astronauts, 11
Hecker, Edward J., 183
Helicopters, 66, 99, 107
*Helios 1*, 46, 47, 61, 212
Helium, 17, 37, 99, 180
Hello, Bastian, 182
HEW. See Health, Education, and Welfare, U.S. Dept. of.
High-altitude research, 13, 49
High-altitude superpressured powered aerostat (HASPA), 219
High Energy Astronomy Observatory (HEAO), 10, 14, 48, 79, 85
High Flight (evangelical foundation), 173, 182
High-temperature core reactors, 30
Highly Maneuverable Aircraft Technology (HIMAT) program, 66
Hinners, Noel W., 73, 143, 158
Hinterman, Hans, 121
HIRS (high-resolution temperature sounder), 109
Hoffman, Hans E. W., 35
Holbrook, Gordon F., 36
Hole in ionosphere, 97
Holloman, Dr. J. Herbert, 68
Holmdel, New Jersey, 166
Homes, heat loss, 130
Honeywell, Inc., 13
Hopi Indians, supernova record, 186
Hosenball, S. Neil, 24, 111, 165
Housing and Urban Development, U.S. Dept. of (HUD), 113, 120, 143, 148, 193
Houston, Tex., ASTP mission control, 132
Houston, Univ. of, 166, 192, 219, 235, 284, 285
Hovis, Warren A., 249
Hsinhua (PRC news agency), 149
Huancayo Observatory, Peru, 90–91
HUD. See Housing and Urban Development, U.S. Dept. of.
Hughes Aircraft Co., 31, 38, 74, 79, 116, 119, 194
Human immunity, 37

Hungary, 153, 229, 241
*Huntsville Times*, 155, 230
Hurricanes, 26, 57, 206
Hurt, Hugh H., 36
Huskie giant aircraft, 24
Hydrochloric acid in stratosphere, 237
Hydrogen, 17, 68, 127, 159, 217, 219, 220
Hypersonic research, 129, 239

## I

IAU (International Astronomical Union), 162, 249–250
IBM (International Business Machines Corporation), 10, 126, 136, 146, 148, 180, 197, 245
*Icarus*, 232
Ice research, 142, 207, 218, 232, 237
Ice-warn (program), 5
IEA (integrated electronic assembly), 7, 181
Iliac 4 (computer), 50
Illinois, Univ. of, 50, 91, 287
Immunity system, human, 37
Immunology, 190
IMOS (Inadvertent Modification of the Stratosphere), Federal Interagency Task Force on, 29
IMP (Explorer Interplanetary Monitoring Platform), 124
Improved TIROS Operational Satellite (ITOS E–2), 2
India, 32, 37, 67, 87, 126, 127, 130, 157, 163, 251
Indians, American, supernova record, 186
Indian Ocean, 89, 180, 195, 207
Indonesia, 4, 31
*Industrial Research*, 193
Industrial Tectonics, Inc., 193
Industry-Federal liaison, 152
Inefficiency, 98
Infections, urinary, 177
Inflation, 21
Infrared observation of upper atmosphere, 100
Infrared Observatory, Airborne, 29, 88
Infrared telescope project, 84
Instituto Geofisico del Peru, 91
Instrumentation of aircraft, 6
Integrated electronic assembly (IEA), 7, 181
Integrated Propulsion Control System (IPCS), 187
Integrated Subsystems Test Bed, 52
Intelligent extraterrestrial life, 164, 231, 232
INTELSAT (International Telecommunications Satellite Organization), 4, 24, 30, 58, 73, 80, 93, 122, 138, 159, 195, 227, 240
*Intelsat I F–1 (Early Bird)*, 89
*Intelsat IV F–1*, 2, 89, 180, 195, 260, 277, 278
*Intelsat IV F–6*, 33, 180
*Intelsat IV F–8*, 30, 180
*Intelsat IVA F–1*, 2, 270, 277, 279

Intelsat V satellites, 138
Intensified-silicon intensified-target (ISIT) camera, 215
Interagency cooperation, 119, 127, 161, 183, 194, 204, 213
Interagency Task Force, 183
Interim Agreements for the Establishment of a Global Commercial Comsat System, 195
Interior, U.S. Dept. of (DOI), 6, 8, 9, 18, 203
Internal combustion engine, 181, 183
Intercosmos (organization), 241
Intercosmos (U.S.S.R. satellites), 251
*Intercosmos 13*, 257
*Intercosmos 14*, 241, 273
International Association of Machinists, 51
International Astronomical Union (IAU), 162, 249–250
International Business Machines Corporation. See IBM.
International communications, ComSat rates for, 230
International cooperation on space, 1, 31, 32, 39, 41, 103, 130, 131, 139, 197, 198, 212
International Explorers Society, 242
International Harvester Company, 214
International Quiet Sun Year, 90
International Space Research Committee, 25
International Telecommunications Satellite Organization. See INTELSAT.
International Telecommunications Union, 93, 216
Interplanetary Monitoring Platform. See IMP.
Interpretation of satellite imagery, 9
Ionosphere, 13, 97, 244
Ion-thruster propulsion system, 93
IPCS (Integrated Propulsion Control System), 187
IR: infrared.
Iran, 3, 4, 203
Irwin, James R., 173, 182
ISIT (intensified-silicon intensified-target) camera, 215
Isolation, biological, 51
Israel, 233
Italy, 44, 60, 78, 203
Itek Corporation, 112
ITOS E–2 (Improved TIROS Operational Satellite), 2
Ivanchenkov, Aleksandr S., 28
*Izvestiya*, 38, 39

## J

Jaguar (close-air-support aircraft), 95
Jakarta, Indonesia, 31
*Jane's Weapons Systems*, 244
Japan, 4, 35, 48, 53, 139, 184, 251
Jastrow, Robert, 175
Javelin, rocket, 284
Jet engines, 32, 215
Jet Propulsion Laboratory, 2, 22, 50, 51–52, 55, 77, 80, 102, 159, 166, 181, 183, 184–185, 202, 203, 205, 208, 223–224, 228

Jet Star (business jet), 190
Jet aircraft, 44
Jetstream, 31, 233
Jever Air Force Base, West Germany, 126
Jicamarca Observatory, Peru, 90–91
Jodrell Bank radio telescope station, England, 178
John C. Lindsay Award, 124
John Jeffries Award, 37
Johns Hopkins University, 31, 63, 79, 166, 176, 179, 226, 285
Johnson, Clarence L., 190
Johnson, R. Tenney, 24, 111
Johnson Space Center, 11–12, 22, 28, 37, 44, 49, 58, 59, 62, 67, 69, 83, 88, 104, 110, 120, 139, 144, 164, 195, 196, 198, 202, 219, 222, 235, 237, 240, 243
Johnston, Richard S., 147, 149
Jones, J. Lloyd, Jr., 88
Jones, Robert T., 115
*Journal of Atmospheric Sciences*, 28
JPL. See Jet Propulsion Laboratory.
JSC. See Johnson Space Center.
Jupiter (planet), 29–30, 35, 42–43, 89, 108, 114–115, 124, 199, 228, 242
Justice, U.S. Dept. of, 29

## K

Kadlec, Paul W., 36
Kaliningrad, U.S.S.R., 66
Kansas, Univ. of, 129
Kapustin Yar, U.S.S.R., 167
*Karjalaimen* (Finnish publication), 153
Kayser, Lutz T., 247
Kazakhstan, 247
Kennedy, Sen. Edward M., 12
Kennedy, President John F., 136, 226
Kennedy Space Center (KSC), 2, 4, 5–8, 11, 12, 22, 23, 28, 29, 32, 44, 61, 67, 69, 73, 77, 80, 85, 92, 93, 94, 96, 100, 110, 112, 117, 121, 123, 132, 137, 143, 146, 157, 161, 167, 168, 171, 183, 195, 206, 207, 213, 214, 218, 219, 240, 242, 243
Kent, Wash., Boeing Space Center, 206
Kenya, San Marco launch platform, 78
Kerguelen Is., launches, 39, 283
Kerwin, Joseph P., 64, 74
Kidney-cell experiment, 4
Kiev, U.S.S.R., 192
*Kiku* (Japanese satellite), 184, 251, 267
Kleppe, Thomas S., 212
Klimuk, Pyotr, 91, 92, 121, 282
Kline, Raymond A., 189
Knights of Columbus, 173
Kompfner, Rudolf, 189
Kostroma Oblast, U.S.S.R., 39
Kotelnikov, Vladimir A., 77, 131, 145
Kourou, French Guiana, 86, 111
Kraft, Christopher C., 235, 243
Kramer, Robert S., 110
Kranzer, Ditmar, 216
Kremlin, 192
KSC. See Kennedy Space Center.
Kubasov, Valery N., 1, 28, 91, 130, 131, 132, 133, 134, 168, 191, 205, 265, 282

Kuiper Airborne Observatory, 88, 138
Kuiper, Gerard P., 88
Kuwait, ground station, 130

## L

Labor and Public Welfare, Senate Committee on, 116
Laboratoire de Meteorologie Dynamique, 207
LACIE. See Large area crop inventory experiment.
LAGEOS. See Laser Geodynamic Satellite.
Lake Tahoe, Nev., 168
Lake Victoria, Africa, 87
Lakeland, Fla., 243
Laminar flow, 54
Lancaster, Malcolm C., 37
Land-use planning by satellite, 9, 183
Landau, Ralph, 68
Landsat program, 5, 9, 74, 168, 208, 212, 232. See ERTS.
 Landsat–B, 5
 Landsat–C, 21
 *Landsat 1*, 5, 9, 33, 38, 163, 179, 185, 202, 222
 *Landsat 2*, 5, 8–9, 33, 38, 53, 90, 157, 185, 202, 237, 242, 253, 277, 278
Langley Research Center (LaRC), 2, 22, 31, 36, 53, 54, 79, 96, 110, 126, 129, 143, 160, 171, 202, 204, 214, 243, 293
LaRC. See Langley Research Center.
Large, Arlen J., 33
Large area crop inventory experiment (LACIE), 5, 9, 44, 59, 172, 224, 237
Large Lift Vehicle (LLV), 154
Large Space Telescope (LST), 14, 105, 113, 127, 144, 147, 148, 194, 230
Laser Geodynamic Satellite (LAGEOS) network, 201
Laser Tracking Subnet, 244
Lasers, 44, 63, 119, 229, 244
Lassen, Herbert A., 37
Launch complex, 39, 61, 112
Launch processing system (LPS), 93, 152
Launch vehicles, 45, 101, 223
Launchings, 1–2, 21, 89, 163, 277, 283
Lawrence Sperry Award, 37
Law, space, 165
Lawrence Livermore Laboratory, 234
Lazar, James B., 190
Lazarev, Vasily, 59, 282
LCMS. See Low-Cost Modular Spacecraft.
Lead-antimony eutectic, 210
Lead-silver alloys, 210
LeBeau, Andre, 66
LeBourget Airport, Paris, 95
Lee, Chester M., 64, 110, 189
Leningrad, 192
Leninsk, U.S.S.R., Baykonur Cosmodrome at, 83
Leonov, Aleksey A., 1, 28, 38, 91, 121, 130, 131, 133, 134, 137, 146, 168, 191, 205, 265, 282
LeRC. See Lewis Research Center.
Leucocytes increase in cosmonauts, 92

Lewis Research Center (LeRC), 2, 22, 33, 42, 49, 50, 67, 82, 93, 111, 139, 146, 171, 193, 195, 209, 213
Lick Observatory, Univ. of California, 126
Life, 35, 164, 176
Life sciences, 33, 165, 197
Life-support system for Space Shuttle, 81
Lifting body (aircraft), 128, 172–173
Light flashes observed by *Skylab 4*, 94
Lightning strikes, 94, 110
Lilly, William E., 22, 226
Limb-radiance inversion radiometer (LRIR), 109
Lindsay, John C., Award, 124
Liquid-metal plasma valve, 38
Liquid mixing in space, 210
Lloyd, Russell P., 168
LLV. See Large Lift Vehicle.
Lockheed Aircraft Corporation, 10, 44, 78, 99, 148, 190, 246, 284
Lockheed Electronics Co., Inc., 202
Lockheed Jet Star aircraft, 65
Lockheed Missiles and Space Co., Inc., 14, 120, 228
London, England, 111, 122, 212
Long Beach, Calif., 178
Lord, Douglas R., 49, 224
Los Alamos Scientific Laboratory, 119, 190
Los Angeles, 112, 160
*Los Angeles Times*, 47, 139, 159, 172, 183, 203, 229, 233
Louis W. Hill Award, 36
Lousma, Jack R., 6, 64, 65, 123
Love, Eugene S., 54
Love, Michael V., 172, 293, 294
Lovell, Sir Bernard, 178
Low, George M., 21, 77, 105, 110, 131, 185, 189
Low-Cost Modular Spacecraft (LCMS), 25
Low-Cost Silicon Array (JPL project), 208
Low-gravity processing, 11
LPS (launch processing system), 152
LRIR (limb-radiance inversion radiometer), 109
LST. See Large Space Telescope.
LTV Aerospace Corporation, 115
Lucas, William P., 125, 203
Luciferase, 176
Luciferin, 176
Ludwig, George H., 212
Luksch, Walter, 8, 66
Luminescence, 176–177
*Luna 22*, 99, 100, 176
Lunar and planetary explorations, 11, 14, 45, 101
Lunar Curatorial Facility, 114
Lunar Orbiter (NASA program), 12
Lunar Roving Vehicle, 125
Lunar samples, 23, 50, 114, 144, 188
Lundin, Bruce T., 213
Lung lesion in astronaut, 171
*Lunik I*, 25
Lunney, Glynn S., 7, 59, 60, 77, 110, 127, 195

**M**

M2 (lifting body), 128
Machinists, International Association of, 51
Madagascar, 130
Madrid, 233
MAF. See Michoud Assembly Facility.
Magellanic Clouds, 217
Magnetic equator, rockets launched at, 90
Magnetic field, 105, 233, 245
Magnetic particle, 166
Magnetic-tape certification facility, 201, 244
Magnetically interlinked points, 39
Magnetohydrodynamic generator, 54
Magnetosphere, 100, 241
Makarov, Oleg, 59, 282
Malagasy Republic, 129–130
Malkin, Myron S., 222, 223
Man, effects of space flight, 214
Management and Budget, U.S. Office of, 96
Manami trees, 138
Manikins used by NASA, 88
Manke, John N., 161, 293, 294
Manned laboratories in space, 96–97
Manned maneuvering unit, 62
"Manned Orbital Systems Concepts" (MSFC study), 239
Manned Space Flight, NASA Office of, 48
Manned space flights, 22, 160, 164, 251, 281
Manned space station, U.S., 105, 222, 239
Mannheimer, Harry, 5
Manufacturing Chemists Association, 240
Mao Tse-tung, 149
Maps, radioastronomy, 217
Marine Environmental Sciences Consortium, 90
Mariner–Jupiter–Saturn mission, 14, 35, 89, 185, 235
Mariner–Jupiter–Uranus mission, 14, 235
Mariner, Ancient, 206
Mariner–Saturn, 199
Mariner missions, 88, 89, 185, 199
*Mariner 5*, 206
*Mariner 6* and *7*, 171
*Mariner 8*, 33
*Mariner 10*, 47, 49, 50, 51–52, 58, 88, 115, 124, 175, 210
Mariner–Venus team award, 77
Marisat (maritime-satellite system), 2, 51, 60, 82, 151, 180
Mark, Hans M., 36, 231
Marots (ESRO's maritime orbital test satellite), 48
Marquardt Corporation, 129
Mars (planet), 1, 8, 14, 31, 55, 78, 96, 110, 111, 115, 143, 160, 162, 164, 169, 170, 171, 172, 175, 176, 184, 205
Marshall, Mrs. George C., 125
Marshall Space Flight Center (MSFC), 4, 7, 8, 10, 12, 13, 14–15, 22, 23, 30, 47, 49, 57, 61, 83, 85, 89, 90, 102, 105, 108, 110, 113, 116, 120, 125, 129, 130, 146, 154, 160, 162, 164, 181, 182, 186, 198, 201, 203, 210, 211, 212, 213, 217, 221, 223, 232, 235, 238, 239, 240
*Marshall Star*, 5, 53, 89, 110, 120, 122, 233
Martin, James S., Jr., 96, 143
Martin Marietta Corporation, 14, 15, 62, 93, 95, 108, 152, 171, 185, 221
Maryland, Univ. of, 231

Massachusetts Institute of Technology (MIT; see also Alcator; Center for Space Research), 79, 215, 219−220, 289
Materials processing, 24, 179, 205, 210
Mathews, Charles W., 5, 224
Matilsky, Terry, 158
Matthoefer, Hans, 43
Mauna Kea, Hawaii, 137
Max-Planck Institut, 212
May, Ellery B., 110, 160, 189
Mayfair Construction Company, 157
McCluney, Ross, 179
McConnell, Joseph H., 82
McDonald Observatory, 205
McDonnell Douglas Astronautics Company, 120, 174, 229, 239
McDonnell Douglas Corporation, 2, 28−29, 32, 44, 51, 60, 78, 79, 80, 95, 97, 178
McElroy, Michael B., 223
MCI Communications Corporation, 10, 148
McLeaish, John E., 104
McLucas, John L., 4
McMurtry, Thomas C., 295
MDM (multiplexers-demultiplexers), 130
Mechanical Engineers, American Society of, 235
Mechanical-thermal simulator, 26−27
Medical isolation of ASTP crew, 120
Medicine and space technology, 51, 177, 219
Megaton strike capacity, U.S., 216
Memory device, 90
Mercanti, Enrico P., 179
Mercury (planet), 47, 49, 52, 106, 115, 250
Mercury propellant, 93
Mercury valve, 38
Mesabi range, Minn., 233
Messerschmitt−Boelkow−Blohm Company, 95, 163, 240
Metals, 111, 205
Meteor (U.S.S.R. meteorological satellite), 167, 251
 *Meteor 21*, 257
 *Meteor 2−1*, 265
 *Meteor 22*, 269
 *Meteor 23*, 275
Meteorology, 2, 9, 24, 26, 27, 41, 71, 117, 142, 168, 207, 225, 244
Metric conversion, 238, 249
mev: million electron volts.
Mexico−Spain satellite link, 122
mhz: megahertz (one million cycles per second).
*Miami Herald*, 244, 247
Miami, Univ. of, 44
Michanowsky, George, 215
Michigan, Environmental Research Institute of, 179
Michigan, Univ. of, 284, 286
Michoud Assembly Facility (MAF), 14, 32
Microbes, 37
Microbiological laboratory, U.S.S.R., 169
Micrometeoroids, 46
Microorganisms, 134
Microprocessor, 215
Microwave energy, 161, 178
Microwave Scanning Beam Landing System, 69
Milan, Univ. of, 212
Mili, M., 93
Militsin, Alberty V., 81
Milky Way (galaxy), 84, 134, 141, 187, 196, 217
Mineral extraction, 8, 203
Mineral resources, 9, 142, 185
Minicomputers for Space Shuttle, 85
Minnesota Public Interest Research Group, 247
Minnesota, Univ. of, 285
Minnows, 197
Mirage (F1−E; French fighter aircraft), 99
Missile (see also individual missiles, kinds of missiles, and missile systems).
 early warning system, 19
 launchings monitored by U.S., 86, 242
 nosetip design, 152
 U.S.S.R. firing, 166−167
Mississippi (state), Landsat applications, 39, 232
Mississippi Air and Water Pollution Control Commission, 90
Missouri, Landsat applications, 232
MIT. See Massachusetts Institute of Technology, 68, 79, 215, 219−220, 289
Mitsubishi Electric Corporation, 53
Mobile Bay, 90
Modular Computer Systems, Inc., 85, 117
Molina, Mario J., 236
Molniya (U.S.S.R. comsat series), 167, 251
 *Molniya 1*, 259, 262, 267
 *Molniya 1−30*, 103, 251
  long-range communications and TV, for, 103
 *Molniya 2*, 254, 265, 267, 274
 *Molniya 3*, 258, 272, 275
 *Molniya 1-S*, 204
Monkeys in space program, 121
Monopole discovery, 166
Montoya, Joseph M., 186
Moog Inc. Controls Division, 223
Moon (see also lunar headings), 14, 23, 63−64, 74, 99−100, 111, 124, 187, 213, 226
Morris, George A., 115
Morrison, James R., 5
Morrison Knudsen Company, 69
Morton, Rogers C. B., 8
Moscow, U.S.S.R., 3, 81, 137, 191, 208, 210, 226, 230, 250
Moscow Mission Control Center, 122, 127
Mosher, Charles A., 43, 49, 108, 151
Moss, Sen. Frank E., 25, 115, 222, 235
Mount Wilson Observatory, 225
MRCA (multirole combat aircraft), 107
MSFC. See Marshall Space Flight Center.
Muir, Warren R., 183
Multiplexers-demultiplexers (MDM), 130
Multirole combat aircraft (MRCA), 107
Munich space operations center, 46
Murcray, David G., 240
Murray, Bruce C., 115
Myers, Dale D., 182
Mylar balloon (Echo 1), 165

N

NAA (National Aeronautic Assn.), 190, 249

NACA. See National Advisory Committee for Aeronautics.
NAE. See National Academy of Engineering.
Names for new planetary features, Moscow meeting on, 250
Nanking Observatory, China, 78
NAS. See National Academy of Sciences.
NASA. See National Aeronautics and Space Administration.
Nash, Edward J., 74
NASTRAN (NASA structural analysis), 7
National Academy of Engineering (NAE), 68, 221, 246
National Academy of Sciences (NAS), 34, 68, 141, 246
National Advisory Committee for Aeronautics (NACA), 69, 202
National Aeronautic Assn. (NAA) 190, 249
National Aeronautics and Space Administration (NASA; see also individual Centers, offices, and programs; Awards; Budget, NASA; Contract, NASA; Employment, NASA; Interagency cooperation; International Cooperation, space; Launchings; and Personnel, NASA), 1, 6, 7, 8, 9, 11, 13, 19, 29, 35, 37, 38, 41, 43, 54, 57, 63, 69, 71, 82, 85, 93, 94, 97, 98, 105, 119, 123, 127, 129, 144, 151, 158, 165, 168, 171, 173, 178–179, 186, 187, 190, 206, 207, 218, 222, 226, 248
National Aviation Club, 74
National Bureau of Standards (NBS), 43, 236, 249
National Business Aircraft Association, 65
National Cancer Institute, 27–28, 60
National Center for Atmospheric Research (NCAR), 89, 207
National Center for Resource Recovery, 61
National Environmental Satellite Service (NESS), NOAA, 211–212, 243, 248–249
National Gallery of Art, 137
National Geodetic Satellite Program (NGSP), 63
National Medal of Science, 189–190
National Oceanic and Atmospheric Administration. See NOAA.
National Radio Astronomy Observatory, 231
National Research Council, 14, 216, 221, 246
National Science Board, 84
National Science Foundation (NSF), 6, 18, 29, 84, 103, 235
National science policy, 151
National Space Club awards, 64
National Space Science Data Center, 207
National Space Technology Laboratories (NSTL), 22, 39, 52, 63, 90, 123, 143
National Weather Service, 71
NATO (North Atlantic Treaty Organization), 59, 108
Natrium (sodium), 244
Natural gas. See Gas, natural.
Natural Resources Defense Council, 247
Natural resources information, 5
*Nature* magazine, 63, 64
Nature, 103
Naugle, John E., 70, 143, 229
*Nauka I Zhin*, 149
Navigation, 116, 205-206
NAVSTAR (navigation system using time and ranging), global positioning system, 19, 206
Navy, U.S. See U.S. Navy.
Nazca plains, 242
NBS. See National Bureau of Standards.
NCAR. See National Center for Atmospheric Research.
Neal, Roy, 64
Neilon, John J., 207
Neptune (planet), 168, 198
*Nepzabadsag*, 153
NESS. See National Environmental Satellite Service.
Ness, Norman F., 47, 124
Netherlands, 8, 99, 107
Netherlands astronomical satellite (*Ans 1*), 196, 208
Neutron stars, 48, 232
New Hampshire, Univ. of, 285
New Mexico Department of Agriculture, 219
New Mexico Institute of Mining and Technology, 94
New Mexico, launchings, 248
New Mexico State Univ., 127, 219
New York City, N.Y., 11, 112
New York State Development Corporation, 86
New York, State Univ. of, 15, 36, 80
*New York Times, The*, 1, 27, 31, 49, 60, 90, 111, 121, 175, 176, 181, 186, 192, 225, 231, 243, 246
Newport News *Times Herald*, 160, 207
*Newsreview* (AFSC publication), 97, 124, 154, 178, 181, 215, 216
*Newsweek*, 99
Newton, Mass., 227
NGSP (National Geodetic Satellite Program), 63
Nicogossian, Arnauld E., 148–149, 171
Nike-Apache, 90, 284, 285, 287, 288
Nike-Cajun, 237, 286, 288, 290
Nike-Hawk (sounding rocket), 187, 289
Nike-Javelin, 290
Nike-Tomahawk, 90, 285, 286, 287, 288, 289, 291
Nimbus (satellites), 13, 26, 142
Nimbus-F experimental meteorological satellite, 1, 51, 60
*Nimbus 1*, 109
*Nimbus 2*, 109
*Nimbus 3*, 109
*Nimbus 4*, 109, 228
*Nimbus 5*, 109
*Nimbus 6*, 108–109, 207, 232, 233, 237, 263, 277, 278
Ninety-Nines, Inc., 249
Nitrogen, 223, 247
Nitrogen-oxide emission standards, 8
Nitrogen tetroxide, 148

Nixon, President Richard M., 90, 104, 142, 143
NOAA (National Oceanic and Atmospheric Administration). (See also National Environmental Satellite Service (NESS)), 2, 19, 24, 25, 37, 43, 57, 59, 71, 94, 179, 206, 211, 212, 231, 243, 246, 248–249
*Noaa* satellites: see ITOS E–2, *Sms*, *Goes*
Nobel Prize, 125, 190
Noggle, Lawrence W., 57, 58
Noise
  galactic, 100
  jet, 32–33, 147
  reduction, 2, 19, 65, 127
Noordwijk, the Netherlands, 93
NORAD. See North American Air Defense Command.
Nordberg, William, 163, 212
Norsk Polarinstitut, Norway, 207
North American Air Defense Command (NORAD), 47
North Atlantic Treaty Organization (NATO), 59, 108
Northrup Corporation, 4, 54, 124, 240
Northrup Services, Inc., 14
Norway, 99, 107
Nosecap on Space Shuttle, 7
Nosetip designs, 152
Novosibirsk, U.S.S.R., 192
NRC (Nuclear Regulatory Commission), 6–7. (See also National Research Council)
NRL (Naval Research Laboratory). See U.S. Navy.
NSF. See National Science Foundation.
NSPL (NASA Standard Parts List), 239
NSTL. See National Space Technology Laboratories.
Nuclear energy, 119, 219, 220
Nuclear fusion, 219–220
Nuclear-powered aircraft, 57, 123
Nuclear-powered lasers, 119
Nuclear reactors, 220
Nuclear Regulatory Commission (NRC), 6–7
*Nuclear Technology*, 30

O

OA (Office of Applications, NASA Hq), 110
*Oao 3* (*Copernicus*, Orbiting Astronomical Laboratory), 73, 113
OAST (Office of Aeronautics and Space Technology, NASA Hq), 7, 88, 293
Oberg, James E., 64
Objects in space, manmade, 47
Observatory, 88, 138, 237
Oceanography, 2, 117, 142, 179, 218
Office of Management and Budget (OMB), 1, 33
Office of Occupational Medicine and Environmental Health, 50
Office of Science and Technology (White House), 90, 116, 151
Offshore oil, 159, 232–233
*Ogo 1*, (Orbiting Geophysical Observatory), 233
Ohia trees, 138
Oil, 107, 151, 159, 185, 232-233
Oja, Heikka, 169
O'Keefe, John A., 111
Okorokov, Boris, 137
OMB (Office of Management and Budget), 1, 33
OMSF (Office of Manned Space Flight, NASA Hq), 48, 123, 198
O'Neill, Gerard K., 250
OPA (Office of Program Assurance), 46
*Opportunities and Choices in Space Science, 1974*, 14
Optical bench, 47
Optical sensations from cosmic rays, 37
Orange County, Calif., 202
Orbita network (U.S.S.R.) 103, 245
Orbital behavior theory, 166
Orbital Test Satellite (OTS), 48
Orbital Transport and Raketen A.G., 247–248
Orbiter, Shuttle, 34, 185, 189, 201
  and *Lander 1* to Mars, 55
  *Orbiter 2*, 73
  *Orbiter 102*, 38
Orbiting Astronomical Observatory (OAO), 73, 113
Orbiting Geophysical Observatory (OGO), 233
Orbiting Solar Observatory. See OSO.
*Orders of Magnitude*, 69
Orion (constellation), 158
Orlando *Sentinel Star*, 165, 168, 229
Orthostatic stability in cosmonauts, 92
Osborne, Weyman A., 166
Osnos, Peter, 230
OSO (Orbiting Solar Observatory), 1, 10, 60, 117–118
*Oso 1*, 51, 118
*Oso 5*, 10
*Oso 7*, 10
*Oso 8*, 117, 118, 208, 264, 277, 278
OSS (Office of Space Science, NASA Hq), 13, 198
OST. See Office of Science and Technology.
OTDA (Office of Tracking and Data Acquisition, NASA Hq), 189
O'Toole, Thomas, 3, 131, 137, 233
OTS (Orbital Test Satellite), 48
Outer solar system, 14
Outer space, 165, 213, 231
"Outlook for Aeronautics," (study), 218
"Outlook for Space," 150
"Outlook for the Space Program," (speech), 221
Ovenden, Michael W., 168
Overmyer, Robert F., 157
Oxygen, 147
Ozone
  air sampling, 42
  depletion, 144, 159, 183, 223, 234, 236–237, 246, 247
  skin cancer, 27–28, 223
  supersonic transports, 8
  stratospheric pollution, 27–28, 70
Ozonesonde balloons, 90

## P

P-38 (fighter aircraft), 190
Pacific Ocean, 180, 195
Page, George F., 207
Paint, inorganic, 148
Paintings following ASTP, 137
Paiute-Tomahawks, 6
Palestine, Tex., 89
Palmdale, Calif., 185
Palmer, Patrick, 231
Pan American World Airways, 42, 224
Parachutes, 100, 160, 186, 190–191, 217
Paris, 95, 106–107, 207
Particle-interface reactions, 210
Particle stream measurements, 39
Particle with one magnetic pole, 166
*Partnership: A History of the Apollo-Soyuz Test Project*, 66, 77, 83, 110, 120, 121, 136
*Pasadena Star News*, 161
Patterson Air Force Base, 58
Patterson, John, 125
Pauling, Linus C., 190
Pearl River buoys, 90
Pecora, William T., Award, 212
Peenemunde, Germany, 235
Pendray, G. Edward, Award, 36
Pennsylvania State Univ., 91, 286, 287, 288
Pentagon, 164, 245, 246
*People's Daily, The* (China), 140, 176
People's Republic of China (PRC). See China, People's Republic of.
Permanent space facility, 151
Personnel, NASA
  appointment, 24–25, 47, 88, 122, 129, 173, 189, 207
  force reduction, 13
Peru, 90, 203, 242, 283
Peterson, Russell W., 29
Petrone, Rocco A., 36, 61, 69, 70, 229
Petrov, Boris N., 7, 191, 241
Ph.D.s, overproduction, 60
*Philadelphia Inquirer*, 86, 142, 157, 158, 164
Phillips, Samuel C., 180, 181
Phillips, William H., 36
Phosphates, water hyacinths absorbing, 39
Photochemical processes caused by solar ultraviolet radiation, 13
Photoelectron distribution measurement, 187
Photography by satellites, 86, 134
Photovoltaic conference, 80
Photovoltaic systems, 209
Physical changes in cosmonauts, 92
*Physical Review Letters*, 141
Physics and astronomy, 45, 101
*Physics Today*, 97
Physiological effects of environment, 214
Phytoplankton *Gymnodinium*, 179
Pickering, William H., 115
Pictographs of supernova, 186
Pilot-astronaut population in 1985, 88
Pilot body-size changes, 88
Pilots of lifting-body flights, 293–295
Pinsky, Lawrence S., 166

*Pioneer II*, 42-43
*Pioneer 6*, 242
*Pioneer 10*, 29, 42, 88, 108, 138, 242
*Pioneer 11*, 29, 242, 245
Pioneer–Jupiter orbiter, 14
*Pioneer–Saturn* (formerly *Pioneer II*), 42, 228
Pioneer–Venus project, 14, 100-101, 113, 127-128, 144, 147, 175, 194
Piper PA–34 Seneca aircraft, 53
Pittsburgh, Univ. of, 91, 187, 284, 285, 287
"Plane for Outer Space," 150
"Planetary Exploration: Earth's New Horizon," 35
Planets, 19, 78, 163, 175, 198, 206
Plant tumor experiment, 197
Plants, aquatic, 39
Plasma processes, 85
Plasma experiment on *Mariner 10*, 52
Platforms, 109, 207, 232
Plesetsk (U.S.S.R. launch site), 229
Plum Brook Station (ERDA), 213
Plume simulation, 14
Pluto (planet), 78, 198
PM1 Facilities Management Corporation, 207
PMR (pressure-modulated radiometer), 109
pndb: perceived noise in decibels
Pogue, William R., 64, 94, 173, 187, 191
Poland, 229
Polar station, Soviet, 218
Polcyn, Fabian, 179
Pollution, 19, 21, 27-28, 39, 106, 154, 181–183, 247.
*Pollux* and *Castor* (French satellites), 86, 251, 260
Polycrystalline solidification, 210
Pope Paul VI, 140
Pope's Easter Message, 173
Powder Puff Derby, 249
Power from orbit, 30–31
"Powersat" (solar power satellite), 161
Pozinsky, Norman, 188-189
PPG Industries, 13
"Practical Applications of Space Systems" (NRC report), 141
Pratt & Whitney Aircraft Division (United Aircraft Corp.), 2
*Pravda*, 210
PRC (People's Republic of China). See China.
Precipitation, 94
Presidential Management Improvement Awards, 77
Pressure groups, 60
Pressure-modulated radiometer (PMR), 109
Price, Paul B., 166
Prichard, Reuben P., 50
Princeton Univ., 250
Productivity, 36, 68
Production in space, 210
*Prognoz 4*, 245, 251, 275
Program Assurance, NASA Office of, 46–47
Propellant research, 154
Propulsion technology, 49, 82, 93, 105, 119, 129, 190

Proxmire, Sen. William, 42, 61, 102, 126, 127, 130, 157, 226
Prudhoe Bay, Alaska, 232
Pruitt, Dow C., 164
Public science policy, 235
Public Service Satellite Users Consortium, 36
Puerto Rico, 231–232
Pulsars, 134, 162, 181, 248
*Puppis A* (supernova remnant), 232
Purple Mountain One and Two, 78

## Q

Qantas Airways, 42
Quasars, 48, 162
Quiet Sun, International Year of, 90

## R

*Rabotnichesko*, 153
Radar, 71, 94
Radiation (see also Cosmic ray, Gamma ray, Solar radiation, Ultraviolet radiation, X-ray), 13, 41, 105, 141, 162, 178
Radio, 39, 140
Radio Astronomy Explorers A and B, 100
Radio Corporation of America (RCA), 1, 2, 74, 112, 116, 215, 241
 Astro-Electronics Division, 185, 208, 231
Radioactivity, 30, 210
Radiosonde, 71
*Raduga (Rainbow)*, (U.S.S.R. satellite), 245, 251, 275
Ranger and Surveyor programs, 88
Ranger (spacecraft), 115
Rango, Albert, 168
Rappaport, Saul A., 158
Rasool, Ichtiaque S., 158
Rats, 121, 198
RATAN–600 antenna, 232
Ratsiraka, Didier, 129
Raytheon Service Company, 154, 201
RBV (return-beam vidicon) cameras, 185
RCA. See Radio Corporation of America.
RCA–*Satcom 1*, 2
Reaction control system (RCS), 135
Reconnaissance satellites, 67, 86, 176
Recovery helicopters, 66
Recurrent nova, 158
Red-cell production in astronauts, 34
Red stars, 17
Red tide (phytoplankton, *Gymnodinium*), 179, 249
Redstone (missile), 125
Reed, Sylvanus Albert, award, 36
Refan program for noise reduction, 32
Reimbursable payloads, foreign, 48
Reinhold Construction Company, 69
Reliability, aircraft, 19
Remote Manipulator System (RMS), 78
Remote-sensing technology, 9–10, 41, 168
Remotely piloted research vehicle (RPRV), 41, 66, 167
Remotely piloted vehicle (RPV), 66, 124
Rescue Coordination Center of U.S. Coast Guard, 55

Research, 25, 45, 50
 appropriations, 19, 21, 45, 46, 49, 159
 composite materials, 24, 179
 energy, 21, 49, 67–68, 82, 119, 154
 hypersonic flight, 239
 meteorology, 41, 71, 117
 solar, 221, 237
 space, 50, 162, 232
 stratosphere, 70, 159
Return-beam vidicon (RBV) cameras for TV, 185
Reusable space vehicle, 149
Reuters news agency, 225
Reynolds, Smith, and Hills Company, 112
Rice Univ., 285
RMS (remote manipulator system), 78
Robert M. Losey Award, 36
Robot vehicles, 222
Rock, William H., 110, 213
Rockefeller, Nelson A., 106
Rockefeller Univ., 226
Rocketdyne Division of Rockwell International Corporation, 78
Rocket technology, 11, 13, 21, 41, 53, 57, 84, 90, 102, 111, 119, 123, 186, 190, 235, 248, 283
Rockwell International Corporation, 7, 25, 37, 52, 66, 70, 78, 81, 90, 92–93, 96, 105, 120, 124, 125, 126, 136, 139, 161, 173, 180, 182, 185, 201, 202, 203, 211, 226–227, 243
Rocky Mountains, 157
Roland, F. Sherwood, 236
Romanenko, Yury V., 28
Romania, 229
Rome, Univ. of, 212
Rosen, George, 36
Ross, Miles, 52
*Roundup* (JSC publication), 188, 196
Rowe, Herbert J., 173
Royal Society, the, London, 111
RPRV (remotely piloted research vehicle), 41, 66, 167
RPV (remotely piloted vehicle), 66, 124
Rukavishnikov, Nikolay N., 28
Rumsfeld, Donald H., 245
Russia. See U.S.S.R.

## S

Saab Viggen (Swedish fighter aircraft), 99
Sabreliner, Rockwell, 65
Safety, aircraft, 19
Safety and Environmental Health, NASA Office of, 50
Sagan, Carl, 231
Sagdeyev, Ronald, 44, 45
SALT (Strategic Arms Limitation Talks), 152, 216
Salyut (U.S.S.R. scientific space station), 225
 *Salyut 2*, 92
 *Salyut 3*, 11
 *Salyut 4*, 3, 59, 91, 92, 126, 145, 191, 225, 253
Samoa, 207, 246
SAMSO. See Space and Missile Systems Organization (USAF).

San Andreas Fault Experiment, 201
San Antonio, Tex., 188
*San Diego Union*, 105, 233
San Fernando Observatory, 237
San Marco platform, 2, 78, 277
San Martin, Robert, 219
Sands, George, 160
Santa Fe Engineers, Inc., 87
Santiago, Chile, 202–203
SAO. See Smithsonian Astrophysical Observatory.
SAS (Small Astronomy Satellite), 1
Sas-C, 277
*Sas 2*, 181
*Sas 3*, 158, 208
Satcom (communications satellites), 241–242
*Satcom 1* (RCA 1), 274, 277, 279
Satellite applications (see also individual satellite names)
 atmospheric, 21
 biological, Soviet, 229
 climate forecasting, 150
 comsats, 44, 121, 165, 229
 crop acreage calculation, and, 224
 earth resources, 23, 33, 74, 106
 educational TV, 87
 fossil fuel searches, 50
 geodynamic, 201
 geostationary environmental, 206
 high-altitude research, 13
 land-use planning, 183
 meteorological, 21, 51, 71, 192, 215
 microwave communications, 44
 mineral location, 185
 offshore communications, 159
 power stations, 131, 250
 printing reproduction, 51
 radiation transfer, 41
 reconnaissance, 67, 86, 242
 snowcover measurement, 249
 solar power, 161
 "spy satellite," 242
 tracking and data relay, 21, 109
 water measurement by, 179
 X-ray research, 196
Satellite Business Systems, 245
Satellite de Recherches et d'Environment Technique (*Sret 2*), 103
Satellite Instructional Television Experiment (SITE), 87, 130
*Satellite Situation Report*, 251
Saturn (launch vehicle), 125
 Saturn I, 125
 Saturn IB, 4–5, 32, 52, 132, 136, 160
 Saturn IVB, 132
 Saturn V, 3, 13, 161, 213
 Saturn Program Office, 160
Saturn (planet), 108, 199, 228, 245
Saudi Arabia, 4, 233
S-band radio receiver on *Viking 2*, 184
SCAMS (scanning microwave spectrometer), 109
Scandinavian Airlines, 42
Scanning microwave spectrometer (SCAMS), 109
Scatterometer, 214

Scherer, Lee R., 12, 67
Schlesinger, James R., 107, 216, 245
Schmitt, Harrison H., 164, 186
Schurmeir, Harris M., 35
Schweikart, Russell L., 179
Science, 103, 104, 106, 151, 190, 235, 242
*Science* magazine, 11, 13, 94, 115, 117, 147, 148, 163, 196, 250
Scientists, 86, 87, 198
Scobee, Francis R., 295
Scott Air Force Base, Ill., 74
Scott Air-Pak 4.5, 196
Scott, David R., 67, 182
Scout-D launch, 238
Sea of Galilee, 243
Sea-ice, 109, 232
Sea satellite (SEASAT), 2, 142, 228
Seamans, Robert C., 6, 49, 213, 220, 221
Search and rescue, 54–55
Sears, William R., 36
SEASAT (specialized experimental applications satellite) mission, 2, 142, 228
Seismic pictures of the earth, 50–51
Selivanov, Arnold, 209
Serv-Air, Inc., 58
Servoactuator, 223
Sevastyanov, Vitaly, 91, 92, 121, 282
Sever–27, (Soviet high latitude expedition), 218
Severny Polyus–22, 218
Severny Polyus–23, 218, 237
Sewage, 39, 202
Seychelles Islands, 207
Shapley, Willis H., 146, 192
Shatalov, Vladimir A., 149, 150, 205, 236
Shearing interferometer, 112
Shell Oil Company, 99
Shepard, Alan B., 28, 146
Ships, 82, 135, 248
Shirk, Edward K., 166
Short takeoff and landing (STOL) aircraft, 54, 116, 167
Shuttle Orbiter, contracts for, 126
Siberia, 103, 238
Siemens AG, West Germany, 130
Silicon devices, 216
Silver Spring, Md., 229
Simonson, S. Christian, 217
Simulator, communications, 43
Singer, S. Fred, 247
Sioux Falls, S.D., 212
SIRIO (Italy's microwave-propagation comsat), 44
Sirius (star), 138
SITE (Satellite Instructional Television Experiment), 87, 130
Skin cancer and stratospheric ozone, 27
*Skylab Earth Resources Catalog*, 58
Skylab (program), 10–11, 15, 58, 94, 122, 141, 143, 191
Skylab Orbital Workshop, 3, 13, 64, 97, 125, 161, 168, 179
*Skylab 1*, 3, 13, 97, 125
*Skylab 2*, 74
*Skylab 4*, 15, 94, 168, 187
Skylark (British sounding rocket), 232
*Skynet IIA* (U.K. military comsat), 23

Skyship dirigible, 99
Slayton, Donald K., 1, 5, 65, 83, 104, 120, 121, 123, 133, 135, 147, 149, 168, 171–172, 176, 179, 189, 191–192, 205, 243, 265, 282
Slidell Computer Complex, MSFC, 210
SLX–14 (Soviet launch vehicles), 67, 149
Small Astronomy Satellite. See SAS.
Small Magellanic Cloud, 134
SMF (space manufacturing facility), 250
Smith, Bruce F., 64
Smith, James W., 29
Smithsonian Astrophysical Observatory (SAO), 283, 285
Smithsonian Center for Earth and Planetary Studies, 243
Smithsonian Institution, 128, 159, 179, 243
SMM. See Solar Maximum Mission.
SMS (Synchronous Meteorological Satellite), 1, 26
SMS–B, 1
*Sms 1*, 206
*Sms 2*, 25, 206, 254, 277, 278
Smylie, Robert E., 54, 88, 222
Snavely, William W., 193
Snickers (galaxy), 217
Snowcover and snow-melt prediction program, 5
Snowcover observations, 168, 248–249
Sochi, U.S.S.R., 192
Socioeconomic benefits from space systems applications, 142
Soehardjono, Maj. Gen., 73
Sol, 111
Solar array, 105, 240
Solar cavity, 85
Solar cells, 80, 108, 208
Solar collectors, 12–13, 42, 193
"Solar eclipse" during ASTP, 133
Solar energy
 agriculture, use in, 250
 bibliography, 127
 collector for, 193
 conversion to electric power, 93
 cooling and heating systems, 13, 29, 116, 211, 219, 233
 demonstration mission, 188
 desalination, 233
 generating system, terrestrial, 31
 heating and cooling systems, 13, 29, 116, 211, 219, 233
 plants, two kinds of, 30
 privy, 211
 propulsion, 105
 recording system, terrestrial, 146
 satellite, 161, 250
 thermal energy, 127, 193
 330-unit installation, 219
 transmitted by microwave, 203
Solar Energy Research Institute, 221
Solar flares, 73
Solar Heating and Cooling Demonstration Act of 1974, 21, 211
Solar heating, 13, 29, 116, 211, 219, 233
Solar maximum mission (SMM), 14
Solar noise bursts, 100
Solar particle problem, 74

Solar power. See Solar energy.
Solar Radiation and Thermospheric Structure Satellite (*Srats*), 35
Solar system, 14, 134, 168, 232
Solar radiation, 13
Solar winds, 52, 85, 105
Solar x-rays, 118
Solid-fueled rocket booster (SRB), 57, 102, 174
Sound barrier, 69
Sound-waves diagnostic system, 102
Sounding rockets, 11, 39, 41, 90, 178, 240, 283
South Africa, 207
South Atlantic Anomaly, 94
South Dakota, 130
South Pole, 207, 246
Southern California, Univ. of, 36
Soviet Academy of Sciences, 131, 145, 165, 226, 230
*Soviet Space Programs, 1971–75*, 251
Soviet Scientific Conference, 44
Soviet Space Research Institute, 44
Soviet State Research Institute of Engineering Studies, 25
Soviet Union. See U.S.S.R.
Soyuz spacecraft (see also ASTP), 59, 65, 140, 251, 257
*Soyuz 10*, 92
*Soyuz 11*, 92
*Soyuz 14*, 11, 92
*Soyuz 15*, 11
*Soyuz 16*, 12
*Soyuz 17*, 3, 91, 92, 227, 253, 282
*Soyuz 18*, 91, 121, 126, 227, 260, 282
*Soyuz 19* (ASTP), 1, 131–132, 134, 167, 251, 265, 281, 282
*Soyuz 20*, 225, 227, 236, 272
Space, 47, 150, 180, 205
Space and Missile Systems Organization (SAMSO, USAF), 154, 229
Space applications, 45, 101, 205
Space biology and medicine, 4, 197–198, 214
Space colony, 250
Space-environment monitor, 212
Space flight operations, 45, 101
Space Launch Complex, 153
Space law, 165, 213
Space manufacturing facility (SMF), 250
Space objectives and the national interest, 150
Space probes, 175, 251
Space Processing Applications Rocket (SPAR), 210, 212, 217
Space program, national, 146, 226
Space research, 25, 45, 50–51
Space Science, NASA Office of, 13, 22, 70, 79
Space Science and Applications, House Subcommittee on, 223, 224
Space Science Award, 37
Space Shuttle (see also AMPS; Apollo-Soyuz Test Project; Employment, NASA; External tank; Launch Processing System; Low Cost Modular Spacecraft; MDM; servoactuator; Space Shuttle main engine; Space

Space Shuttle—continued
  Shuttle orbiter, Space Tug, Spacelab; SRB), 11, 13, 34, 43, 53, 67, 85, 96, 113, 135, 137, 154, 168, 173, 180, 185, 188
  appropriations FY 1976, 45, 46
  budget, 21, 42, 101, 224
  construction of facilities, 61, 69, 87–88, 153–154, 214
  contracts, 7, 14, 34, 69, 84, 85, 93, 105, 108, 112, 113, 130, 152, 202, 213, 226, 240
  development status, 19, 34, 222
  economics effects, 78
  fuel, 52
  payloads, 48–49, 78, 142, 204
  press comment, 146
  testing, 15, 63, 189, 198, 213
  tracking and data acquisition, 11, 31, 96–97, 113, 142, 151
  use, 48, 142
Space Shuttle main engine (SSME), 105, 120, 160
Space Shuttle orbiter, 14, 34, 38, 69, 70, 78, 81, 87, 92, 185, 227
Space station, 149, 150, 151, 203, 236, 239–240, 250
Space Telescope (project), 236
Space Transportation System (STS), 222
"Space truck," 248
Space Tug, 14, 139, 201
Space vehicle, reusable, 149
Spacecraft, 6, 19, 86, 110, 133, 163
Spacecraft Design Award (AIAA), 37
*Spaceflight* (BIS publication), 249
Spaceflight Tracking and Data Network (STDN), 244
Spacelab (ESA payload for Shuttle), 21, 27, 35–36, 48, 49, 57, 67, 100, 113, 123, 160, 165, 182
*Spacelab Newsletter*, 123
Spaceplane, 149
*Spaceport News*, 5, 73, 92, 94, 123, 146, 182, 191, 219
"Spaceship Earth" (lecture), 182
Spacesuit leak, 121
Spacetrack coverage, 215
Spain–Mexico satellite link, 122
SPAR (Space Processing Applications Rocket), 210, 217
Spectrometry, 102
Spectrophotometer, 228
Spence, Roderick W., 190
Sperry Flight Systems, 116, 167
Sperry Rand Corporation, 69, 130
Spinning flight, 167
Spinrad, Hyron, 126
*Sputnik* (magazine), 149, 150
*Sputnik 1*, 25, 165, 192
Sputter-deposited materials, 210
Spy satellites, U.S., 242, 244
*Srats* (solar radiation and thermospheric structure) Japanese satellite, 35
SRB (solid-fueled rocket booster), 57, 102, 174
*Sret 2* (Satellite de Recherches et d'Environment Technique), 103, 251

SS (U.S.S.R. missiles) (SSN–6, SS–7, SS–8, SS–11, SS–17, SS–19), 167
SSME. See Space Shuttle main engine.
SSN–6 (U.S.S.R. submarine-based rocket), 167
SST (supersonic transport), 8, 60, 82, 107, 218, 247
St. Louis, Mo., 191
*St. Louis Post-Dispatch*, 229
Stafford, Thomas P., 1, 5, 65, 83, 120, 123, 132, 133, 135, 147, 149, 168, 178, 189, 191, 192, 265, 282
Standardization of weapons in NATO countries, 108
Standard Parts List, NASA, 239
Stanford Solar Observatory, 225
Star City, U.S.S.R., 191–192
Star Tracker for Economical Long-Life Attitude Reference (STELLAR), 184–185
*Starlette* (passive geodetic satellite), 27, 254
Stars, 48, 73, 196
State-Federal liaison, 152
STDN. See Spaceflight Tracking and Data Network.
Steam plants, 49
Stecker, Floyd W., 141
Steering by gyro, 216
STELLAR (Star Tracker for Economical Long-Life Attitude Reference), 184–185
Stellar fires, 17
Sterilization, 110, 112, 143
Stermer, R. L., 243
Stern, Richard J., 80
Stever, H. Guyford, 29, 84, 103
Stirling external combustion engine, 181–182
STOL (short takeoff and landing) aircraft, 54, 116, 167
Storms. See Meteorology.
Strategic Arms Limitation Talks (SALT), 152, 216
Stratosphere, 13, 21, 29, 70, 144, 159, 183–184, 237
Streamflow observations, 168
Strike of aircraft workers, 28–29
Strom, Robert G., 58
Structural surveys, 9
STS. See Space Transportation System.
Stuhlinger, Ernst, 235
Stullken, Donald E., 104
Submarines, 19, 242
Sullivan, Walter, 1, 27, 158, 231
Sumerians, 225
Sun-earth relationship, 118
Sun (see also solar headings), 10, 19, 118, 224–225, 245
Sundic, Milika, 152
Sung Dynasty, 186
Super Arcas (sounding rocket), 90, 284, 285, 286, 287, 289, 290, 291
Super Loki (sounding rocket), 90, 237
Supernovae, 48, 162, 181, 186–187, 232
Supersonic transport (SST), 8, 60, 82, 107, 154, 247

*Surveyor* (spacecraft), 115
Suslov, Mikhail, 230
Sweden, 60, 99
Symphonie–B, (French–German experimental comsat), 2
*Symphonie 1*, 23, 177
*Symphonie 2*, 177, 251, 267, 277, 279
Symphonie Project Operations Group, 177
Synchronous Meteorological Satellite (SMS), 1, 25
Synthetic fuels, 67–68
Syromyatnikov, Vladimir, 139, 140

T

T–500S (Paymover tractor), 214
T&DR (tracking and data relay), 109
*Taiyo*, 251, 255
Talley, Wilson K., 183
Tanegashima launch site, Japan, 184
Tape recorder on Landsat, 222
Target visibility, 6
Tass news agency, 91, 99, 100, 105, 218, 226, 230, 241, 244
Tau Ceti (star), 73
Taurus, (constellation), 186
Tbilisi, U.S.S.R., 192
TDRSS (Tracking and Data Relay Satellite System), 27, 113, 233
Teague, Olin E., 8, 25, 43, 61, 108, 116, 141, 151, 219
Technical Univ. of Vienna, 216
Technology, 22, 36, 45, 50–51, 77, 101
Teleoperator system, earth-orbiting, 14
Telephone, 4, 230, 241
Telesat–C (Anik 3), 2, 29, 51
Telescope, 14, 30, 84, 105, 112–113, 127, 137, 144, 147, 148, 165, 169, 185, 194, 214, 215, 230
Television, 6, 24, 86, 87, 245
Television Infrared Operational Satellite *(Tiros–N)*, 215
Temperature, 24, 26, 27
Temperature/humidity infrared radiometer (THIR), 109
Tennessee, Landsat applications, 232
Tenth planet, 168
ter Horst, J. F., 172
Terrier Malamute (rocket), 288
Texas Medical Center, 176, 179
Texas, Univ. of, 205, 289
Thermal energy, 31, 127, 185
Thermionic conversion, 131
Thiokol Corporation, 84
THIR (temperature/humidity infrared radiometer), 109
Thole, John M., 64
Thom, Karl, 119
Thomas, Garland L., 183
Thor-Delta launch vehicles, 2, 8, 23, 25, 44, 60, 62, 63, 73, 79, 108, 110, 119, 162–163, 177
Thoria-dispersed magnesium, 210
*Tiros–N* (Television Infrared Operational Satellite) spacecraft, 26, 208, 211, 215, 231
*Tiros 1*, 57

Titan-Centaur launch vehicles, 2, 87, 143, 164, 171, 173, 223
Titan (Saturn moon), 245
*Today* editorial, 172
Toilets, solar-powered, 211
Tokamak (nuclear reactor), 220
Tokyo Univ., Institute of Space and Aeronautics, 35
Toronto, 173
Toxic chemicals, 39
Tracking and data acquisition, 22, 45, 101, 188–189
Tracking and data relay (T&DR), 109
Tracking and Data Relay Satellite System (TDRSS), 27, 113, 233
Tracking stations, 21, 130, 201
Tractor, towing, 214
Transit (U.S. Navy navigation system), 179
Transonic flight, 161, 172
Transport aircraft, 8, 82, 107
Transportation, U.S. Department of, 8, 27, 29, 55, 80
Trella, Massimo, 8, 66
Trendelenburg, Ernst, 66
*Triad 2* (U.S. Navy satellite), 271
Trident missile, 19, 23
Tripler Hospital, Hawaii, 147, 149
Trimethylaluminum, 6
Tropical wind-energy conversion and reference level experiment (TWERLE), 109
Truman, President Harry S, 84, 226
TRW Company, 43
TRW Systems Group, 37, 43, 136, 204, 221
Tsibin, Sergei, 140
Tsiolkovsky, Konstantin, 137
Tu–144 (U.S.S.R. supersonic aircraft), 44
Tulinius, Jan R., 37
Tunney, Sen. John V., 32, 127, 128, 161
Tupolev, Aleksey A., 44
Tupolev–144 airliner, supersonic, 44, 247
Turbine technology, 50
Turbofans, 82
Turbojet, 82
TWERLE (tropical wind-energy conversion and reference level experiment), 109
Tyuratam, U.S.S.R., 67

U

U–2 (high altitude aircraft), 13, 70, 138, 190, 237
UAL. See United Air Lines.
Ubangi River, Africa, 233
Uchinoura Space Center, Japan, 35
*Uhuru (Explorer 42)*, 79
U.K. (United Kingdom), 23, 60, 93
Ultrasound diagnostic systems, 102
Ultraviolet radiation, 13, 68, 105, 134, 223
U.N. See United Nations.
Unemployment of scientists and engineers, 86
Union of Soviet Socialist Republics. See U.S.S.R.
United Air Lines, Inc. (UAL), 2, 42, 218
United Aircraft Corporation, 4, 36
United Kingdom. See U.K.

United Nations (U.N.)
  ad hoc committee on uses of outer space, 165
  General Assembly, 231
  Political and Security Committee, 213
  Scientific and Technical Subcommittee for the Peaceful Use of Space, 25
  World Meteorological Organization, 139
United Nations Public Registry of Space Flights, 251
United States (see U.S.), 68, 121, 158
United States Air Force (USAF)
  Aero Propulsion Laboratory, 67
  AFSC. See Air Force Systems Command.
  airborne comsat terminal, 29
  Airborne Warning and Control System (AWACS), 59
  Avionics Laboratory, 81
  Cambridge Research Laboratories, 6
  comsat failures, 87
  dirigibles, 99
  Flight Dynamics Laboratory, 152
  Flight Test Center, Calif., 10, 24, 27
  hypersonic aircraft, 239
  manned orbiting laboratory, 153
  –NASA joint programs, 293
  nuclear-powered aircraft, 57
  reconnaissance satellites (Big Bird), 67
  satellite communications (AFSATCOM), 43
  Space and Missile Systems Organization (SAMSO), 154, 229
  upper atmosphere experiments, 6
United States Army, 71
  Ballistic Missile Agency, 125
  Corps of Engineers, 9, 14, 213
United States Navy (USN), 71
  Auxiliary Landing Field, Calif., 167
  combat fighter development, 23
  Naval Observatory, 168
  Naval Research Laboratory (NRL), 94
  sixth fleet, 59
  Transit navigation system, 179
United Technologies Corporation, 81, 162
Universal Construction Company, 213
Universe, 19, 178
Upper atmosphere, 6, 100
Uranium, 119, 233
Uranus (planet), 168, 198–199, 235
Urey, Harold C., 111
Urinary infections detection, 177
Urokinase, 4
Uruguay, 203
USAF. See United States Air Force.
USDA. See Agriculture, U.S. Department of.
U.S.–European joint venture, Spacelab's first mission, 35
USGS. See Geological Survey, U.S.
U.S.–Malagasy agreement on tracking station, 129
USN. See United States Navy.
U.S.–U.K. cooperative satellite *(Ariel 5)*, 158
U.S.–U.S.S.R. cooperation (see also Apollo-Soyuz Test Project), 31–32, 41, 131, 140, 191–192

U.S.–West German solar probe *(Helios I)*, 46
U.S.S.R. (Union of Soviet Socialist Republics), 24, 28, 42, 44, 53, 54, 70, 77, 107, 121, 127, 146, 168, 204, 222, 234
  missile (see also individual missiles, such as SS–19, SSN–6), 166–167
  satellites and spacecraft (see also individual satellites and spacecraft, such as Cosmos, Intercosmos, Luna, Mars, Meteor, Molniya, Salyut, and Soyuz), 4, 47, 67, 99, 100, 103, 149, 169, 176, 188, 204, 208, 209
  space program, 81, 102, 130, 149, 218, 231
Ute-Tomahawk rockets, 6
UV: ultraviolet (radiation).

V

VAB: Vehicle Assembly Building (KSC).
Van Allen radiation belts, 42, 94, 103
Van Flandern, Thomas C., 168
Van Reeth, George, 66
Vandenberg Air Force Base, 29, 64, 153, 154
*Vanguard 1*, 47
Vatican satellite coverage, 173
Vecchietti, George J., 59
Vela (constellation), 181, 225
Venera spacecraft, 106, 162, 209, 210
  *Venera 9*, 105, 209, 210, 251, 263
  *Venera 10*, 105, 209, 210, 251, 263
Venus (planet; see also Mariner 10, Pioneer–Venus, *Venera 9*, and *Venera 10*), 49, 52, 100, 101, 105, 106, 113, 115, 127, 144, 162, 183, 209–210, 226
Vertical Motion Simulator (VMS), 159, 196
Vertical takeoff and landing aircraft (VTOL), 159, 196
Vibration testing of Space Shuttle, 15
Video inertial pointing (VIP), 214
Video sensors, 215
Video service by Westar (Western Union), 112
Vidicon tube, 215
Viking Data Review Management Center, 182
*Viking: Mission to Mars*, 171
Viking (program), 8, 21, 113, 143, 160, 172–173, 174–175
Viking (spacecraft), 1, 55, 73, 95, 96, 112, 153, 164, 169–170, 277
Viking–A spacecraft, 153, 166
Viking–B spacecraft, 143, 153, 166, 182
*Viking 1*, 110, 169, 170–171, 217, 223, 266, 279
*Viking 1* lander, 110, 111, 170–171
*Viking 2* lander, 73, 94, 184, 217, 220, 267, 279
Vinogradov, Alexander P., 226
VIP (video inertial pointing), 214
Virginia, Univ. of, 247
VMS (Vertical Motion Simulator), 159
Volcanoes on Mars, 175
Volgograd, U.S.S.R., 192
Vologda Oblast, U.S.S.R., 39

von Braun, Wernher, 31, 125, 166
von Karman lecture, 35
Vortices, wake, 80
VSTOL: vertical or short takeoff and landing.
VTOL. See Vertical takeoff and landing aircraft.

## W

Wake vortices, 80
*Wall Street Journal*, 33, 51, 228, 236
Wallops Flight Center (WFC), 2, 22, 91, 237
Wallops Island, Va., 41, 206
Walverton, British Columbia, 39
Washington, D.C., 128, 159, 195
*Washington Post, The*, 3, 4, 42, 54, 77, 81, 96, 102, 126, 127, 131, 136, 137, 157, 166, 173, 174, 186, 190, 219, 230
*Washington Star, The*, 10, 57, 58, 84, 116, 137, 163, 175, 231, 246
Water, 37, 44, 90, 142, 177, 179
Water hyacinths, 39
Waterman, Alan T., Award, 84
Wave radiation measurements, 39
Weapons standardization in NATO, 108
Weather. See Meteorology.
Weber Metals and Supply Company, 129
Weightlessness, 3, 135, 197, 213, 229
Weinberger, Caspar W., 36
Weinreb, Marius B., 187
Weitz, Paul J., 64, 74
West Germany, 2, 4, 35, 46, 59, 60, 67, 126, 152, 177, 247–248, 277
Westar (Western Union comsat), 4, 136, 159, 241
  *Westar 1*, 112, 136, 164, 228–229
  *Westar 2*, 112, 136
Western European Union, 235
Western Test Range (WTR; see also Vandenberg AFB, Calif.), 2, 8, 60, 62, 108, 162, 238
Western Union Telegraph Company, 112, 116, 136, 159, 164, 241
WFC. See Wallops Flight Center.
Wheat, 9, 264
White, George C., 47
White House, 60, 103, 190, 205
White Sands Missile Range (WSMR), 215, 240
*White Sun of the Desert* (film), 130
White, Thomas D., Space Trophy, 187
Whitlow, John B., 82
Whitten, Les, 77
Wicks, Alden, 137
Wide-bodied jets, 78
Williams, Walter C., 201–202

Wilmington, Del., 249
Wilson, Robert G., 25
"Wind Energy Utilization" (bibliography), 127
Wind tunnel, 69, 73, 152
Winfrey, Lee, 142
Wing-sweep control on B–1, 10
Wingless lifting body, X–24B, 161
Wingless vehicle, M2–F3, 128
Winter, David L., 33, 34, 96, 197
Wischnia, Herbert F., 113
Wisconsin, Univ. of, 284, 289
Witkin, Richard, 181
Woodfin, Kenneth L., 59
Woomera Rocket Range, Australia, 232
Worden, Alfred M., 182
World Meteorological Organization (U.N.), 139
World War II, 69, 235
Wren, Christopher S., 225
Wright Brothers Memorial Trophy, 190
WSMR. See White Sands Missile Range.
WTR. See Western Test Range.
Wu, Edward K., 149, 236
Wyeth, Jamie, 137
Wyld, James H., Propulsion Award, 190
Wyoming, 168, 187
Wyoming, Univ. of, 84

## XYZ

X–1 (NACA–USAF aircraft), 69
X–24 (lifting body), 161, 172, 194, 293
X-ray
  astronomy, 208
  medical, 51
  Orion constellation, 158
  sensors, 187
  sources, 84, 118, 134, 158, 196
  telescopes, 10, 48, 91
Yak (U.S.S.R. airliner), 53
Yakushin, Anatoly, 137
Yardley, John F., 61, 189, 223, 224
YC–15 (short-takeoff-and-landing aircraft), 178
Yeliseyev, Aleksey S., 81, 222
Yoshikawa, Tsuneo, 216
Young, A. Thomas, 110, 111
Young, John W., 188
Young, Kenneth A., 37
Yugoslavia, 152
Zaire, 38, 203
Zhukov, G. P., 165
Zodiacal light, 15, 46
Zuckerman, Benjamin M., 231
Zvenigorod, U.S.S.R., 44
Zvezdny Gorodok, U.S.S.R., 121

# ERRATA

## In Earlier Volumes of Astronautics and Aeronautics

*Astronautics and Aeronautics, 1967*

p. 65: In second item on March 7, "Impact of Jan. 27 Apollo accident...," add three ellipsis dots at end of first paragraph. Substitute for source at end of item: "(NASA Auth. Hearings transcript, pp. 151–2, 186–7)"

p. 442: Index: Under "Apollo (program)" in left column, substitute "Apollo 6" for "Apollo 5" in parentheses after "AS–502." In right column after "Apollo 5," substitute "AS–204, originally AS–206" for "AS–502" in parentheses.

*Astronautics and Aeronautics, 1968*

p. 316: In 4th paragraph, 1st line, change "Sonnett" to "Sonett." Make corresponding change in Index, p. 417, left column.

p. 355: In Appendix A, 1st column, add two asterisks after Sept. 16, for *Zond V*.

*Astronautics and Aeronautics, 1969*

p. 64: In 2nd paragraph from bottom, change time of *Apollo 9* splashdown to 12:00:53 pm EST.

p. 376: In 8th line, change "July 21" to "July 20" [using Eastern Standard Time rather than Greenwich Mean Time].
In paragraph 3, line 4, for "contingency sample" substitute "contingency and selected samples."

*Astronautics and Aeronautics, 1972*

p. 345: In 13th line from bottom, change identification of Dr. Robert R. Gilruth to "NASA Director of Key Personnel Development and former MSFC Director."

p. 148: In paragraph 3, line 3, change liftoff date to April 23. [It was 1:26 am April 24 by Greenwich Mean Time.]

*Astronautics and Aeronautics, 1973*

p. 432: Index: After Gilliland, Whitney, change "5" to "4."

# NASA HISTORICAL PUBLICATIONS

## HISTORIES

- Anderson, Frank W., Jr., *Orders of Magnitude: A History of NACA and NASA, 1915–1976*, NASA SP–4403, 1976, GPO.*
- Benson, Charles D., and William Barnaby Faherty, *Moonport: A History of Apollo Launch Facilities and Operations*, NASA SP–4204, 1978, GPO.
- Brooks, Courtney G., James M. Grimwood, and Loyd S. Swenson, *Chariots for Apollo: A History of Manned Lunar Spacecraft*, NASA SP–4205, 1979, GPO.
- Byers, Bruce K., *Destination Moon: A History of the Lunar Orbiter Program*, NASA TM X–3487, 1977, NTIS.†
- Corliss, William R., *NASA Sounding Rockets, 1958–1968: A Historical Summary*, NASA SP–4401, 1971, NTIS.
- Ezell, Edward C. and Linda N., *The Partnership: A History of the Apollo-Soyuz Test Project*, NASA SP–4209, 1978, GPO.
- Green, Constance McL., and Milton Lomask, *Vanguard: A History*, NASA SP–4202, 1970, GPO; also, Washington: Smithsonian Institution Press, 1971.
- Hacker, Barton C., and James M. Grimwood, *On the Shoulders of Titans: A History of Project Gemini*, NASA SP–4203, 1977, GPO.
- Hall, R. Cargill, *Lunar Impact: A History of Project Ranger*, NASA SP–4210, 1977, GPO.
- Hartman, Edwin P., *Adventures in Research: A History of the Ames Research Center, 1940-1965*, NASA SP–4302, 1970, NTIS.
- Link, Mae Mills, *Space Medicine in Project Mercury*, NASA SP–4003, 1965, NTIS.
- Rosholt, Robert L., *An Administrative History of NASA, 1958-1963*, NASA SP–4101, 1966, NTIS.
- Sloop, John L., *Liquid Hydrogen as a Propulsion Fuel, 1945–1959*, NASA SP–4404, 1978, GPO.
- Swenson, Loyd S., James M. Grimwood, and Charles C. Alexander, *This New Ocean: A History of Project Mercury*, NASA SP–4201, 1966, GPO.

## REFERENCE WORKS

- *Aeronautics and Space Report of the President*, annual volumes for 1975–1978, GPO.
- *The Apollo Spacecraft; A Chronology*, NASA SP–4009, vol. 1, 1969, NTIS; vol. 2, 1973, GPO; vol. 3, 1976, GPO; vol. 4, 1978, GPO.
- *Astronautics and Aeronautics: A Chronology of Science, Technology, and Policy*, annual volumes 1961–1974, with an earlier summary volume, *Aeronautics and Astronautics, 1915–1960*. Early volumes available from NTIS, recent ones from GPO.
- Hall, R. Cargill, ed., *Essays on the History of Rocketry and Astronautics: Proceedings of the Third through the Sixth History Symposia of the International Academy of Astronautics*, NASA CP–2014, 2 vol., 1977, GPO.
- Roland, Alex F., *A Guide to Research in NASA History*, HHR-50, 3rd ed., Feb. 1979, looseleaf, available from NASA History Office, Washington, DC 20546.
- *Skylab: A Chronology*, NASA SP–4011, 1977, GPO.
- Van Nimmen, Jane, and Leonard C. Bruno, *NASA Historical Data Book, 1958–1968*, NASA SP–4012, vol. 1, *NASA Resources*, 1976, NTIS.
- Wells, Helen T., Susan H. Whiteley, and Carrie E. Karegeannes, *Origins of NASA Names*, NASA SP–4402, 1976, GPO.

---

*GPO: Order from Superintendent of Documents, Government Printing Office, Washington, DC 20402.
†NTIS: Order from National Technical Information Service, Springfield, VA 22161.

www.ingramcontent.com/pod-product-compliance
Lightning Source LLC
Chambersburg PA
CBHW081718170526
45167CB00009B/3616